Springer Tracts in Modern Physics
Volume 178

Managing Editor: G. Höhler, Karlsruhe

Editors: H. Fukuyama, Kashiwa
J. Kühn, Karlsruhe
Th. Müller, Karlsruhe
A. Ruckenstein, New Jersey
F. Steiner, Ulm
J. Trümper, Garching
P. Wölfle, Karlsruhe

Honorary Editor: E. A. Niekisch, Jülich

Now also Available Online

Starting with Volume 165, Springer Tracts in Modern Physics is part of the Springer LINK service. For all customers with standing orders for Springer Tracts in Modern Physics we offer the full text in electronic form via LINK free of charge. Please contact your librarian who can receive a password for free access to the full articles by registration at:

http://link.springer.de/series/stmp/reg_form.htm

If you do not have a standing order you can nevertheless browse through the table of contents of the volumes and the abstracts of each article at:

http://link.springer.de/series/stmp/

There you will also find more information about the series.

Springer
Berlin
Heidelberg
New York
Barcelona
Hong Kong
London
Milan
Paris
Tokyo

Springer Tracts in Modern Physics

Springer Tracts in Modern Physics provides comprehensive and critical reviews of topics of current interest in physics. The following fields are emphasized: elementary particle physics, solid-state physics, complex systems, and fundamental astrophysics.
Suitable reviews of other fields can also be accepted. The editors encourage prospective authors to correspond with them in advance of submitting an article. For reviews of topics belonging to the above mentioned fields, they should address the responsible editor, otherwise the managing editor.
See also http://www.springer.de/phys/books/stmp.html

Managing Editor

Gerhard Höhler
Institut für Theoretische Teilchenphysik
Universität Karlsruhe
Postfach 69 80
76128 Karlsruhe, Germany
Phone: +49 (7 21) 6 08 33 75
Fax: +49 (7 21) 37 07 26
Email: gerhard.hoehler@physik.uni-karlsruhe.de
http://www-ttp.physik.uni-karlsruhe.de/

Elementary Particle Physics, Editors

Johann H. Kühn
Institut für Theoretische Teilchenphysik
Universität Karlsruhe
Postfach 69 80
76128 Karlsruhe, Germany
Phone: +49 (7 21) 6 08 33 72
Fax: +49 (7 21) 37 07 26
Email: johann.kuehn@physik.uni-karlsruhe.de
http://www-ttp.physik.uni-karlsruhe.de/~jk

Thomas Müller
Institut für Experimentelle Kernphysik
Fakultät für Physik
Universität Karlsruhe
Postfach 69 80
76128 Karlsruhe, Germany
Phone: +49 (7 21) 6 08 35 24
Fax: +49 (7 21) 6 07 26 21
Email: thomas.muller@physik.uni-karlsruhe.de
http://www-ekp.physik.uni-karlsruhe.de

Fundamental Astrophysics, Editor

Joachim Trümper
Max-Planck-Institut für Extraterrestrische Physik
Postfach 16 03
85740 Garching, Germany
Phone: +49 (89) 32 99 35 59
Fax: +49 (89) 32 99 35 69
Email: jtrumper@mpe-garching.mpg.de
http://www.mpe-garching.mpg.de/index.html

Solid-State Physics, Editors

Hidetoshi Fukuyama
Editor for The Pacific Rim
University of Tokyo
Institute for Solid State Physics
5-1-5 Kashiwanoha, Kashiwa-shi
Chiba-ken 277-8581, Japan
Phone: +81 (471) 36 3201
Fax: +81 (471) 36 3217
Email: fukuyama@issp.u-tokyo.ac.jp
http://www.issp.u-tokyo.ac.jp/index_e.html

Andrei Ruckenstein
Editor for The Americas
Department of Physics and Astronomy
Rutgers, The State University of New Jersey
136 Frelinghuysen Road
Piscataway, NJ 08854-8019, USA
Phone: +1 (732) 445 43 29
Fax: +1 (732) 445-43 43
Email: andreir@physics.rutgers.edu
http://www.physics.rutgers.edu/people/pips/Ruckenstein.html

Peter Wölfle
Institut für Theorie der Kondensierten Materie
Universität Karlsruhe
Postfach 69 80
76128 Karlsruhe, Germany
Phone: +49 (7 21) 6 08 35 90
Fax: +49 (7 21) 69 81 50
Email: woelfle@tkm.physik.uni-karlsruhe.de
http://www-tkm.physik.uni-karlsruhe.de

Complex Systems, Editor

Frank Steiner
Abteilung Theoretische Physik
Universität Ulm
Albert-Einstein-Allee 11
89069 Ulm, Germany
Phone: +49 (7 31) 5 02 29 10
Fax: +49 (7 31) 5 02 29 24
Email: steiner@physik.uni-ulm.de
http://www.physik.uni-ulm.de/theo/theophys.html

Dieter Langbein

Capillary Surfaces

Shape – Stability – Dynamics,
in Particular Under Weightlessness

With a Foreword by Ulf Merbold
With 182 Figures

 Springer

Prof. Dr. Dieter Langbein
Universität Bremen
Am Fallturm
28359 Bremen, Germany
E-mail: d.langbein@t-online.de

Library of Congress Cataloging-in-Publication Data applied for

Die Deutsche Bibliothek - CIP-Einheitsaufnahme

Langbein, Dieter:
Capillary surfaces: shape – stability – dynamics, in particular under
weightlessness/Dieter Langbein. – Berlin; Heidelberg; New York;
Barcelona; Hong Kong; London; Milan; Paris; Tokyo: Springer, 2002
(Springer tracts in modern physics; Vol. 178)
(Physics and astronomy online library)
ISBN 3-540-41815-6

Physics and Astronomy Classification Scheme (PACS): 36, 47, 68, 81,

ISSN print edition: 0081-3869
ISSN electronic edition: 1615-0430
ISBN 3-540-41815-6 Springer-Verlag Berlin Heidelberg New York

This work is subject to copyright. All rights are reserved, whether the whole or part of the material is concerned, specifically the rights of translation, reprinting, reuse of illustrations, recitation, broadcasting, reproduction on microfilm or in any other way, and storage in data banks. Duplication of this publication or parts thereof is permitted only under the provisions of the German Copyright Law of September 9, 1965, in its current version, and permission for use must always be obtained from Springer-Verlag. Violations are liable for prosecution under the German Copyright Law.

Springer-Verlag Berlin Heidelberg New York
a member of BertelsmannSpringer Science+Business Media GmbH

http://www.springer.de

© Springer-Verlag Berlin Heidelberg 2002
Printed in Germany

The use of general descriptive names, registered names, trademarks, etc. in this publication does not imply, even in the absence of a specific statement, that such names are exempt from the relevant protective laws and regulations and therefore free for general use.

Typesetting: Data conversion by Steingraeber Satztechnik GmbH, Heidelberg
Cover design: *design & production* GmbH, Heidelberg

Printed on acid-free paper   SPIN: 10732617   56/3141/tr   5 4 3 2 1 0

Foreword

Even today flying into space is an undertaking at the borderline of what is currently feasible. A very complex assembly of sophisticated machinery is needed in order to reach orbit and return from there to Earth safely. An astronaut knows that only if the rocket engines, inertial navigation platforms, hydraulic systems, generators of electrical power, attitude control systems, life support systems, computers, data transmission equipment, heat shields and many other components work flawlessly can he or she accomplish a mission in good health. In other words there is risk, but the odds are better than even. How else could a man or a woman see the incredible beauty of our planet Earth from a distance? Take the combination of its colours, white and blue. It is so breathtakingly stunning, that most likely not even a poet could find words to adequately describe it. There is the curved line of the horizon with an incredibly black sky above it. At first sight it is evident how endless and empty the universe actually is. The Earth's horizon is fringed with a royal blue seam, our atmosphere. Its beauty and fragility are thrilling, and seen from orbit it is instantly apparent that there is not much air around us.

Since it takes only 90 minutes to accomplish one full orbit, an astronaut's perception changes. Within this rather short time he or she passes over the day side of Earth, where he or she can easily find and observe all the features and landmarks of the area on the ground rolling through his or her field of view. He or she not only sees oceans, coastlines, continents and the world's big mountain areas, but he or she can accurately pinpoint the fine structures of the terrain underneath. As he or she is moving swiftly, the Sun sets 16 times faster than we experience on the ground. On the night side of the orbit, all stars shine on him with unprecedented splendour. When he gets close to one of the magnetic poles, the spectacular displays of the Northern or Southern Lights will not only catch his eyes, but most likely also capture his imagination. Of course there is not much time for contemplation. The Sun will rise again in next to no time and will shed its light on other wonders of nature.

As it takes so little time to travel once around Earth, an astronaut views our globe with different eyes. Suddenly it loses its mighty size and its apparent robustness. Evidently we do not live on an indestructible planet. An astronaut, trapped into the small artificial world of a spacecraft, sooner or later also realizes that in the long run there is no place that offers the same quality of life as Earth does. For these reasons astronauts and cosmonauts consistently share the view that the Earth ought to be conserved for the coming generations more vigorously. Sustainability is a term frequently heard in discussions amongst space flyers.

Manned space missions are not only extremely rewarding due to the striking view from orbit and its deep emotional impact but also because of the intellectual dimension of the science involved. It is the responsibility of the scientist astronaut to perform a series of experiments in flight. They are conceived by an international group of scientists working in various fields. The multidisciplinary mix of typically several dozen scientific experiments selected for one specific flight constitutes a major challenge on the one hand, but on the other hand provides a singular chance to the astronaut to widen his horizon. Today scientists are usually forced to focus on rather narrow fields of research, otherwise they will not be able to keep up with the ever-accelerating progress in knowledge. In line with the Latin proverb variatio delectat a scientist astronaut finds himself in a more fortunate situation. He gets involved in many scientific disciplines like astronomy, atmospheric physics, Earth observation, materials science, human physiology, biology, plasma physics, and, as here, fluid physics, to name a few. A good experiment per definition attempts to move the borderline of knowledge further into the unknown domain. In other words the scientist astronaut needs not only to comprehend the basics of all disciplines represented by the experiments of his mission, he is forced to advance to the transition area between the known and the unknown. He of course knows that the so-called principal investigators, the originators of the various experiments, have spent years of work developing their science to flight maturity. The cost of the mission, which puts a high burden of responsibility on his shoulders, and also the enthusiasm and dedication of the many scientists involved in the flight concern him a lot. On the one hand, there are their high expectations of obtaining data of high quality and sufficient quantity; on the other hand, there is a rigid constraint: The flight time is limited to a few days only. There is normally only a single chance to perform an experiment. Under these circumstances it is obvious that an astronaut has to spend a number of years of his life in training in order to build up adequate proficiency and to become a credible substitute for all the scientists who have to stay on the ground. It happens frequently that the intense interaction between the scientists and the astronauts results in a trustful relationship far beyond the working one. The international framework around a flight apparently stimulates a cooperative spirit amongst people originating from different countries, professions and affiliations. Indeed teamwork is indispensable, but it is encouraging to see how much different players involved in a mission learn from each other, not only as regards science.

Fluid physics is a very special discipline amongst those utilizing the microgravity environment aboard a spacecraft. I remember a NASA movie showing

Ed Gibson aboard Skylab, the first US space station. During the three months Ed spent in orbit fluids like orange juice and tomato juice obviously excited him progressively. The movie shows him in his spare time playing with free-floating big droplets. Ed makes them collide. In most cases they fuse and start to oscillate in very interesting modes. Ed also rotates them such that they get deformed from the perfect spherical shape and finally break apart. The movie also presents an apparatus he devised in which he creates a liquid column suspended between two metallic plates. He varies the length of the column as well as its shape. As far as I remember he even manages to rotate it. By now Ed's experiments have been repeated under much more controlled conditions. Using liquids of different viscosities and surface tensions we have expanded our knowledge further. The questions related to interface or Marangoni convection also became of great scientific interest.

In short, fluid physics evolved into one of the most dynamic disciplines as regards research under microgravity. One of the reasons for the rapid progress in the field of fluid physics is most likely the combined employment of theory and experimental work. The availability of a rather large number of equations precisely describing capillary forces and their effects has definitely contributed to the impressive development. Dieter Langbein is one of the scientists who got involved in the field at the pioneering stage, as a theoretician as well as an experimenter. In my view there is nobody else who could condense the results obtained by the many members of the community so far into a monograph as comprehensively. His book is being published at the right time, because the International Space Station (ISS) is in orbit. Although it will take a few more years to finish its assembly, it is already available as a new scientific platform. The most relevant improvement in comparison to former systems like Spacelab and Spacehab is its permanent availability, and in comparison to Mir its superior characteristics relate to power, cooling, data acquisition, communication and teleoperation.

Noordwijk, The Netherlands   *Ulf Merbold*
2001

Preface

This book is devoted to interfaces between two fluids, that is, between a liquid and a gas (e.g. water and air) or between two liquids (e.g. water and oil). The main motivation for the book is twenty-five years of cooperation between materials scientists and life scientists in research under the low-gravity conditions available in sounding rockets and orbiting spacecraft. This unique environment has made possible numerous qualitative and quantitative observations of effects that are masked by gravity on earth. Large liquid surfaces have been created and their stability and dynamics have been studied. The experimental insights gained have, in turn, strongly stimulated further systematic theoretical and mathematical investigations.

The initial idea of producing high-quality materials in space had to give way, after the first missions, to the insight that it is much more realistic to acquire basic knowledge in space and to apply that knowledge to materials production on earth. Examples are provided by the control of convection during crystal growth, the active exploitation of thermocapillary migration during the production of finely dispersed samples of monotectic materials, and the observation of considerably lower diffusion coefficients in molten metals.

Working in the interdisciplinary field of microgravity is a great experience. The unusual experimental conditions strongly unite researchers. Cooperation is absolutely necessary. And the astronauts make a strong contribution, too. I had experience of this during training sessions in the classroom, with the Plateau tank, at the various facilities, during parabolic flights and during the performance of experiments with air-to-ground contact. In particular, let me mention Reinhard Furrer, Ulf Merbold, Ernst Messerschmid, Chiaki Mukai, Wubbo Ockels, Hans Schlegel, Gerhard Thiele and Ulrich Walter. Let me also acknowledge my colleagues from the DLR (formerly DARA, and before that DFVLR) and from the ESA for their support for the preparation of the experiments and for providing the flight opportunities. Numerous colleagues, notably Robert Finn, Paul Concus, Masamichi Ishikawa, Armin de Lazzer, Robert Naumann and Mark Weislogel, provided their results openly, sent their preprints, had very open and provocative discussions with me, read parts of the manuscript, etc. And, last but not least, it is a pleasure to thank my wife Walburga for her great patience during the preparation of this manuscript.

Bremen, November 2001 *D. Langbein*

Contents

Notation XV

1. **Introduction** 1
   1.1 Space Missions 1
   1.2 Interdisciplinary Stimuli 6
   1.3 Problems of Fluid Physics 9
   1.4 Zero Mass Acceleration or Weightlessness 13
   1.5 Flight Selection and Simulation 15
   References 18

2. **Interface Tension and Contact Angle** 21
   2.1 Molecular Attraction and Condensation 21
   2.2 The Interface Tension 24
       2.2.1 Theoretical Aspects 24
       2.2.2 Experimental Methods 25
       2.2.3 Qualitative Rules for the Interface Energy 27
   2.3 The Static Contact Angle 30
   2.4 The Dynamic Contact Angle 31
   2.5 Merging of Drops and Bubbles 35
   2.6 Adhesion Forces in Liquid Films 37
   References 38

3. **Capillary Shape and Stability** 41
   3.1 Balance of Forces 41
   3.2 Minimization of Energy 44
   3.3 Analytical Solutions of the Capillary Equation 47
       3.3.1 Rise of Liquid in a Tube 47
       3.3.2 Spherical Surfaces 49
       3.3.3 Rise of a Liquid in Contact with an Infinite Plane 51
   3.4 Axisymmetric Surfaces 52
   3.5 Container Shape and Wetting 57
   3.6 Drops at Low Bond Numbers 59

XII   Contents

    3.7   Representations of the Capillary Equation 61
           3.7.1   Cartesian Coordinates $z(x,y)$ 62
           3.7.2   Polar Coordinates $r(\vartheta,\varphi)$ 62
           3.7.3   Cylindrical Coordinates $r(\varphi,z)$ 63
           3.7.4   Cylindrical Coordinates $z(r,\varphi)$ 63
           3.7.5   Axisymmetry 64
    References 65

4. Stability Criteria 67
    4.1   Stability of Capillary Surfaces 67
    4.2   Breakage of Cylindrical Surfaces 69
    4.3   Second Variation of Energy 73
    4.4   Normal Deformations of Liquid Zones 75
           4.4.1   Instabilities of Periodic Surfaces 75
           4.4.2   Normal Deformations of a Circular Cylinder 76
           4.4.3   The Symmetric Instability of the Catenoid 77
    4.5   Nonaxisymmetric Instabilities 79
           4.5.1   Lateral Deformations of the Center Line 79
           4.5.2   Liquid Rings 81
    4.6   The Minimum-Volume Condition 83
    4.7   Linear Stability Analysis 85
    References 87

5. Axisymmetric Liquid Columns at Rest and Under Rotation 89
    5.1   Introduction 89
    5.2   The Normal Deformations 90
           5.2.1   The Symmetric Mode $D\{2,0\}$ 92
           5.2.2   The Antimetric Mode $D\{1,0\}$ 94
           5.2.3   The Lateral Instability $D\{0,1\}$ 96
           5.2.4   Stability of a Liquid Ring 98
    5.3   Nearly Cylindrical Surfaces 101
           5.3.1   Fourier Expansion of an Axisymmetric Surface 101
           5.3.2   The Symmetric Instability $D\{2,0\}$ 102
           5.3.3   The Antimetric Instability $D\{1,0\}$ 102
           5.3.4   The Lateral Mode $D\{0,1\}$ 103
           5.3.5   Nonzero Bond Number 104
    5.4   Rotating Free Drops 106
           5.4.1   Motivation 106
           5.4.2   Shape of Rotating Drops 107
           5.4.3   Stability 110
           5.4.4   Conservation of Angular Momentum 113
           5.4.5   Finite-Element Analysis 114
    References 117

Contents

6. **Liquid Zones** 119
   - 6.1 Liquid Bridges Between Parallel Plates 119
     - 6.1.1 Introduction 119
     - 6.1.2 Branches of Solutions of the Capillary Equation ... 120
     - 6.1.3 Properties of the Inflection Point 123
     - 6.1.4 The Instability Due to the Bifurcation (Due to $D\{1,0\}$) 125
     - 6.1.5 The Instability Due to the Minimum Volume (Due to $D\{2,0\}$) 127
     - 6.1.6 Differing Contact Angles 129
     - 6.1.7 Gravity 129
     - 6.1.8 Key Points 134
   - 6.2 Double Float Zones 135
     - 6.2.1 Introduction 135
     - 6.2.2 Unduloids and Nodoids 137
     - 6.2.3 Branches of Solutions 138
     - 6.2.4 Results of the Spacelab Experiments 141
     - 6.2.5 The Stability Diagram 143
     - 6.2.6 Key Points 145
   - References 147

7. **Canthotaxis/Wetting Barriers/Pinning Lines** 149
   - 7.1 Introduction 149
   - 7.2 Straight Wetting Barriers 151
     - 7.2.1 The Wetting Tile 151
     - 7.2.2 The Wetting Stripe 153
     - 7.2.3 The Wetting Cross 154
     - 7.2.4 Circular Tubes 155
     - 7.2.5 Large Liquid Volumes 157
   - 7.3 Liquid Surfaces in Wedges 158
   - 7.4 Taylor Expansions at Small Radii 162
     - 7.4.1 Alternative Winding Functions 162
   - 7.5 Liquid Surfaces in Square Cylinders, $\cos\gamma_1 + \cos\gamma_2 = 0$ 164
   - 7.6 Towards Modeling Canthotaxis 169
     - 7.6.1 Helicoid and Catenoid 169
     - 7.6.2 Winding Rates $[\partial z(\varphi)/\partial\varphi]_{r=0} \propto [\cos(s\varphi)]^k$ 170
     - 7.6.3 Winding Rate of Infinity 171
     - 7.6.4 Circular Tube with Complementary Contact Angles . 172
   - References 177

8. **Cylindrical Containers** 179
   - 8.1 Introduction 179
     - 8.1.1 Fields of Application 179
     - 8.1.2 Liquids in Edges 180

	8.2	The Integral Theorem for Cylindrical Vessels 182	
		8.2.1 Application of Divergence Theorem 182	
		8.2.2 Minimization of Energy with Respect to Height 183	
		8.2.3 Evaluation of Wedge Contributions 185	
	8.3	Examples 186	
		8.3.1 Ice Cream Cone 186	
		8.3.2 Rhombic Cylinder 188	
		8.3.3 Regular Polygon 190	
		8.3.4 Liquid in a Rotating Wedge 192	
		8.3.5 No Wetting of Wedge 193	
		8.3.6 Liquid Volume Pressed into a Wedge 195	
	8.4	Stability of Convex Cylindrical Surfaces 199	
		8.4.1 Longitudinal Normal Deformations 199	
		8.4.2 Axially Periodic Meniscus Shapes 200	
		8.4.3 Adjustment to Fit Solid Edges 201	
		8.4.4 Volume and Energy 203	
		8.4.5 Rotating Wedges 205	
	8.5	The MAXUS Experiment DYLCO 205	
	References 211	

9. Liquid Surfaces in Polyhedral Containers 213
   9.1 Spherical Surfaces at Edges and Corners 213
       9.1.1 Nonwetting Drops 213
       9.1.2 Drops in Planar Wedges........................... 214
       9.1.3 Drops in Spherical Wedges........................ 216
       9.1.4 Liquid Drops in a Tripod 217
       9.1.5 Regular N-Pods................................. 217
   9.2 Transition Between the Corner and the Wedge 222
       9.2.1 Liquid Volumes in Polyhedra...................... 222
       9.2.2 Exponential Piling-Up in Corners.................. 223
       9.2.3 Numerical Calculation of Corner Volume 225
       9.2.4 Similarity of Corner Volumes 228
       9.2.5 Finite Wedge Length............................. 229
       9.2.6 Accuracy of the Present Approach 231
       9.2.7 Prospects.. 232
   References 233

10. Playing with Stability 235
   10.1 Proboscides 235
       10.1.1 Finite Rhombic Prisms 235
       10.1.2 Canonical Proboscides 238
       10.1.3 Interface Configuration Experiment 241
   10.2 Exotic Containers 246
       10.2.1 Circular Tubes with Unusual Properties 246
       10.2.2 Adjustment of Container Shape 249
       10.2.3 Integration of Container Shape 251

	10.2.4	Mismatch of Volume and/or Contact Angle 253
	10.2.5	Residual Gravity 254
	10.2.6	Drop Tower Tests 256

References 258

11. Liquid Penetration into Tubes and Wedges 259

11.1 About the Momentum, or Navier–Stokes, Equation ... 259
11.2 Penetration into Capillaries 261
    11.2.1 Cylindrical Vessels 261
    11.2.2 Liquid Rise in Capillaries 263
    11.2.3 Liquid Penetration into Wedges 264
    11.2.4 Similarity Solutions for Long Times 266
    11.2.5 Numerical Solution 269
11.3 Dynamics of Liquids in Edges and Corners 272
    11.3.1 The DYLCO Experimental Module 272
    11.3.2 Drop Towers Tests for DYLCO 273
    11.3.3 Conduct of the IML-2 Experiment 274
    11.3.4 Results of the DYLCO IML-2 Experiment 276
11.4 The Geometric Friction Coefficient Φ 278
    11.4.1 Flow in Rectangular Tubes 278
    11.4.2 Flow in Parallelograms 283
References 285

12. Oscillations of Liquid Columns 287

12.1 Introduction 287
12.2 Theory 288
    12.2.1 Infinite Liquid Columns 290
    12.2.2 The Free Fluid Surface 291
    12.2.3 Natural Frequencies 292
    12.2.4 Finite Liquid Columns 292
    12.2.5 Axially Damped Oscillations 294
    12.2.6 Symmetric and Antimetric Oscillations 295
    12.2.7 Resonance Detection and Flow Patterns 297
12.3 Experiments 302
    12.3.1 Short Liquid Columns 302
    12.3.2 Plateau Simulation 303
    12.3.3 Automatic Resonance Detection 305
    12.3.4 The LICOR Runs 308
12.4 Lateral Oscillations of Liquid Bridges 312
    12.4.1 Damped Harmonic Oscillations 312
    12.4.2 Periodic Lateral Deformations 314
    12.4.3 Coupled Damped Oscillations 316
References 321

13. Microgravity Experiments in Sounding Rockets, Spacelab and EURECA 323
   13.1   TEXUS 1–39 323
   13.2   MAXUS 1–4.. 336
   13.3   MiniTEXUS 1–6 337
   13.4   MASER 1–8 338
   13.5   SPAR I–X... 340
   13.6   TR-IA 1–7 343
   13.7   Skylab, May 1973 345
   13.8   Apollo–Soyuz Test Project (ASTP) 346
   13.9   Spacelab 1 (STS-9) 347
   13.10 Spacelab 3 (STS-51B) 348
   13.11 Spacelab D-1 (STS-61A) 349
   13.12 Spacelab D-2 (STS-55) 351
   13.13 IML-1 (STS-42) 353
   13.14 Spacelab J (STS-47) 354
   13.15 IML-2 (STS-65) 355
   13.16 EURECA 357
   13.17 MIR and FOTON 357
   Bibliography 358

Subject Index 361

Notation

$\tilde{}$	superscript denoting dimensionless quantities
A, A_1, A_s	surface area
$A_c = A/(2\pi RL)$	area/area of cylindrical column
A_H	Hamaker constant
$Bo = \Delta\rho g L^2/\sigma$	Bond number, also defined by means of disk radius R
$C, c, c_1, c_2, \ldots$	constants
$Ca = v\eta/\sigma$	capillary number
const	constant without further definition
$D\{1,0\}$	antimetric longitudinal deformation with one node
$D\{2,0\}$	symmetric longitudinal deformation with two nodes
$D\{m,0\}$	longitudinal normal deformation with m nodes
$D\{0,n\}$	lateral normal deformation with $2n$ nodes according to $\cos(n\varphi)$
d	waviness parameter
E_{kin}	kinetic energy
E_{pot}	potential energy
E_{rot}	energy of rotation
E_{sur}	surface energy
E_{vcs}	viscous energy loss
I	moment of inertia
L	length of column (spacing of disks)
$L_c = L/2R$	aspect ratio = length of column/disk diameter
$\mathbf{n}$	normal to fluid surface
$l = [\sigma/(g\rho)]^{0.5}$	capillary length
$\log(z)$	natural logarithm of z
$Oh = [\rho\nu^2/(\sigma R)]^{0.5}$	Ohnesorge number
p	pressure
$\widehat{p} = p/\sigma$	curvature
$P_c = pR/\sigma$	dimensionless pressure or curvature
q	radial wavenumber
R	(or R_1 and R_2) radius of disks
R_c	neck or belly radius
$R_{\text{crn}}, V_{\text{crn}}, A_{\text{crn}}$	parameters related to a liquid surface in a corner
$Rn = \Delta\rho\omega^2 R^3/\sigma$	rotation number

XVIII   Notation

$R_{\text{wdg}}, V_{\text{wdg}}, A_{\text{wdg}}$	parameters related to cylindrical liquid surface in a wedge
s	azimuthal wavenumber
Sc	Schmidt number
T	temperature
t	time
U_0, U_1, U_2	potentials
$\boldsymbol{U}(\boldsymbol{r})$	vector potential of flow
$\boldsymbol{u}(\boldsymbol{r}) = (u_x, u_y, u_z)$ $= (u_r, u_\varphi, u_z)$	flow velocity
V	liquid volume
$\boldsymbol{V}(\boldsymbol{r})$	vorticity
$V_{\text{c}} = V/(\pi R^2 L)$	volume/(volume of cylindrical column)
v	velocity
$We = \rho \omega^2 R^3 / \sigma$	Weber number
α	half dihedral angle, half wedge angle
γ	(or $\gamma_1, \gamma_2, \gamma_i$) contact angle
δ	$= \pi/2 - \alpha - \gamma$, arc of liquid meniscus in wedge
$\Delta \rho$	density difference
η	dynamic viscosity
κ_1, κ_2	principal curvatures
λ_1, λ_2	Lagrange parameters
ν	kinematic viscosity
ρ	density
Σ	periphery of cylindrical vessel
σ	(or $\sigma_1, \sigma_2, \sigma_i$) surface or interface tension
Φ	geometric friction parameter
Ψ	(or Ψ_1 and Ψ_2) angle of inclination at supporting disks
$\Psi_{\text{c}} = \Psi/(\pi/2)$	reduced angle of inclination at supporting disks
Ω	cross section of cylindrical vessel
ω	circular frequency of oscillation or of rotation
$\tilde{\omega} = (\rho \omega^2 R^3 / \sigma)^{0.5}$	dimensionless circular frequency of oscillation or of rotation

1. Introduction

The advantages of performing experiments under conditions of weightlessness are discussed in this chapter; the inherent handicaps are not ignored. In any experimental study or procedure in materials science or in the life sciences which may gain from performance under microgravity conditions, at least one fluid phase is involved. A fluid means a liquid or a gas. The most obvious effects of microgravity are the constancy of fluid-static pressure, the absence of free convection and the absence of sedimentation.

The relevant missions of Skylab and of the Space Shuttle (in particular, Spacelab), and experiments in the space station MIR, and also in the national and international sounding-rocket programs SPAR, TEXUS, MASER, TR-1A, etc. are quoted. The essential selection criteria are microgravity quality, microgravity time, cost, availability and repeatability. Easy access and the use of laboratory equipment are limited to drop towers and parabolic flights of aircraft.

1.1 Space Missions

For about a century, interest in the shape of liquid surfaces in differently shaped containers at low gravity was mainly academic. Since zero gravity considerably simplifies the relevant equations, these surfaces have been favored topics of mathematical and theoretical investigations. The everyday experience that liquids rest at the bottom of their containers is a rare exception under low-gravity conditions. Limited modeling of such surfaces has been possible by means of soap bubbles. On those, gravity has only little impact, and quite a number of papers on their fascinating shapes have been published [Boys 1959]. Explicit applications of microgravity conditions, however, have been rare. We all know that water running from a tap forms a jet, which becomes thinner and thinner and breaks into droplets. The droplets, after some initial oscillations, form spheres [McDonald 1954], as do raindrops and their solidified equivalents, hailstones. The same effect applies to lead drops. In the late 18th and in the 19th century, free fall of lead drops in drop towers was used for producing lead pellets all over the world [Minchinton 1995, Kuschnigg & Sprenger 1996, Sprenger 1996].

The attitude towards liquid surfaces at zero gravity changed strongly in the early 1970s when, following the race to the moon, space flights were made available for scientific and technical purposes. It was obvious from the beginning that fluid physics would become a central topic. In the solid phase gravity gives rise to marginal effects only. It is always in the fluid phase that reducing gravity causes numerous changes. Following various exploratory activities in the Apollo program, systematic research into materials and fluid sciences started in Skylab during 1973–74. In addition to an intensive medical program, the astronauts performed 14 experiments on crystal growth and composite materials in two furnace facilities, the "Material Processing Facility" and the "Multipurpose Furnace System". Also, nine "Science Demonstration Experiments" were largely created and run from onboard components and liquids such as water, salt water and soapy water [Snyder 1976, Vreeburg 1986].

Table 1.1. Active Skylab periods

	Crew	Arrival	Departure
Skylab	Unmanned	14 May 73	11 Jul. 79
Skylab 2	Conrad, Kerwin, Weitz	25 May 73	22 Jun. 73
Skylab 3	Bean, Garriot, Lousma	28 Jul. 73	25 Sep. 73
Skylab 4	Gibson, Pogue	16 Nov. 73	08 Feb. 74

During the operational time of Skylab, the European Spacelab and its facilities were planned and designed for flight on one of the early Space Shuttle missions. The respective Memorandum of Understanding between NASA and ESRO was signed in Washington on 24 September 1973. The time interval until its launch was bridged by sounding-rocket programs, the SPAR program in the United States (*Sp*ace *P*rocessing *A*pplication *R*ockets, first flight 11 Dec. 1975, 10 flights until June 1983) and the TEXUS program in Europe (*T*echnologische *Ex*perimente *u*nter *S*chwerelosigkeit, first flight 13 Dec. 1977, 39 flights until May 2001).

The European Spacelab had its maiden flight on the Space Shuttle Columbia on 28 November 1983. Many further Spacelab missions with considerably improved equipment and control of experiments from the ground by methods suchs as telecommanding followed. On the other hand, the continuity of these flights was severely interrupted by the Challenger disaster on 29 January 1986. The last but one flight of the Space Shuttle had carried the German Spacelab mission D-1. The Challenger disaster, in addition to causing considerable delay to all further Spacelab missions, also had a strong impact on the further planning. The ambitious plans of developing a smaller, independent European Shuttle Hermes, which was suggested to be launched by Ariane 5 and would serve a Columbus Freeflying Laboratory, had to be abandoned when the Hermes payload turned negative owing to the additional

Table 1.2. Major microgravity flights

STS[a]		Shuttle	Launch	Landing	Payload Specialists
09	Spacelab 1	Columbia	28 Nov 83	08 Dec 83	Merbold, Lichtenberg
51B	Spacelab 3	Challenger	29 Apr 85	06 May 85	van den Berg, Wang
51F	Spacelab 2	Challenger	29 Jul 85	06 Aug 85	Acton, Bartoe
61A	D-1	Challenger	30 Oct 85	06 Nov 85	Furrer, Messerschmid, Ockels
38	Astro-1	Columbia	02 Dec 90	11 Dec 90	Durrance, Parise
40	SLS-1	Columbia	05 Jun 91	14 Jun 91	Gaffney, Fulford
42	IML-1	Discovery	22 Jan 92	30 Jan 92	Bondar, Merbold
	MIR 92		17 Mar 92	25 Mar 92	
50	USML-1	Columbia	25 Jun 92	09 Jul 92	deLucas, Trinh
46	EURECA	Atlantis	31 Jul 92	Deploy	Malerba
47	Spacelab J	Endeavour	12 Sep 92	20 Sep 92	Jemison, Mohri
52	USMP-1	Columbia	22 Oct 92	01 Nov 92	MacLean
55	D-2	Columbia	26 Apr 93	06 May 93	Schlegel, Walter
57	EURECA	Endeavour	Retrieval	01 Jul 93	Also: Spacehab-1
58	SLS-2	Columbia	18 Oct 93	01 Nov 93	Fettman
	MIR 92E		10 Oct 93	04 Jan 94	
	FOTON 9		10 Dec 93	16 days	
60	Spacehab-2	Discovery	03 Feb 94	11 Feb 94	
62	USMP-2	Columbia	04 Mar 94	18 Mar 94	
65	IML-2	Columbia	08 Jul 94	23 Jul 94	Mukai, Thomas
	EUROMIR 94		Aug 1994	30 days	
	FOTON 10		Oct 94	16 days	
66	Atlas-3	Atlantis	03 Nov 94	14 Nov 94	
	EUROMIR 95		1995	135 days	
63	Spacehab-3	Discovery	03 Feb 95	11 Feb 95	
73	USML-2	Columbia	20 Oct 95	05 Nov 95	Leslie, Sacco
75	USMP-3	Columbia	22 Feb 96	09 Mar 96	Guidoni
77	Spacehab-4	Endeavour	19 May 96	29 May 96	
78	LMS	Columbia	20 Jun 96	07 Jul 96	Favier, Thirsk
83	MSL-1	Columbia	04 Apr 97	08 Apr 97	Crouch, Linteris
94	MSL-1A	Columbia	01 Jul 97	17 Jul 97	Crouch, Linteris
87	USMP-4	Columbia	19 Nov 97	05 Dec 97	Kadenyuk
90	Neurolab	Columbia	17 Apr 98	03 May 98	Buckey, Pawelczyk
95	Spacehab-5	Discovery	29 Oct 98	07 Nov 98	Mukai, Glenn

[a]STS = *S*pace *T*ransportation *S*ystem

safety measures required. The name Columbus Laboratory had been selected because it was intended to be launched on the occasion of the semimillennium of Columbus's landing in America in 1992!

Parallel to the Spacelab flights, the sounding-rocket programs were continued and new ones were initiated: the ESA programs MASER (*Ma*terial *S*cience *E*xperiments *R*ockets, first flight 19 Mar. 1987, 8 flights until May 1999) and MAXUS (an acronym formed from MASER and TEXUS, first flight 8 May 1991, 4 flights until April 2001) and the Japanese program TR-IA (first flight 19 Sep 1991, 7 flights until summer 1998). Research will

4   1. Introduction

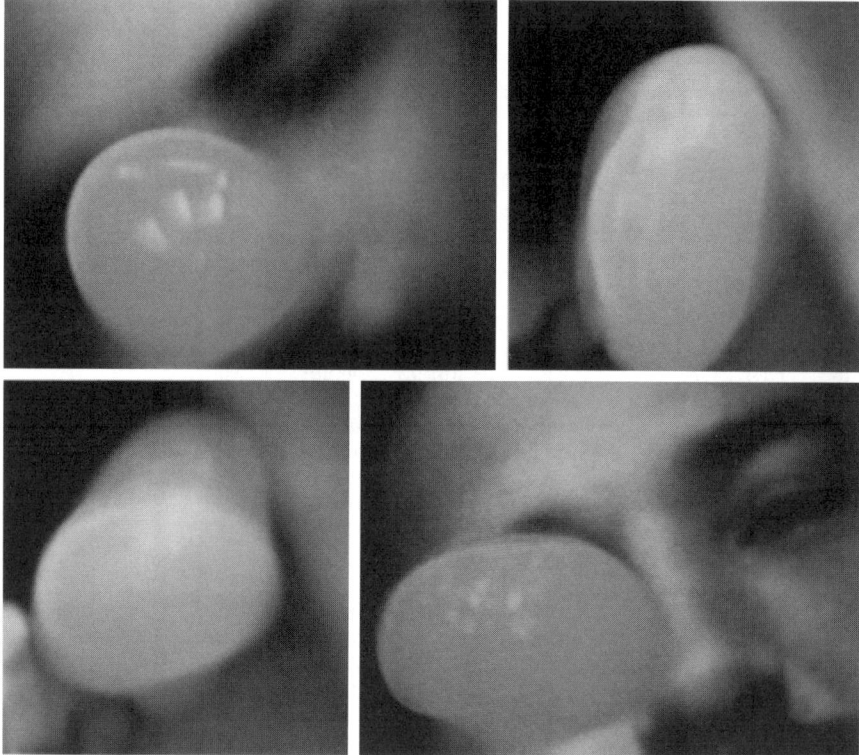

Fig. 1.1. Wubbo Ockels playing with orange juice in Spacelab D-1

be continued in the forthcoming International Space Station (ISS). One must not forget the short-time microgravity environments available at drop towers and during parabolic maneuvers of aircraft, which are heavily used for original research and for the preparation of longer and more expensive missions as well.

A microgravity environment can be achieved, with strongly differing available time intervals and costs, on any of the following platforms:

- drop tower (Cleveland, OH; CNES, Grenoble; ZARM, Bremen; JAMIC, Sapporo)
- aircraft (KC-135, Caravelle, Airbus, MU-300 (automatic))
- sounding rockets (SPAR, TEXUS, MASER, MAXUS, TR-IA)
- satellites (EURECA)
- orbiting space platforms (ISS Freedom).

It also may be simulated, in a few limited cases and to a certain extent, in ground-based laboratories.

Whether an experiment is best performed in the Spacelab or the International Space Station, in a sounding rocket, in a drop tower or during a

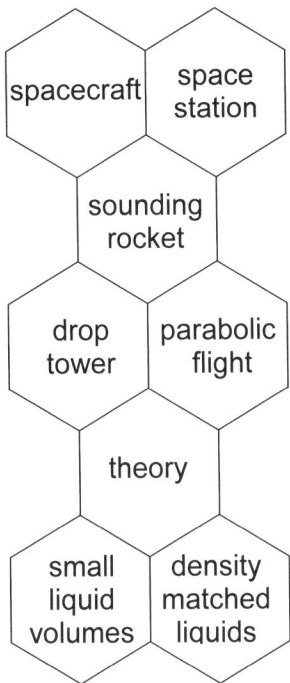

Fig. 1.2. Alternative methods of achieving low Bond numbers

parabolic flight of an aircraft depends on time, required microgravity quality and cost. Drop towers and parabolic flights are low in cost, they provide microgravity times up to 10 and 30 seconds, respectively. The microgravity quality in parabolic flights is not the very best. It depends on the skill and experience of the pilot. The flights, however, are cheap and help effectively in eliminating basic mistakes. Usually they serve more for testing ideas than for quantitative measurements. Sounding rockets allow between six and fifteen minutes of microgravity; they are flown regularly. The Space Station provides plenty of time, but brings along high costs, and regular access for experimenters is still difficult and needs plenty of time and effort.

The attainable g-levels are influenced by absolute rotation and angular accelerations of the space platform, by external forces acting on it and, in particular, by non-uniformities of these forces. Optimum conditions are achieved if

- there are no maneuvers of the spacecraft
- the platform does not rotate
- the experimental apparatus is located as close as possible to the trajectory of the platform's center of mass.

Experimenters appreciate the skill and flexibility of the crew, on the one hand, and oppose their movements, on the other hand.

1.2 Interdisciplinary Stimuli

From the very beginning, research under microgravity conditions has been strongly multidisciplinary. In any experimental study or procedure in materials science or in the life sciences which may gain from performance under microgravity conditions, at least one fluid is involved. In the solid state, gravity may give rise to marginal effects only. Thus, in all disciplines research focuses on fluid-dynamic effects, including temperature and mass diffusion, phase diagrams and nucleation. And clearly, for fluid physicists microgravity means an excellent chance to reconsider many of their basic concepts.

The most obvious effects are the constancy of fluid-static pressure, the absence of free convection and the absence of sedimentation. The constancy of fluid-static pressure allows the establishment of large liquid surfaces. The absence of free convection affects heat and mass transport and enables accurate investigations into diffusion. The absence of sedimentation strongly influences the separation of monotectic alloys.

Crystal growers follow two basic concepts, namely

- control of convection
- growth without the use of cartridges.

The aim is to grow large, high-quality crystals, which usually requires high homogeneity. In addition, the crystals should grow fast. Growth may be from vapor, from a melt or from a solution, i.e. from a fluid in some way. Growth requires a temperature gradient, the crystal grows at the melting temperature corresponding to the local concentration. The fluid phase, hence, is susceptible to convection. Convection must be perfectly steady in order not to provoke local temperature and/or concentration fluctuations and thus cause crystal inhomogeneities. This becomes more critical the faster the crystal is grown, i.e. the stronger the temperature gradient applied. These obstacles may be reduced by applying zero gravity.

A second severe cause of inhomogeneity is nucleation of secondary crystals at the cartridge wall. In order to let such secondary crystals grow outwards, growth with a convex solidification front is preferred. The alternative of growing without a cartridge by using a freely floating zone once more requires microgravity. This allows one to establish large, freely floating liquid zones. The temperature gradient along a floating surface, convection (Marangoni convection). The surface tension between a liquid and a gas and the interface tension between two liquids usually depend on temperature. The regions exhibiting high interface tension pull on the regions exhibiting low interface tension, such that an interface convection, usually directed from hot (= low interface tension) to cold (= high interface tension), results. Regarding this effect, the experimenter indeed falls between two stools.

A further field of strong activity is the development of new materials, in particular finely dispersed samples of monotectic alloys. Monotectic alloys exhibit a miscibility gap, i.e. the melts of the two metals separate during

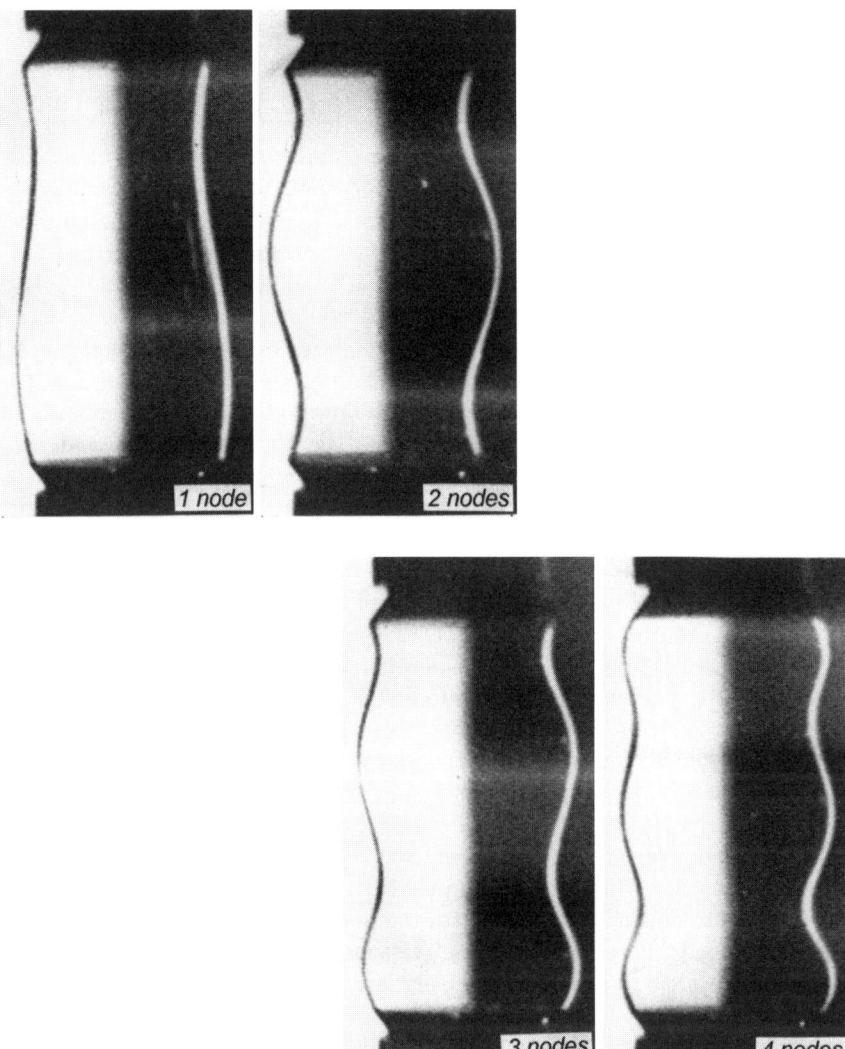

Fig. 1.3. Resonance modes of a cylindrical liquid column observed in Plateau simulation

cooling down before the solidification temperature is reached. There exist several hundreds of such monotectic metal pairs, and many of them hold out prospects of practical application, for instance their use as bearings for car engines. In this respect, compounds based on aluminum containing lead inclusions are considered most appropriate, since the former metal is light, whereas the latter metal is slippery. If one attempts to produce such dispersions on the ground the precipitates of the heavier component are going to sink to the bottom prior to solidification. In the first samples of aluminum/indium processed in space in the SPAR II and SPAR V missions [Ahlborn & Löhberg 1976, 1984, Gelles & Markworth 1977, 1980], indium has been found on the outside and aluminum on the inside, which is not considered a great success.

A large number of mechanisms contributing to that separation and the final species distribution under low-gravity conditions have been identified. The most important are the wetting of the crucible by the two components and thermocapillary migration of the precipitates. Indium wets the alumina crucible used in the corresponding studies better than aluminum. By changing the crucible material to silicon carbide, it is possible to make aluminum the outside component [Potard 1984]. And cooling is necessarily initiated from the outside, such that the resulting temperature gradient brings about an outward migration of the precipitates. Numerous further experiments in sounding rockets and in the Spacelab have followed [ESA SP-219 1984, Langbein 1990, Ratke & Diefenbach 1995]. From these experiments, researchers have learned to handle wetting conditions and to exploit directed thermocapillary migration of droplets by optimizing the temperature gradients, sample thickness and cooling time.

Another promising research topic is the engulfment of solid particles by a solidification front. Engulfment is due to intermolecular interactions. Crystal growers want to eliminate foreign particles, i.e. to clean their crystals by zone refining. In contrast, metallurgists want to produce compound materials, i.e. to keep the foreign particles finely distributed inside the material. Here again the wetting conditions between the particles, the bulk material and the crucible are crucial. If the interface energy of the particles is lower in the melt (or liquid) than in the solid, $\sigma_{pl} < \sigma_{ps}$, they will try to stay outside the solid. In the opposite case, for $\sigma_{pl} > \sigma_{ps}$, they are going to be integrated. The reason is the short-ranged van der Waals repulsion, which pushes the particles ahead of the solidification front. Successful pushing, however, requires that the melt is continuously flowing into the narrowing gap between the particle and the solid. This needs more energy against viscous friction of the melt the faster the solid grows, the higher the viscosity η of the melt and the smaller the spacing d between the particles and the growth front; $v\eta < f(d/R)(\sigma_{ps} - \sigma_{pl})$ (v = growth rate, R = particle radius). All particles are more easily captured at high growth rates v, and large particles are more easily captured than small particles. There exists an equilibrium spacing $d_0(v, R)$ [Langbein 1981, Körber 1988, Pötschke & Rogge 1989]. The theoretical relations between van der Waals forces, interface tension and particle repulsion certainly deserve further efforts. Quantitative experiments suffer particularly from the lack of sphericity of the tiny particles required, from the difficulties in measuring their interface tension and, very often, from a residual gas content, for which the particles constitute nucleation centers.

Finally, let us stress the effects of wetting and convection on measurements of diffusion coefficients. Diffusion is very slow and therefore particularly susceptible to other causes of mass transport in fluids. Various experiments on self-diffusion and interdiffusion in liquid metals have been performed during the FSLP, D-1 and D-2 Spacelab missions [Frohberg et al. 1984, 1987a,b]. The diffusion coefficients found are generally considerably lower than those obtained on the ground. At the same time, their accuracy is much higher. This is clearly related to the absence of free convection, which even in thin capillaries cannot be fully suppressed. Diffusion coefficients D usually have an order of magnitude of 10^{-3} mm^2/s, whereas mass transport due to convection is governed by the liquid's kinematic viscosity ν, of the order of magnitude of 10 mm^2/s. The ratio of these experimental parameters, the Schmidt number $Sc = \nu/D$, quite generally has an order of magnitude of 10^4.

The corresponding microgravity experiments, which cover a wide temperature region, indicate that the diffusion coefficients do not follow the anticipated exponential law (Arrhenius law), but exhibit a T^2 dependence on temperature [Frohberg et al. 1995a,b]. This allows discrimination between the different theories in existence. However, the same authors reported quite different results in the case of monotectic alloys. In interdiffusion experiments with the systems Al/In and Al/Bi, complete mixing was obtained during the Spacelab D-2 mission, whereas the ground reference experiments showed partial mixing only. A stronger mixing under microgravity conditions than on the ground has to be ascribed to an additional convective effect. In the case of monotectic alloys this effect is most likely due to the changing wetting conditions. At compositions near the critical point, where the two consolute melts differ in concentration only slightly, one of the melts may easily creep along the cartridge wall, thereby fully enclosing the other.

Microgravity offers advantages, and disadvantages as well. Failure often is just due to lack of experience. In the preparation of one of the first exploratory experiments on reduced convection under low gravity it was overlooked that convection in the light bulb used for illumination would also be reduced. The bulb burned out rapidly.

1.3 Problems of Fluid Physics

The topics presented below roughly outline the variety of research areas in materials science in which fluid physics is being investigated in a microgravity environment. Many more questions arise in physical chemistry and the life sciences. One must not forget other problems of fluid physics such as fluid handling and spacecraft stability. The term "fluid" in this context comprises liquids and gases, for example water, a metallic melt used for production of composite materials, a drop created to study its oscillations or an unexpected gas bubble hampering the growth of a single crystal.

Putting all the research areas together, we can identify the following problems of fluid physics:

- to study the behavior and handling of fluids
- basic research on fluids
- to test computer codes for the shape, stability and oscillations
- to learn about freshly formed interfaces
- to learn about thin liquid layers
- to study edge effects
- to investigate wetting and spreading of liquids
- to study coalescence
- to observe thermal and solutal interface convection
- to investigate oscillatory and turbulent flows
- to study heat and mass transfer
- to accurately measure diffusion coefficients
- to perform transparent model experiments
- to assist in the production of finely dispersed emulsions of metals
- to assist in experiments on crystal growth
- to study microconvection during cellular growth, dendritic growth, etc.
- to assist in experiments in the life sciences
- to analyse g-level tolerability.

It is always the fluid phase on which reduced gravity has a strong impact, i.e. where the phenomena caused by gravity are strongly diminished. The main effects regularly mentioned in this context are

- the constancy of fluid-static pressure
- the prevalence of wetting conditions
- the absence of free convection
- the absence of sedimentation.

The effects of pressure are very well known. The fluid-static (often referred to as "hydrostatic") pressure is higher by about 0.17 bar in our feet than in our head. A diver is aware that the hydrostatic pressure increases by 1 bar every 10 m. The altimeters used in aircraft and by mountain climbers likewise rely on such a pressure difference. This pressure drop is greatly reduced in a microgravity environment. Pressure differences only arise because of residual accelerations, due to so-called g-jitters. The present requirements on the tolerable residual accelerations in the International Space Station (ISS), namely less than $10^{-6}\,g$ at frequencies below 1 Hz and a quadratic increase with frequency above 1 Hz, were originally formulated in the field of fluid physics.

It is everyday experience that liquids rest at the bottom of the container used. This situation is a rare exception under low-gravity conditions. The shape of a liquid surface is basically determined by the container shape, the liquid volume and the liquid's wettability with the container material. Usually, special measures must be taken to keep a liquid resting on the bottom. Knowledge of all liquid configurations which may arise is hence a prerequisite for performing scientific research under microgravity conditions.

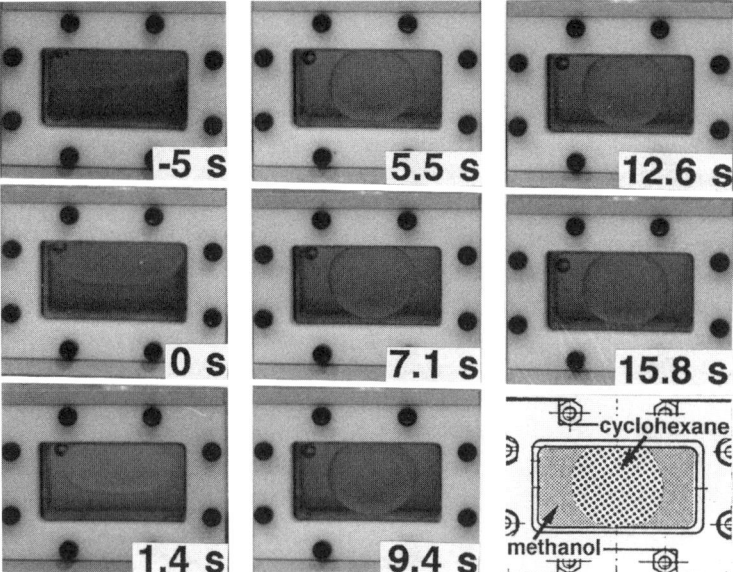

Fig. 1.4. Spreading of methanol around cyclohexane during a parabolic flight of the KC-135 aircraft. The size of the fluid cell used is 40 mm × 20 mm × 10 mm. The time inset shows that spreading takes only about five seconds. A spherical cyclohexane volume touching the front and rear windows is formed [Langbein 1990, 1993]

The effects of free convection are obvious as well. The updraft which a glider exploits when flying long distances is caused by the atmosphere being heated differently over light ground and dark ground. Air, as do all other gases, expands and thus becomes lighter with increasing temperature, such that the more strongly heated warm air starts rising and the less strongly heated cold air starts sinking. Similar effects arise if a liquid in a closed container is heated differently at different positions. A central heating system works without a circulation pump if the heater is placed in the basement; hot water rises anyway.

Sedimentation has the same origin as convection, i.e. buoyancy. The term "sedimentation" is used for the migration of solid or fluid particles, whereas the term "convection" is applied to the motion of different regions of the same fluid. A stone falls quickly, small dirt particles sediment only slowly and a bubble rises rather than sinks. Sedimentation depends on the density and shape of the particles considered and on the density and viscosity of the fluid in which they are embedded. In the absence of gravity, no differences in weight exist, i.e. in a space laboratory an astronaut experiences a neutral equilibrium. It is indeed more correct to speak about near-weightlessness than about microgravity.

12   1. Introduction

Fluid physics differs from other fields of research under microgravity by having numerous exact equations at hand. Convection may be calculated from the so-called momentum equation relating flow velocity, external forces, local pressure differences and viscous friction of the fluid. This equation represents the extension of Newton's law of acceleration to fluids. And there are well-known equations describing energy and mass transport. An extensive literature on analytical and numerical solutions of these equations is available. Nevertheless, a lot of questions remain unsolved. This is due to the fact that these equations include various nonlinear terms, which allow unique solutions at low flow rates and low transport rates only. At high flow rates, when convection strongly affects the force driving it, strongly different flow patterns may come into existence. These patterns depend on the initial conditions and on the method by which they were established, and often are not reproducible at all. Qualitative investigations may be based on dimensionless numbers which appear in the relevant differential equations when all physical parameters are normalized by means of characteristic experimental parameters.

By using large fluid volumes and surfaces, it is possible to study

- volume and surface oscillations
- surface stability
- nonlinear oscillatory effects
- wetting dynamics and advancing and receding contact angles
- convective stability
- thermocapillary and solutocapillary convection (Marangoni convection).

Water in a glass usually exhibits a flat surface. If water wets the container, the surface is slightly bent upward at the periphery. On the other hand, if one cautiously overfills the glass, one observes a liquid cap which may be slightly bent outwards at the periphery. A single additional drop of water breaks the situation, such that water spills down outside the glass. If, instead, a glass coated with Teflon is used, the surface bends down slightly inside the glass. And, as already mentioned, water running from a tap forms a jet, which becomes thinner and thinner and breaks into droplets. The droplets, after some initial oscillations, form spheres. If one reduces the flow rate, the droplets form earlier. On the other hand, if honey is running down from a spoon, one observes little or no tendency to breakage. Honey, in contrast to water, has a high viscosity.

On a NASA website one finds the following:

Fluid physics is vital to understanding, controlling, and improving all of our industrial and natural processes. The engines used to propel a car or an airplane, the shape of the wings of an airplane that allow it to fly, the operations of boilers that generate steam used to produce over 90% of the world's electric power, the understanding of how cholesterol is transported in our bloodstream and effects heart disease, and how pollutants are transported

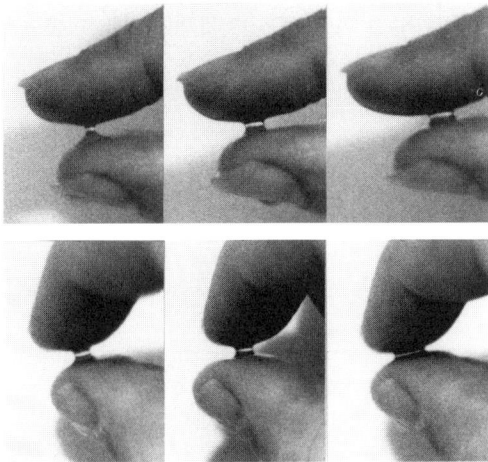

Fig. 1.5. Water bridges between fingers

and dispersed in air and water are just a few examples of how fluid physics affects our everyday life and forms the very basis for an industrial society.

1.4 Zero Mass Acceleration or Weightlessness

If one stands on a balance on a catapult and is thrown upwards, one experiences a much higher weight than one is normally used to. Very likely one is forced onto one's knees. However, as soon as one has left the catapult, one floats upwards together with the balance, without exerting any force on the latter. The balance shows a weight of zero! This state is maintained during the free rise, at the peak of the free flight and during the subsequent free fall. Then one must hope to be caught gently, since otherwise the impact may cost several bones. Mountain climbers keep saying that there is no problem with falling down a steep mountainside. The real problem is the final impact. The same is true for bungee jumping. It is said to be pure fun (if the rope is not too long).

Weightlessness results if all masses follow the forces acting on them in like manner, such that in a coordinate system moving with the masses one gets the impression of zero acceleration. The relevant forces are, primarily, gravity and centrifugal force. (An equivalent electrostatic attraction, for example, requires homogeneous charging of all masses.) In a spacecraft there exists weightlessness, and astronauts or cosmonauts can float around, since the spacecraft, the cosmonauts and all the equipment follow the attraction of the earth and the centrifugal force of the orbit around the earth. These are the same forces as those which keep the planets orbiting around the sun and give rise to Kepler's three famous laws. Attraction by the earth decreases in inverse proportion to the square of the distance r from the earth's center.

Equilibrium between gravity and centrifugal force in an orbit with radius r and circular frequency ω is achieved for

$$g\frac{R^2}{r^2} = \omega^2 r, \tag{1.1}$$

where $g = 9.87\,\text{m}\,\text{s}^{-2}$ is the value of the earth's gravity at the surface and R is the earth's radius, $R = 6366\,\text{km}$. In a low orbit of about $300\,\text{km}$ above the earth's surface, gravity still has 91% of its value on the ground. One finds a circular frequency of $\omega = 1.16 \times 10^{-3}\,\text{s}^{-1}$ and an orbital period of 5407 seconds or 90 minutes.

Weightlessness is experienced not only at the center of mass, but everywhere along the trajectory of the center of mass. The use of the term "microgravity", where in fact one means weightlessness, is, strictly, incorrect. Gravity is just compensated, a balance in which any body force counts. Gravity is used in this context as an unofficial unit of mass acceleration. If an aircraft flies in a sharp curve, the force on the aircraft and the pilot is said to increase to several g. A military pilot must be able to withstand $6g$. If a stone is rotated on a rope, one has to counteract the centrifugal force acting on it; this may amount to several g as well.

In positions closer to the earth than the center of mass, gravity prevails; one experiences attraction. Farther from the earth, centrifugal force prevails; one experiences repulsion. The gravity gradient

$$\frac{\mathrm{d}}{\mathrm{d}r}\left(g\frac{R^2}{r^2} - \omega^2 r\right) = -\left(2g\frac{R^2}{r^3} + \omega^2\right) \approx -3g\frac{R^2}{r^3} \tag{1.2}$$

equals $4 \times 10^{-6}\,\text{s}^{-2}$, which leads to a residual gravity of $4 \times 10^{-7}g$ at a radial distance of $1\,\text{m}$ from the trajectory of the spacecraft's center of mass.

An everyday experience of a gravity gradient is seen in the tides traveling around our planet earth according to the position of the sun and moon. The earth runs in its orbit around the sun under the balance of gravity and centrifugal force. In the trajectory of the earth's center of mass, gravity is fully compensated. In the regions closer to the sun a residual acceleration is directed to the sun; in the regions more distant from the sun a residual acceleration is directed from the sun. In consequence, flood waves arise on both sides of the earth. The earth's orbit, additionally, is affected by the moon. Once again this occurs in such a manner that at the earth's center of mass the moon's gravity is fully compensated. This gives rise to further, even larger flood waves directed towards and away from the moon. The pairs of flood waves caused by the sun and by the moon basically run independently. Each pair of waves is nearly equal in strength on the front side and on the rear side.

1.5 Flight Selection and Simulation

Before undertaking the tasks of performing an experiment in a sounding rocket, in a spacecraft or in a space station, it is most advisable to carefully check all terrestrial alternatives. Accessibility, cost, repeatability and microgravity quality must be taken into account. Easy access and the use of laboratory equipment are limited to drop towers and parabolic flights of aircraft.

In an ESA brochure, *Challenges and Prospectives of Microgravity Research in Space* [Malméjac et al. 1981], we read:

In planning materials science experiments in space one must take into account the constraints associated with the type of space platform to be used, each one having its own level of microgravity, its own limitations as regards time, power, energy, cooling, data acquisition and safety regulations, and its own equipment. This set of imposed conditions will necessarily influence all aspects of the microgravity experiments from the earliest works to the postflight analysis philosophy, the only common point being an increase of both the global complexity and the individual difficulties, even in the case of experiments already perfectly known in their classical terrestrial environment. It is important that any new space investigator be aware of such problems and of their possible solutions before entering this very intriguing but not so simple field.

When one is aware of the main effects of gravity as presented in the preceding section, it follows that one should consider methods of reducing or even completely avoiding them. First of all, one is reminded of the Archimedean principle: Archimedes observed that the weight of a body in water is lower than in air by the weight of water displaced. It is said that he used this principle to confirm to his king that some precious gifts were made from pure gold. If the body immersed has the same density as water, it appears weightless. The body may be made of any solid or fluid and the outer medium may be any fluid not miscible with this solid or fluid. This is one of the aspects of weightlessness: in density-matched, or isopyknic, media there is no sedimentation. The potential energy of an immersed body is independent of its height.

The alternatives to going to a microgravity environment are

- theoretical and numerical modeling: reduces   all gravitational effects
- to apply density-matched liquids: reduces   sedimentation, pressure difference
- to use small liquid volumes: reduces   fluid-static pressure, convection
- to work in a klinostate: reduces   sedimentation.

Common to all methods of simulation is a low Bond number, i.e. a small ratio of the effects of gravity to the effects of surface tension. If an experiment involves two immiscible fluids 1 and 2, which differ in density by $\Delta\rho = \rho_2 - \rho_1$, the fluid-static pressure difference increases from bottom to top in proportion to $\Delta\rho$ and to gravity, $\Delta p = g\,\Delta\rho\,L$, where L is the vertical extension of the experiment. The fluid interface reacts to that pressure difference with an increase in capillary pressure given by the interface tension σ times the local curvature. From dimensional considerations, that curvature may be equated to $1/L$, such that for the ratio of the gravitational to the capillary pressure one obtains the Bond number

$$Bo = \frac{g\,\Delta\rho\,L^2}{\sigma}. \tag{1.3}$$

An alternative way of judging the influence of gravity is to introduce the capillary length l,

$$l = \sqrt{\frac{\sigma}{g\,\Delta\rho}}, \tag{1.4}$$

at which the Bond number equals 1, and to compare l with the experimental dimensions.

Working with density-adjusted fluids was heavily exploited by the blind Belgian physicist Plateau in the mid 19th century [Plateau 1843, 1873]. He showed by means of this technique that the maximum stable length L of a uniform cylinder of liquid, contained at its ends by nonwetting solid surfaces, roughly equals three diameters. In another celebrated experiment conducted by Plateau, a spherical oil drop was rotated in a mixture of water and alcohol that had the same density. The drop first appeared ellipsoidal, then became oblate (as is true for our rotating planet earth) and eventually a ring separated at the equator. Subsequently this ring disintegrated into separate parts [Minkowski 1921, Myshkis et al. 1976]. Actually, the rotating drops must have been slightly more dense or else have had a lower kinematic viscosity, since otherwise no pressure difference along their surface could have resulted during spinning up.

The dominance of surface tension at low heights L can be conveniently demonstrated by taking a small volume of water between the thumb and forefinger and slowly increasing the spacing of the fingers in the vertical direction: the liquid bridge becomes longer and narrower and eventually breaks. The maximum height achievable is roughly three to four millimeters.

The use of a klinostat, which avoids sedimentation by slowly turning an experimental apparatus around a horizontal axis, has been reported with respect to solidification experiments and biological investigations [Otto & Lorenz 1978].

The easily accessible microgravity opportunities, namely drop towers and aircraft, clearly allow short times of microgravity only. On the other hand, the experiments may easily be repeated and modified. The most obvious advantages of drop towers and aircraft maneuvers are

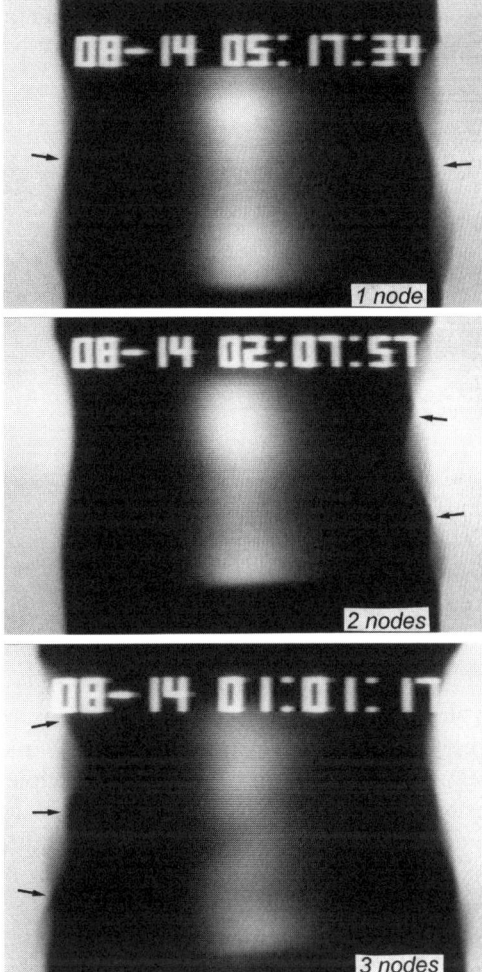

Fig. 1.6. The first three resonance modes of a small water column 3 mm in diameter and 2 mm in height

- easy and frequent accessibility
- easy repetition of experiments
- the ground reference experiment and flight experiment are identical
- slicing of experiments
- experiment history at $1g$ is conserved (drop tower only)
- very high microgravity quality (drop tower only)
- no limitations on weight
- low safety requirements
- low cost.

Drop towers and parabolic flights allow tests of technical or experimental details such as

- where is the liquid?
- is the speed of liquid injection adequate?
- does the bubble detach from the injection needle?
- do the wetting barriers work?
- do the camera, video recorder, etc. work correctly?

Among the successful drop tower experiments, let us mention

- settling and choking of fluid surfaces
- capillary flows in tubes with different cross sections
- thermocapillary drop migration
- instability of thermocapillary flows
- ultrasonic absorption
- nucleation, condensation and bubble growth
- flow patterns in ferromagnetic fluids
- combustion research
- formation of metal foams
- ignition limits and velocities of meagre flames
- interferometric analysis of flame structure
- graviperception and gravikinesis of ciliates.

The most obvious advantage of the drop tower is the direct switch from $1g$ to microgravity. Various long-lasting parts of an experiment may be prepared on top of the tower, under resting conditions before the drop. Examples are experiments on heating, diffusion, etc. This definitely distinguishes the drop tower from all other possibilities for obtaining microgravity. A quite typical example is thermocapillary drop and bubble migration. On top of the drop tower, sufficient time is available for establishing a temperature or concentration field. After release of the capsule, a very short time (< 0.2 seconds) is required for the creation of bubbles, and 4.7 seconds are left for observation of the migration.

Parabolic flights are less accurate by far with respect to the g-level. However, they allow for direct participation and observation by the experimenter. He/she may change the parameters in each subsequent parabola, and there can be thirty or forty of them. This means very high flexibility.

References

1. Ahlborn H, Löhberg K: Ergebnisse von Raketenversuchen zur Entmischung flüssiger Aluminium-Indium-Legierungen. Statusseminar Spacelab-Nutzung des BMFT (1976) Paper 12.1
2. Ahlborn H, Löhberg K: Influences affecting separation of monotectic alloys under microgravity. ESA SP-222 (1984) 55–62

3. Boys CV: Seifenblasen und die Kräfte, die sie formen. (Soap bubbles and the forces which mould them.) Desch-Taschenbuch: Natur und Wissen: 70–73 (1959)
4. Frohberg G, Kraatz KH, Wever H: Selfdiffusion of Sn112 and Sn124 in liquid tin. ESA SP-222 (1984) 201–205
5. Frohberg G, Kraatz KH, Wever H: Atomic diffusion and transport in liquids. In: *Scientific Results of the German Spacelab Mission D1*. P.R. Sahm, R. Jansen, M.H. Keller (eds.), Cologne (1987a) 144–151
6. Frohberg G, Kraatz KH, Wever H: Transport kinetics and structure of metallic melts, self-, impurity- and interdiffusion. In: *Research Program of the German Spacelab Mission D2*. P.R. Sahm, R. Jansen, M.H. Keller (eds.), Cologne (1987b) 144–151
7. Frohberg G: Diffusion in liquids. In: *Scientific Results of the German Spacelab Mission D2*. P.R. Sahm, M.H. Keller, B. Schiewe (eds.), Cologne (1995a) 275–287
8. Frohberg G, Kraatz KH, Griesche A, Wever H: Diffusion in liquid metals and alloys: self- and impurity diffusion. In: *Scientific Results of the German Spacelab Mission D2*. P.R. Sahm, M.H. Keller, B. Schiewe (eds.) Cologne (1995b) 288–294
9. Gelles SH, Markworth AJ: Agglomeration in immiscible liquids. Final postflight report on SPAR II experiment 74-30, NASA TM-78125 (1977)
10. Gelles SH, Markworth AJ: Agglomeration in immiscible liquids. Final postflight report on SPAR V experiment 74-30, NASA TM-78275 (1980)
11. Körber C: Phenomena at the advancing ice–liquid interface: solutes, particles and biological cells. Q. Rev. Biophys. **21** (1988) 229–298
12. Kuschnigg I, Sprenger HJ: Shot towers – predecessors of low gravity utilization. Low G, INTOSPACE 7.1, Hannover (1996) 10–11
13. Langbein D: The motion of particles ahead of a solidification front. In: *Intermolecular Forces*. B. Pullman (ed.), Reidel (1981) 547–562
14. Langbein D: Separation of binary alloys with miscibility gap in the melt. In: *Progress in Low-Gravity Fluid Dynamics and Transport Phenomena*. J.N. Koster, R.L. Sani (eds.), IAA Series (1990) 631–659
15. Langbein D: Fluid physics. In: *Research in Space – The German Spacelab Missions*. P.R. Sahm, M.H. Keller, B. Schiewe (eds.), WPF, Cologne (1993) 91–114
16. Malméjac Y, Bewersdorff A, Da Riva I, Napolitano LG: Challenges and prospectives of microgravity research in space. ESA BR 05, October 1981
17. McDonald JE: The shape of raindrops. Sci. Am., February 1954, p. 64
18. Minchinton W: Shot towers – precursors of modern low gravity drop facilities. Low G, INTOSPACE 6.3, Hannover (1995) 3–5
19. Minkowski H: Kapillarität. In: *Encyklopädie der Mathematischen Wissenschaften*. A. Sommerfeld (ed.), B.G. Teubner, Leipzig, Vol. 9 (1921) 558–613
20. Myshkis AD, Babskii VG, Kopachevskii ND, Slobozhanin LA, Tyuptsov AD: *Low-Gravity Fluid Mechanics* [translated by R.S. Wadhwa]. Springer, Berlin, Heidelberg (1976)
21. Otto GH, Lorenz H: Simulation of low gravity conditions by rotation. AIAA 16th Aerospace Sciences Meeting, Huntsville, AL (1978), AIAA Paper 78–273
22. Pötschke J, Rogge V: On the behavior of foreign particles at an advancing solid–liquid interface. J. Cryst. Growth **94** (1989) 726–738
23. Plateau J: Statique expérimentale et théorique des liquides. Mém. de l'Acad. de Belgique **16** (1843)
24. Plateau J: *Statique expérimentale et théoretique des liquids soumis aux seules forces moléculaires*. Vol. 2. Gauthier-Villars, Paris (1873)

25. Ratke L, Diefenbach S: Liquid immiscible alloys. Mater. Sci. Eng. **R15** (1995) 263–347
26. Potard C: Filtration-theory approach to immiscible alloys solidification. Proceedings of the RIT/ESA/SSC Workshop, Järva Krog, Sweden, 18–20 January 1984. ESA SP-219, 79–82
27. Snyder RS: Summary of Pre-ASTP Results. Proceedings of the Second European Symposium on Material Sciences in Space, Frascati, 6–8 April 1976. ESA SP-114, 19–26
28. Sprenger HJ: Low G production in shot towers. Low G, INTOSPACE 7.2, Hannover (1996) 11–13
29. Vreeburg JPB: Summary review of microgravity fluid science experiments. ESA Report, Nov. 1986
30. European Space Agency: The effect of gravity on the solidification of immiscible alloys. Proceedings of the RIT/ESA/SSC Workshop, Järva Krog, Sweden, 18–20 January 1984. ESA SP-219

2. Interface Tension and Contact Angle

Wherever liquids occur, the effects of surface tension are obvious. The theory of surface tension, however, is still in a phenomenological state. Several microscopic arguments, in particular those based on van der Waals attraction, are available. They yield qualitative results at best.

This situation becomes even worse if dynamic effects are considered, e.g. the creation of fresh surfaces during surface oscillations, and the motion of a solid/liquid contact line. An advancing contact angle is steeper than the stationary contact angle, a receding one is flatter. The main aspects of these dynamic effects are discussed here.

2.1 Molecular Attraction and Condensation

Water drops which fall from a tap or which fall from the sky as raindrops generally assume a spherical shape. They may oscillate owing to the preceding formation process and may adopt a slightly oval form owing to atmospheric resistance. The primary cause of their spherical shape is molecular attraction. And, from the fact that a spherical shape exhibits the minimum surface area for a given liquid volume, it may be concluded that molecular attraction is isotropic and a surface energy exists. A further repeatedly cited effect of surface tension is the fact that many insects are able to walk on a water surface.

There is an attractive force not only between water molecules, but also between any two molecules of matter. Molecules may lower their joint potential energy by contracting into a sphere. The first quantitative description of this attraction between molecules was given by van der Waals [1881], when he formulated his equation of state of real gases. He found that the behavior of these gases can be accurately described by introducing into the equation of state for ideal gases

$$pv = R_\mathrm{B} T \tag{2.1}$$

- a minimum volume b, which can be interpreted as the volume occupied by the close-packed molecules, and

- a contribution $a/(\text{volume})^2$ to the pressure accounting for molecular attraction,

$$\text{leading to} \quad \left(p + \frac{a}{v^2}\right)(v - b) = R_\text{B} T \,, \tag{2.2}$$

where R_B is the molar gas constant, $R_\text{B} = 8.3144 \,\text{J/K/mol}$. The distribution of the molecules' energy between the thermal (kinetic) energy and the potential energy of attraction lowers the kinetic energy by a term proportional to $(\text{volume})^{-2}$, which is equivalent to $(\text{mean distance})^{-6}$. The change in kinetic energy corresponds to a change in molecular momentum and in the pressure exerted by the molecules on the wall. Unfortunately, the observed attractive potential, proportional to $(\text{mean distance})^{-6}$, cannot be extrapolated to zero molecular spacing. Neither does the van der Waals energy at near-zero spacing satisfy the usual $(\text{mean distance})^{-6}$ relation, nor may the other attractive and repulsive contributions be neglected. Otherwise it would be possible to extrapolate, for example to the heat of evaporation. Equation (2.2) contains the first terms of a virial expansion of the pressure p with respect to the volume v.

Van der Waals attraction is due to a correlated motion of electrons in neighboring molecules. The electrons orbiting around their nuclei "note" the motion of the electrons in the nearby molecules (nuclei) and adapt their motion accordingly. This lowers their joint energy by an amount that increases the closer the molecules approach each other, such that an attractive potential results. This attractive potential is usually stronger between molecules of the same species than between molecules of different species. In consequence, molecules of the same species push aside molecules of different species, thus causing a separation of the two phases. (This effect is similar to the Archimedean principle. All media experience gravity. However, the denser species drives the less dense species off the bottom, and an effective rise rather than sinking of the lighter species results.) Further contributions to attraction may result from electronic overlap, hydrogen bonds, etc. Molecular attraction turns out to be particularly strong in metals, where the nearly free electrons optimize their mutual overlap. Therefore, the surface tension of metals is higher by an order of magnitude than in insulators.

Analytic calculations of van der Waals attraction are commonly based on dipole–dipole interactions. The electrons in their orbitals around the nuclei may undergo transitions to other orbitals by photon absorption or emission. The resultant electromagnetic dipole field is "noted" by the neighboring electrons. They react accordingly, i.e. they may emit or absorb a photon themselves. The probability of such transitions is governed by the quantum mechanical zero-point energy, which is paralleled macroscopically by the fluctuation–dissipation theorem. Dipoles which fluctuate dissipate energy according to the imaginary part of their dielectric polarizability. The coefficient a in (2.2) therefore equals an infinite frequency integral over the product of the molecular polarizabilities [Langbein 1974]:

$$\Delta E = -\frac{\hbar}{4\pi} \int_{-\infty}^{\infty} d\omega \, Tr\left[X_i(i\omega)T_{ij}X_j(i\omega)T_{ji}\right] . \qquad (2.3)$$

Here $\hbar$ is Planck's constant, $X_i(i\omega)$ is the polarizability of molecule i at circular frequency ω, and T_{ij} is the dipole interaction tensor

$$T_{ij} = -\boldsymbol{\nabla}_i \boldsymbol{\nabla}_j \frac{1}{|r_i - r_j|} . \qquad (2.4)$$

A liquid surface is not static, but is stationary. A few surface molecules always gain sufficient thermal energy to evaporate, while others are captured by the surrounding gas phase. Whether molecular attraction suffices to enable condensation from the gas to the liquid phase clearly depends on pressure and temperature:

- A low gas pressure involves a large mean spacing of the molecules, such that the formation of clusters is unlikely. When the pressure is increased, the molecules come into contact more often, and clusters become more likely.
- The decrease in potential energy corresponding to zero molecular spacing must exceed the energy of thermal motion. Lowering the temperature thus makes clustering more likely.
- The more molecules are already contained in a cluster, the more an additional molecule joining it lowers the potential energy. The energy gain per additional molecule increases with increasing cluster size.
- Depending on the species, the molecules may lower their energy by spreading over a solid surface. In the case of good wetting, clustering at the surface becomes more likely.
- The molecules in a liquid experience attraction by the neighboring molecules, without already being in the state of lowest energy. The second phase transition, to crystallization, is once more due to cluster formation.

When a cluster is formed, most molecules may be considered as surface molecules. The gain in potential energy per molecule still is smaller than the thermal kinetic energy. However, with increasing cluster size the relative number of bulk molecules rises and the gain in potential energy per molecule can exceed the kinetic energy. The Gibbs model of nucleation states that a cluster of molecules needs a surface energy proportional to $4\pi r^2 \sigma$ and gains a bulk energy proportional to $(4\pi/3)r^3 h_c$, i.e.

$$\Delta E = 4\pi r^2 \sigma - \frac{4\pi}{3} r^3 h_c , \qquad (2.5)$$

where h_c is the heat of condensation. According to this very simple model, the net free energy becomes negative, i.e. the cluster becomes a nucleus, for $r_n \geq 3\sigma/h_c$. (A cluster is called a nucleus if it grows thermodynamically.)

2.2 The Interface Tension

2.2.1 Theoretical Aspects

A liquid volume attempts to minimize its surface area, i.e. between any two fluids (two liquids or a liquid and a gas), a surface energy or interface energy exists. A molecule near the surface rather than in the bulk of a liquid volume is missing about one half of its partners of attraction. Energy must be supplied against the attraction by the other molecules to move an additional molecule from the bulk to the surface.

A liquid interface, on the other hand, is not as static as the common assumption of an interface energy suggests. Two immiscible phases are in fact separated by a diffuse interface layer, the thickness of which is usually a few molecular diameters, and in which the properties vary smoothly from one phase to the other. The two phases continuously exchange molecules, momentum and energy. The thermodynamic state of equilibrium is not the one that exhibits lowest energy, but the one that is the most likely under the given intensive parameters of temperature and pressure. The interface layer is governed by the requirement of minimum entropy of the liquid system. Modeling the interface layer at either the molecular (microscopic) or phenomenological (macroscopic) level is still, to a large extent, an open problem. Even less well established is the bridging of the gap between the microscopic and macroscopic descriptions.

Surface tension is a component of the surface stress tensor. It is usually assumed that under stationary conditions this tensor is isotropic. The limits of applicability of this assumption have not been well investigated. The reversible part of the stress tensor may even be nonsymmetric. Calculations of the surface energy thus are subject to many imponderabilities. Neither the arrangement of the molecules adjacent to the surface nor the forces of attraction at small distances are properly known. In addition to the van der Waals attraction electronic overlap, hydrogen bonds, etc. have to be taken into account.

Investigations of nucleation indicate that the value of the interface tension between two media even depends on the curvature of the interface. This dependence becomes stronger the smaller the radius of curvature is. Since reliable measurements and a generally accepted theory are missing, one describes the dependence of the interface tension on the radius of curvature by empirical or semiempirical formulae [Tolman 1949]. The number of missing attraction partners increases with increasing curvature. This, however, does not mean that the surface energy increases as well.

In spite of the many theoretical objections, let us calculate the surface energy by integrating the van der Waals potential between a sphere of radius R_1 and the outer liquid. Convergence of this integration requires that a small spacing ΔR between the sphere and the outer liquid is maintained,

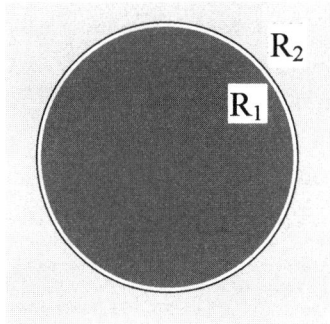

Fig. 2.1. Sketch of the integration of the van der Waals potential

see Fig. 2.1. From the potential between two molecules (points) at distance r apart,

$$U_0 = \frac{A_H}{r^6}, \tag{2.6}$$

we find, for the potential between the inner sphere and a single molecule at distance R_2 from its center,

$$U_1 = \frac{4\pi A_H}{3} \frac{R_1^3}{(R_2 - R_1)^3 (R_2 + R_1)^3}. \tag{2.7}$$

Integration over the concentric outer liquid with a void radius $R_2 = R_1 + \Delta R$ leads to

$$U_2 = \frac{\pi^2 A_H}{3} \left[\frac{R_2 R_1}{(R_2 - R_1)^2} + \frac{R_2 R_1}{(R_2 + R_1)^2} + \log\left(\frac{R_2 + R_1}{R_2 - R_1}\right) \right],$$

$$\sigma \propto \frac{U_2}{4\pi R^2}. \tag{2.8}$$

In (2.6) to (2.8), A_H is known as the Hamaker constant. From (2.8) we find the energy σ per unit surface area to increase with decreasing particle radius if the surface area is calculated at the inner radius R_1, but to decrease if the surface area is calculated at the outer radius R_2, and to slightly increase if the surface area is calculated at the mean radius $(R_1 + R_2)/2$ (Fig. 2.2). This arbitrariness adds further objections to such integrations.

2.2.2 Experimental Methods

On the experimental side, several of the well-established terrestrial methods of measuring surface or interface tension make explicit use of gravity. It is actually the Bond number defined by (1.3), which is determined. Among these methods, let us mention the following [Gebhardt 1981]:

- Measurement of liquid rise in a thin capillary.

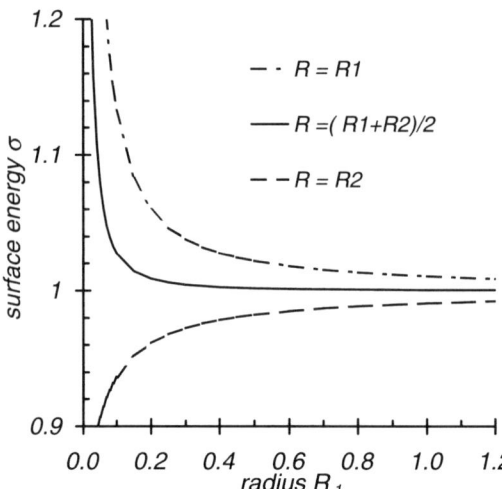

Fig. 2.2. The surface energy resulting from the van der Waals potential (2.8)

- The sessile- or hanging-drop method, i.e. observation of the shape of a drop sitting on a horizontal plane and numerical comparison of that shape, pixel by pixel, with shapes computed for different Bond numbers.
- The measurement of the maximum volume of a hanging drop by counting the number of falling drops and measuring their total volume. This method has lost its former attractiveness. Calculation of the maximum volume does not represent a problem. The volume actually dropping, however, depends on the shape and wettability of the liquid outlet, the viscosity of the liquid, etc.
- The Wilhelmy method, i.e. measuring the force required to pull a plate out of a bath. This method presumes a plane liquid surface and a horizontal contact line. These presumptions are no longer valid under weightlessness. Instead, one may measure the force required to maintain (or pull apart) a liquid bridge between two coaxial circular disks.
- The force required to pull a thin film out of a bath. This method works also under weightlessness.
- The difference in vapor pressure between a curved and a plane interface. This requires long times and has poor accuracy.
- The measurement of the pressure inside a drop. This pressure equals the surface tension times the curvature. The curvature is taken as positive if the interface is convex when viewed from the liquid side of the surface. This method is fast and has recently been applied by Liggieri et al. [1996] in the TEXUS 33 mission.
- Observation of the shape of a bubble or drop in a spinning container. In this case surface tension works against centrifugal force rather than against gravity. Substituting $\omega^2 R$ for g in the Bond number leads to the rotation number

$$Rn = \frac{\Delta\rho\,\omega^2 R^3}{\sigma}\ . \tag{2.9}$$

2.2.3 Qualitative Rules for the Interface Energy

It needs about one half the energy to bring a molecule from the bulk of a liquid volume to the surface than it needs to bring it to the gas phase. The resultant surface energy therefore equals about half the heat of condensation (or evaporation). Likewise, the interface energy between a liquid phase and the corresponding solid is about one-half of the heat of melting. Many diagrams of this kind have been reported, see Figs. 2.3 and 2.4 [Chalmers 1959, 1964].

The interface tension changes with temperature, with concentration and with several other external parameters. As a rule, the surface energy decreases with increasing temperature. It is roughly proportional to the density. This gives rise to thermal interface convection (also referred to as thermocapillary convection or Marangoni convection). That effect has been known since the 19th century and is attributed to Lord Rayleigh and to Marangoni [1865, 1871]. If for any reason the interface tension along a fluid interface varies, a shear force from regions exhibiting low interface tension to regions exhibiting high interface tension results, which gives rise to a corresponding fluid flow. Marangoni convection is the reason why dirt particles move rapidly from the

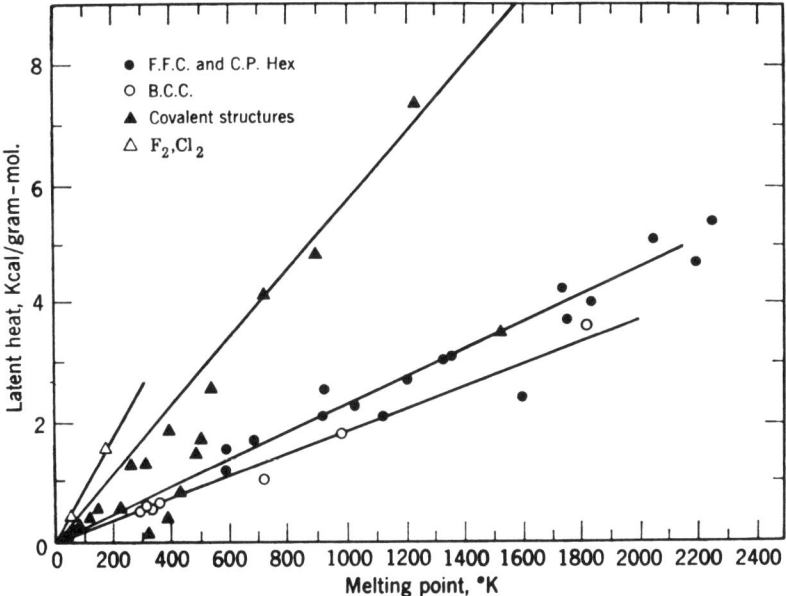

Fig. 2.3. Heat of melting versus melting point [Chalmers 1959]

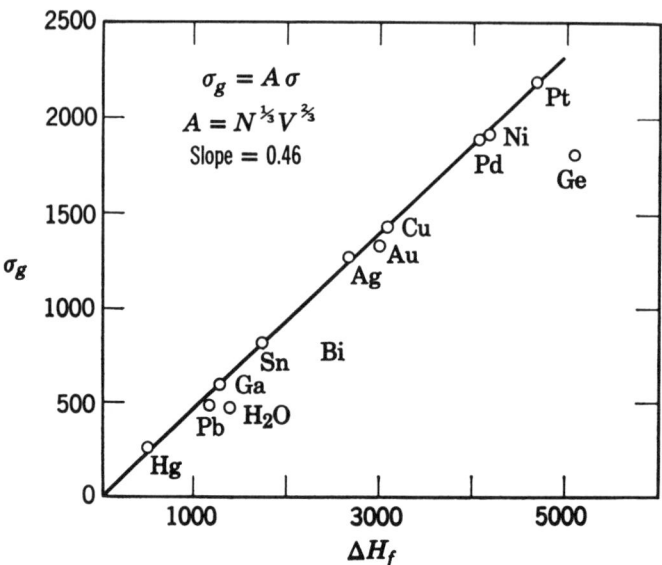

Fig. 2.4. Surface tension versus heat of melting [Chalmers 1959]

center to the periphery of a burning candle and the reason why cognac creeps up a suitably designed glass.

- *Candle:* The molten wax (stearin) is hotter near the wick than at the periphery, which leads a fluid flow to the periphery at the surface and to the center at the bottom. The soot particles act as tracers for observation of this motion.
- *Cognac glass:* The more volatile alcohol evaporates at a higher rate than water at the periphery. There, the decreasing alcohol concentration increases the interface tension. A flow up the glass results. The rising cognac film, however, breaks into droplets, which run back down to the normal liquid level.

Numerous investigations of this natural convection, which is masked by buoyancy on the ground, have been performed under weightlessness. Particular attention has been paid to drop and bubble migration. The surface energy of drops and bubbles decreases if they move from regions of high to low interface tension, i.e. from cold to hot regions. Marangoni migration of droplets plays an important role during the separation of monotectic alloys, as mentioned in Sect. 1.2. Other fields of research are interface convection in liquid zones (bridges between coaxial circular disks) and in plane liquid layers. Both geometries exist in crystal growth experiments (floating-zone growth and Bridgman growth).

During investigations of interface convection, one has to keep in mind the following:

- Freshly formed interfaces, as long as surfactant molecules have not reached their final arrangement, usually exhibit an increased interface tension. The interface tension, generally, is sensitive to impurities.
- The surface tension of highly pure distilled water, which equals $0.07\,\text{N/m} = 70\,\text{erg/cm}^2$, changes strongly with contamination.
- Water, therefore, is not suited at all to quantitative investigations of Marangoni convection. Experimenters prefer silicone fluids, which are available in a wide range of viscosities and show only little sensitivity to contamination.

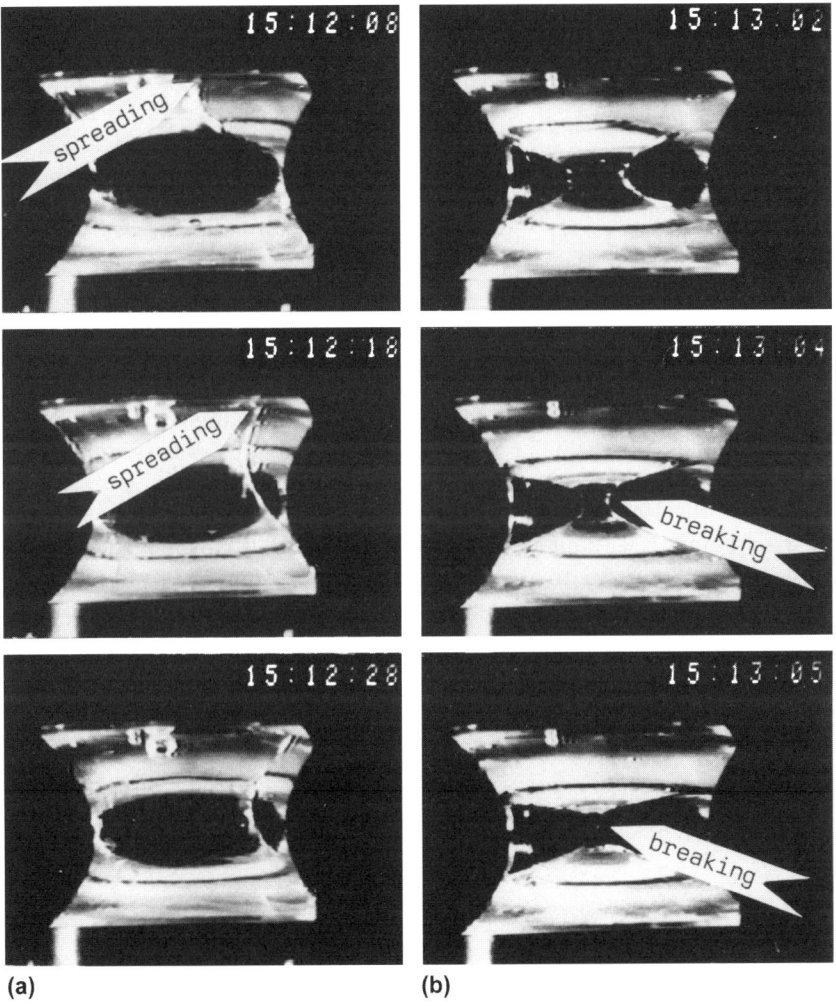

Fig. 2.5. (a) Spreading of benzyl benzoate along the heated upper disk in the D1 experiment "Mixing and Demixing of Transparent Liquids"; (b) breakage of the inner column of benzyl benzoate in the outer column of paraffin oil

- Aging of liquids and of solid surfaces often poses problems to space experiments. Usually the samples have to be delivered three months before the flight.
- Another effective possibility for changing the surface tension is application of electromagnetic fields.
- A stable interface is not static; it is stationary. Even if an interface is dissolving owing to a previous change of the parameters of thermodynamic equilibrium, it still is an interface and acts as such (Fig. 2.5).

2.3 The Static Contact Angle

Two phases 1 and 2 which contact each other form an surface or interface 12 exhibiting a tension σ_{12}. The interface actually represents an interface layer. If three phases 1, 2 and 3 contact each other, three interfaces 12, 13 and 23 arise. They meet in a contact line. Minimization of the energy entails that the stresses along the contact line must balance each other. In the case of three fluid phases, i.e. liquid 1 + liquid 2 + gas 3 or liquid 1 + liquid 2 + liquid 3, Neumann's boundary condition along the contact line results. Vectorial equilibrium of the three interface tensions is required (see Fig. 2.6):

$$\boldsymbol{\sigma}_{12} + \boldsymbol{\sigma}_{23} + \boldsymbol{\sigma}_{31} = 0 \, . \tag{2.10}$$

According to observations by Marangoni [1871], the interface tension between two liquids 1 and 2 is in all cases smaller than the difference between their surface tensions in air (3), $\sigma_{12} < |\sigma_{13} - \sigma_{23}|$.

If one of the three phases considered is solid, i.e. liquid 1 + gas 2 + solid 3 or liquid 1 + liquid 2 + solid 3, the solid is able to compensate any shear stress normal to its surface, such that only the residual forces tangential to the solid surface must balance each other. This leads to Young's boundary condition along a contact line at a solid surface (= equilibrium of the tangential components of the interface tensions) [Young 1805] (Fig. 2.7):

$$\sigma_{13} + \sigma_{12} \cos \gamma_1 = \sigma_{23} \, . \tag{2.11}$$

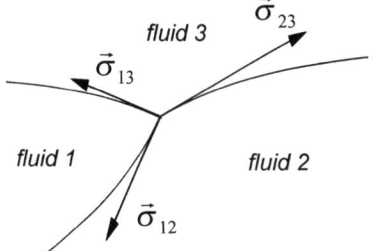

Fig. 2.6. Fluid contact line: the vector sum of the three interface tensions acting at the contact line vanishes, $\boldsymbol{\sigma}_{12} + \boldsymbol{\sigma}_{23} + \boldsymbol{\sigma}_{31} = 0$ (Neumann's boundary condition)

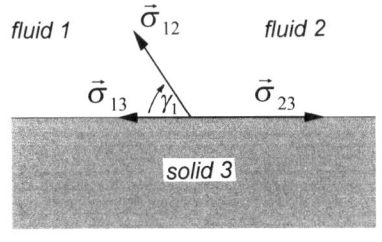

Fig. 2.7. Solid contact line: the interface tension σ_{23} between fluid 2 and solid 3 balances the sum of the interface tension σ_{13} between fluid 1 and solid 3 and the tangential component $\sigma_{12}\cos\gamma_1$ of the interface tension between fluids 1 and 2: $\sigma_{12} = \sigma_{13} + \sigma_{23}\cos\gamma_1$ (Young's boundary condition)

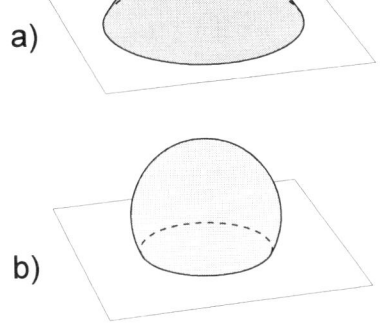

Fig. 2.8. A drop (spherical cap) on a solid surface, (**a**) with small contact angle (e.g. water on glass), (**b**) with large contact angle (e.g. mercury on glass, water on Teflon)

Wetting of solid 3 by fluid 1 releases energy if $\gamma_1 < \pi/2$ or $\cos\gamma_1 > 0$; it needs energy if $\gamma_1 > \pi/2$ or $\cos\gamma_1 < 0$. On the other hand, wetting of a solid surface usually results in an increase of the fluid surface. The effective change in energy is given by $\sigma_{12}(A_\mathrm{f} - A_\mathrm{s}\cos\gamma)$, where A_l and A_s are the surface areas of fluid and the solid, respectively. The interface energies σ_{13} and σ_{23} of solid 3 with fluids 1 and 2 are theoretical parameters and cannot be measured absolutely. And at present it is impossible to calculate them. Experimentally, only the difference $\sigma_{13} - \sigma_{23}$ can be determined by means of Young's boundary condition (2.11).

Young's boundary condition on the contact angle is easily formulated as a balance of forces or as a minimum in energy. It is a static condition, i.e. it presumes that the fluids have plenty of time to assume a minimum in energy. This time includes the exchange of molecules across the interface due to Brownian motion, i.e. evaporation, mixing or solution. The contact line is actually a contact region with a large extension along the line and an extension of molecular dimensions perpendicular to it.

2.4 The Dynamic Contact Angle

If a vertical plate is moved up and down through a liquid interface, one observes obvious variations of the contact angle. When the plate is pushed in, the contact angle becomes steeper; when the plate is pulled out, the contact

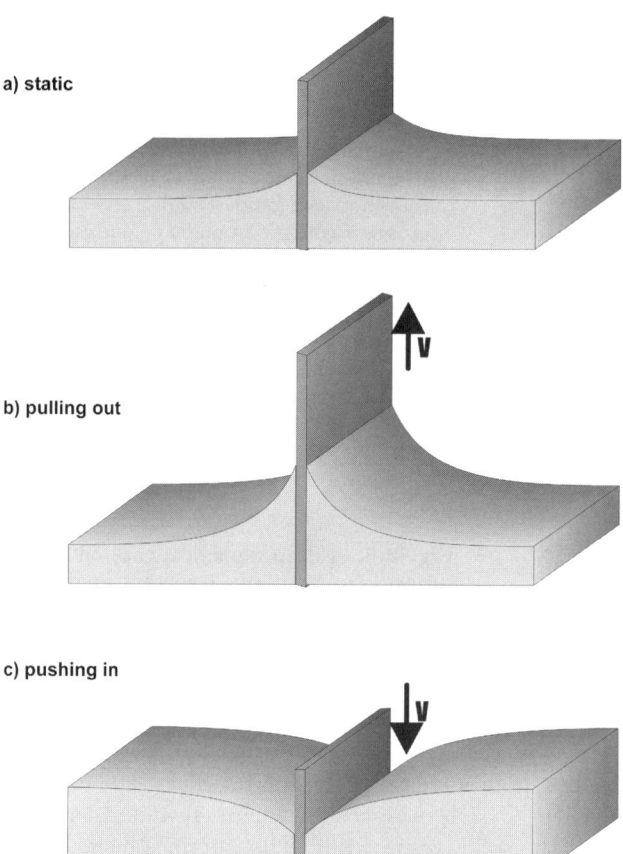

Fig. 2.9a–c. When the plate is pushed in, the contact angle becomes steeper; when the plate is pulled out, the contact angle becomes flatter

angle flattens; see Fig. 2.9. An equivalent effect is observed if a capillary containing a finite liquid plug is moved up and down over a piston, or if a drop is squeezed between parallel plates. Experimentally, one usually finds two regions: at low rates of motion v of the contact line, the contact angle increases linearly with the rate, whereas at high rates of motion, the contact angle approaches a constant value (Fig. 2.10).

The fluid flow resulting from the presence of a moving contact line requires a force proportional to the rate v of motion and to the liquid viscosity η. Substitution of this force into Young's boundary condition (2.9) leads to

$$\sigma_{13} + \sigma_{12}\cos\gamma_1 + cv\eta = \sigma_{13} + \sigma_{12}\cos\gamma_1 + c\sigma_{12}\,Ca = \sigma_{23}\;, \qquad (2.12)$$

where Ca is the capillary number,

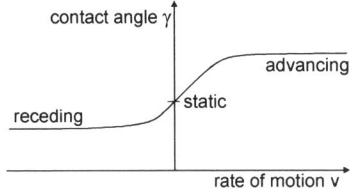

Fig. 2.10. At low rates v of motion of the contact line, the contact angle increases linearly with the rate, whereas at high rates of motion, the contact angle approaches a constant value

$$Ca = \frac{v\eta}{\sigma} \qquad (2.13)$$

Ca describes the ratio of the fluid-dynamic pressure to the capillary pressure, i.e. the relative deformation of the liquid interface due the fluid flow. According to (2.12), $\cos\gamma$ decreases linearly with Ca. The front angle γ_1 steepens, the rear angle γ_2 flattens. The viscous force has to be provided by the fluid, as it wets the surface, i.e. by the energy gain of the advancing fluid, or else by an external force.

In attempting to quantitatively calculate the force required for contact line motion, one faces a problem of principle: the contact line cannot really move! The motion of a liquid near a solid surface is expected to satisfy the no-slip condition. Most common liquids, the so-called Newtonian liquids, show zero flow velocity at solid surfaces. This assumption is inconsistent with the motion of a contact line. Modeling the resulting fluid flow by means of the usual flow equations (the momentum and continuity equations, see Chap. 11) leads to an infinite shear stress at the contact line. This cannot be traced back to any kinematic incompatability between spreading and the no-slip condition [Dussan & Davis 1974]. It is inherent in using the macroscopic flow equations together with the no-slip boundary condition.

In computer codes, one may weaken or even abandon the no-slip condition close to the contact line. Several slip boundary conditions could, in principle, be imposed. Dussan [1976] has shown, however, that the flow distant from the contact line, where almost all fluid mechanical measurements are made, is quite insensitive to the form of these boundary conditions. This is a rather discouraging conclusion, since it indicates that the mechanism of motion of the contact line cannot be elucidated by macroscopic measurements.

Here again, it is necessary to return to the microscopic, molecular description and to throw overboard the familiar concept that the fluid interface is a sharp interface. It is an interface layer which is subject to a continuous exchange between the molecules of the two fluids. A few molecules, owing to their thermal motion, always intrude into the other fluid. A macroscopic interface is actually a Gibbs dividing interface. Likewise, the contact line is not a line but a contact region. Within that contact region the interfacial concentration profile of fluids 1 and 2 is modified by the attraction by solid 3.

In Fig. 2.11a it is assumed that fluid 1 is the better-wetting species, such that the contact angle γ_1 of this fluid is smaller than the contact angle γ_2 of fluid 2. The angles γ_1 and γ_2 are complementary, i.e. $\gamma_1 + \gamma_2 = \pi$. The

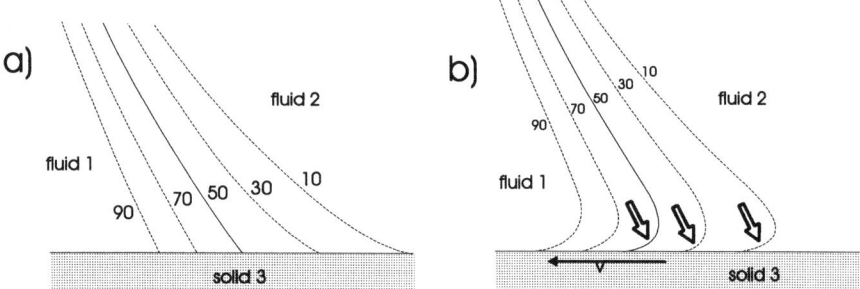

Fig. 2.11. Microscopic view of the isopleths (the lines of equal concentration) in the contact region of fluid 1, fluid 2 and solid 3: (**a**) static; (**b**) bending of the isopleths and diffusive transport during motion of the contact line to the right

molecules of fluid 1 are thus likely to be spread out at the interface with the solid. Let us assume that the contact line moves to the right; then, in the coordinate system moving with the contact line, the solid surface is moving to the left. The isopleths (the lines of equal concentration) are bent to the left accordingly; see Fig. 2.11b. A portion of the receding fluid 2 becomes trapped between the advancing fluid 1 and the solid 3. The more strongly attracted molecules of fluid 1 attempt to balance this situation by strong diffusion to solid 3. Thereby they gain energy of wetting, to an extent that decreases with increasing rate v of motion owing to irreversible diffusive losses. Both the convective flow of fluid 1 parallel to the solid surface and the diffusive flow perpendicular to it are proportional to the dynamic viscosity (as expressed in Einstein's relation between the viscosity and the diffusion coefficient). The prewetting of solid 3 by fluid 1, which is considered necessary for wetting, is supplemented by post-dewetting of fluid 2.

Let us repeat that reasoning with some slight modifications. At low rates of motion, the contact region moves uniformly. It rolls over the solid surface through near-equilibrium thermodynamic states. This is assisted by prewetting, which, however, requires sufficient time. With increasing rate v of motion, the isopleths increasingly bend over at the solid surface. Eventually, fluid 1 assumes a convex interface which is no longer stable, but must break up according to a Rayleigh instability (see Chap. 4). The contact region now does not move uniformly any more, but "hobbles" (tumbles) along the solid surface and shows fingering. The interface just splashes onto the solid surface. The receding fluid becomes trapped below the advancing fluid where it sticks to grooves and cracks and forms puddles there. Even if, owing to diffusion, the molecules of the receding fluid 2 succeed in reaching the appropriate side, they succeed too late, at the cost of a large portion of the wetting energy. There is post-dewetting of the receding fluid 2. Surface heterogeneities cause additional hysteresis.

Fingering and puddle formation mean that only a portion of the wetted solid area counts in the energy and force balance. It is no longer $\sigma_{12} - \sigma_{23}$ which enters the modified boundary condition (2.12). With increasing rate v of motion, clearly, nonequilibrium thermodynamic situations become involved more and more. Whatever the detailed shape of the interface, only a portion of the wetting energy contributes to the energy balance. The advancing contact angle thus assumes a limiting value.

The fast motion of a contact line invites us to reflect on several analogies:

- The motion of a contact line has much in common with the removal of a piece of Scotch tape from a sheet of paper. If the removal is very slow and is done at a very flat angle, it is possible to remove the glue together with the tape and to leave the paper largely undamaged. Work has to be done only against the glue. However, if one pulls the tape too fast, the paper will certainly be damaged. A steep angle of removal further increases the risk of rupture.
- Another comparable process is crystal growth from a melt. If one chooses too high a growth rate v, the planar growth front breaks up into dendrites and eventually a mushy zone arises. The allowed rate for planar growth has an order of magnitude of a few micrometers per second; the molecules need time to find their appropriate sites. If the temperature gradient is increased, the growth rate may be increased also. The resulting crystal is three-dimensional and stable, such that one is able to draw conclusions about the processes involved.
- A fast advance of a winning army causes a disorderly retreat of the losing army. This usually involves high losses on the latter side. Scattered troops are taken prisoner; to escape from the camps and to return to the correct side causes difficult logistical problems and is risky as well.

Among the early studies of the dynamic contact angle, let us mention the papers by Friz [1921] dealing with complete wetting, by Hansen & Toong [1971] on the relationship to hydrodynamic forces, and several experimental papers by Dussan [1976, 1979], Dussan et al. [1974, 1982, 1991] and Chen et al. [1997], see Fig. 2.12. A general overview of the dynamics of spreading and of the models and assumptions has been given by deGennes [1985]. Since then, an increasing number of papers have been published [Kröner 1990, Shikhmurzaev 1997, Cox 1998].

2.5 Merging of Drops and Bubbles

The fact that a fluid interface represents an interface layer leads to further consequences, for example to Ostwald ripening and merging of drops and bubbles. Two bubbles within a liquid may merge if they contact each other. The surface energy of the single, merged bubble is lower than that of the two separated bubbles. Terrestrially, two bubbles have a good chance to come into

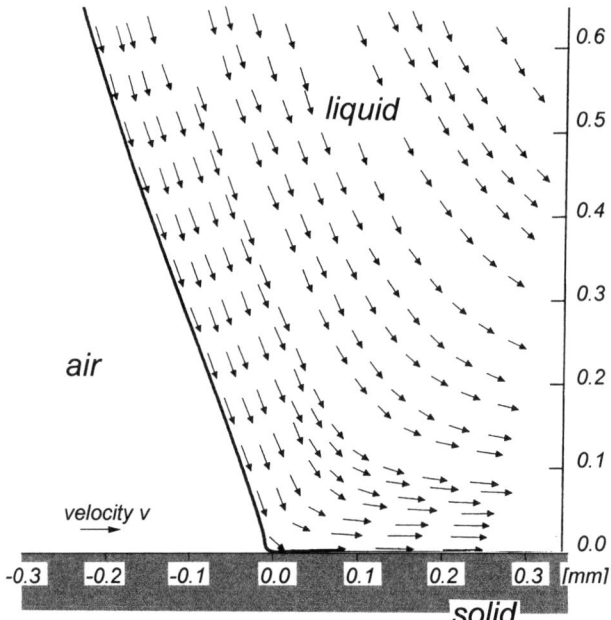

Fig. 2.12. A typical flow field near a moving contact line at $Ca = 0.1$. The *solid line* represents the liquid/air interface; $v = 35\,\mu\text{m/s}$ is the tube immersion velocity. Only every fifth data point is shown [from Chen et al. 1997]

contact owing to buoyancy and convection. In a microgravity environment they may stay apart for long times, i.e. merging is considerably reduced. Nevertheless, thermodynamics finds a way to reach the absolute minimum of free energy in the absence of forces. In the case of bubbles, it is the solubility of the gas in the surrounding liquid which does the job. The solubility generally increases with decreasing bubble radius (increasing bubble pressure). A small bubble thus tries to raise the gas concentration in the liquid, whereas a large bubble tries to lower it. Diffusive gas transport from a small bubble to a large bubble results, i.e. the latter grows at the expense of the former. This thermodynamic instability of a small bubble, in the absence of direct contact, is termed the Kelvin instability.

When several drops, bubbles or solid particles are involved and all surface regions with low curvature grow at the expense of those with high curvature, one speaks about Ostwald ripening instead. In analogy to the increase of the solubility of a gas in a liquid with increasing bubble curvature, there is an increase in vapor pressure of a liquid in a gas with increasing drop curvature.

When the two drops or bubbles considered are a short distance apart, they may encounter another effect: each particle affects the solubility and concentration in its neighborhood. At small spacing each particle "notes" the change in concentration caused by the other. This may give rise to direct attraction and subsequent merging. The absence of convection and sedimentation under microgravity thus enables direct investigations of the range of intermolecular attraction, the effective thickness of diffuse interfaces, and the mechanisms of coagulation between drops and between bubbles.

2.6 Adhesion Forces in Liquid Films

The D1 experiment "Adhesion Forces in Liquid Films" [Padday 1983, 1987] was aimed at studying the forces exerted on a thin liquid layer by a solid surface. Here, liquid columns with a catenoid shape were established between two disks of unequal radii. The catenoid is distinguished among the axisymmetric surfaces by zero capillary pressure and zero mean curvature; see Sect. 3.4. The liquid used was silicone oil. The larger disk had a conical shape with a slope chosen to match that of the intended catenoid. Thus, a very thin liquid

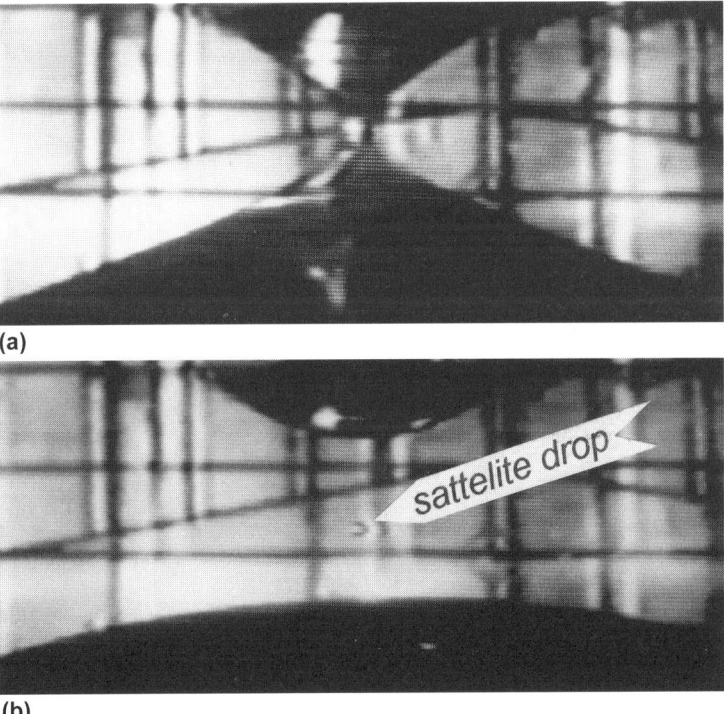

Fig. 2.13. A nearly catenoid shape before breakage (**a**) and a satellite drop after breakage (**b**) of a column of silicone oil during the D1 experiment "Adhesion Forces in Liquid Films". The diameter of the larger disk is 100 mm. The satellite drop moved forth and back along the symmetry axis with decreasing speed. It did not pick up speed in the radial direction, but rather stayed in a stable position during the six minutes during which recording was performed

layer should be formed by the catenoid at the perimeter. The interaction of the solid with the liquid surface was expected to bring about a strong change of the column's shape, which, in order to further increase the sensitivity, was chosen close to a stability limit.

In several trials the wetting film was not flat but was found to be much thicker than anticipated. The increased volume of the film caused an early breakage of the liquid bridge. During several breakages, the formation of a satellite drop was observed. The satellite drop moved forth and back along the symmetry axis and was repelled by the two liquid volumes on the supporting disks. In the run shown in Fig. 2.13 the drop rested in a stable position between the disks for about six minutes.

The axial repulsion may be explained by opposite electric charging of the spherical liquid volumes on the supporting disks. Initially, the drop had the same sign of charge as the liquid volume on the larger disk, from which it was ejected. It was attracted by the opposite charge of the liquid volume on the smaller disk, but changed its charge when it touched that volume, and then was axially attracted by the liquid volume on the large disk. There it once more changed charge, and so on. Eventually, there were electrical charges left at the perimeter of the liquid volumes only. The drop could not pick up charge any more and, owing to dielectric polarization, moved to the position of maximum strength of the electric field. This was an unintended, but effective electrostatic positioner.

References

1. Chalmers B: *Physical Metallurgy*. Wiley, New York (1959) p. 85
2. Chalmers B: *Principles of Solidification*. Wiley, New York (1964)
3. Chen Q, Ramé E, Garoff S: The velocity field near moving contact lines. J. Fluid Mech. **337** (1997) 49–66
4. Cox RG: Inertial and viscous effects on dynamic contact angles. J. Fluid Mech. **357** (1998) 249–278
5. Dussan V. EB: The moving contact line: the slip boundary condition. J. Fluid Mech. **77** (1976) 665–685
6. Dussan V. EB: On the spreading of liquids on solid surfaces: static and dynamic contact lines. Ann. Rev. Fluid Mech. **11** (1979) 371–400
7. Dussan V. EB, Davis SH: On the motion of a fluid–fluid interface along a solid surface. J. Fluid Mech. **65** (1974) 71–95
8. Dussan V. EB, Ngan CG: On the nature of the dynamic contact angle: an experimental study. J. Fluid Mech. **118** (1982) 27–40
9. Dussan V. EB, Ramé E, Garoff S: On identifying the appropriate boundary conditions at a moving contact line: an experimental investigation. J. Fluid Mech. **230** (1991) 97–116
10. deGennes PG: Wetting: statics and dynamics. Rev. Mod. Phys. **57** (1985) 827–863
11. Friz G: Über den dynamischen Randwinkel im Fall der vollständigen Benetzung. Z. angew. Phys. **19** (1921) 374–378

12. Gebhardt KF: The Gauss–Laplace equation, methods to determine interfacial tension, and the preparation of fluid–fluid interfaces under microgravity. In: *Proceedings of the Workshop: Flüssigkeitsgrenzflächen und Benetzung*. D. Langbein (ed.), Frankfurt am Main (1981) 33–44
13. Hansen RJ, Toong TY: Dynamic contact angle. J. Colloid Interf. Sci. **37** (1971) 196
14. Kröner D: Asymptotic expansions for a flow with a dynamic contact angle. In: The Navier–Stokes Equations. Lecture Notes in Mathematics, Vol. 1431. J.G. Heywood, K. Masuda, R. Rautmann, V.A. Solonnikov (eds.). Springer, Berlin, Heidelberg (1990) 49–59
15. Langbein D: *Theory of van der Waals Attraction*. Springer Tracts in Modern Physics Vol. 72, Springer, Berlin, Heidelberg (1974) 1–139
16. Marangoni C: Sull' expansione delle goccie di liquido gallegiante sulla superficie di altro liquido. Pavia (1865)
17. Marangoni CGM: Über die Ausbreitung einer Flüssigkeit auf der Oberfläche einer anderen. Ann. Phys. (Poggendorf) **143** (1871) 337
18. Liggieri L, Ravera F, Passerone A: Scientific results of the Mite-2 TEXUS 33 sounding rocket experiment. In: *Proceedings of the Second European Symposium on Fluids in Space*. A. Viviani (ed.). Naples (1996) 135–143
19. Padday JF: *Fluid Physics in Space – The Kodak Ltd Experiment Aboard Spacelab-1*. Kodak Ltd, Harrow UK (1983)
20. Padday JF: Capillary forces in low gravity. ESA SP-256 (1987) 251–256
21. Shikhmurzaev YD: Moving contact lines in liquid/liquid/solid systems. J. Fluid Mech. **334** (1997) 211–249
22. Tolman RC: The effect of drop size on surface tension. J. Chem. Phys. **17**, (1949) 333–337
23. van der Waals JD: *Die Continuität des gasförmigen und flüssigen Zustandes*. Leipzig (1881)
24. Young T: An essay on the cohesion of fluids. Phil. Trans. Roy. Soc. London **95** (1805) 65–87

3. Capillary Shape and Stability

Two equivalent methods of deriving the capillary equation are considered here. They are based on

- the balance of forces on each element of the liquid surface. Each surface element experiences fluid-static pressure on its area and surface tension along its perimeter. This view was promoted by Laplace [1805, 1806] and Young [1805].
- the minimization of the total energy of the liquid under the constraint of constant liquid volume. This aspect was promoted by Gauss [1830].

The capillary equation is also termed the Gauss–Laplace equation for this reason.

3.1 Balance of Forces

Interest in the capillary behavior of fluids started rather early. In 1712, Taylor reported:

The following Experiment seeming to be of use, in discovering the Proportions of the Attractions of Fluids, I shall not forbear giving an Account of it; tho' I have not here Conveniences to make it in so successful a manner, as I could wish. – I fasten'd two pieces of Glass together, as flat as I could get; so that they were inclined in an Angle of about 2 Degrees and a half. Then I set them in Water, with the contiguous Edges perpendicular. The upper part of the Water, by rising between them, made this Hyperbola; which is as I copied it from the Glass. – I have examined it as well as I can, and it seems to approach very near to the common Hyperbola. But my Apparatus was not nice enough to discover this exactly. – The perpendicular Assymptote was exactly determined by the Edge of the Glass; but the Horizontal I could not so well discover.

Taylor's result was confirmed soon afterwards by Hauksbee [1712, 1713a, 1713b]. Hauksbee showed additionally that the hyperbolic character of the meniscus is independent of the inclination of the wedge relative to the reservoir. Hauksbee repeated the experiments using "spirit of wine".

About one century later Laplace [1805] postulated, on the basis of his observations on the rise and depression of a meniscus in a capillary, a proportional relation between the curvature of a meniscus and the resulting force on a fluid volume:

We are amply justified in concluding, that all the phenomena of capillary action may be accurately explained and mathematically demonstrated from the general law of the equable tension of the surface of a fluid, together with the consideration of the angle of contact appropriate to every combination of a fluid with a solid.

The concept of a surface tension σ of a liquid had been introduced by Segner [1751]. It is the work ∂W required to increase the surface are by ∂A, i.e. $\sigma = \partial W / \partial A$, and gives rise to a pressure difference Δp between two fluids.

Up to terms of second order, a continuous, differentiable curve at any reference point may be described by its local tangent and its curvature $\kappa = 1/r$. Likewise, a smooth surface may be described at any reference point by its tangential plane and the two principal curvatures $\kappa_1 = 1/r_1$ and $\kappa_2 = 1/r_2$. The principal (maximum and minimum) curvatures κ_1 and κ_2 arise in planes perpendicular to the tangential plane and perpendicular to each other, i.e. their intersection contains the normal to the surface considered. The curvature in an arbitrary plane through the normal forming an angle φ with the plane exhibiting the maximum or minimum curvature κ_1 is given by

$$\kappa = \kappa_1 \cos^2 \varphi + \kappa_2 \sin^2 \varphi . \tag{3.1}$$

Figure 3.1 shows an element of a fluid interface together with the planes of principal curvature. Normal to the area $r_1 \, d\varphi_1 \times r_2 \, d\varphi_2$, the interface element withstands the local pressure difference Δp between the two fluids considered. Tangential to the periphery, it experiences the surface tension σ. A vectorial summation along the periphery $r_1 \, d\varphi_1$ with an angular difference $d\varphi_2$ and along the periphery $r_2 \, d\varphi_2$ with an angular difference $d\varphi_1$ leads to a normal force $\sigma(r_1 + r_2) \, d\varphi_1 \, d\varphi_2$. Balance of forces is thus achieved for

$$\Delta p = \sigma \left(\kappa_1 + \kappa_2 \right) = \sigma \left(\frac{1}{r_1} + \frac{1}{r_2} \right) . \tag{3.2}$$

The local pressure difference Δp in (3.2) includes the fluid-static pressure, which may be caused by gravity and/or by rotation or may be applied from outside. It also includes fluid-dynamic contributions, for instance the force required for acceleration and deceleration of the liquid (= inertia forces) or forces generated during oscillation of a drop against viscous friction.

The interface tension tries to reduce areas with convex curvature; the pressure counteracts this tendency. Figure 3.2 shows the typical shape of a liquid column between two coaxial circular disks. Under the earth's gravity

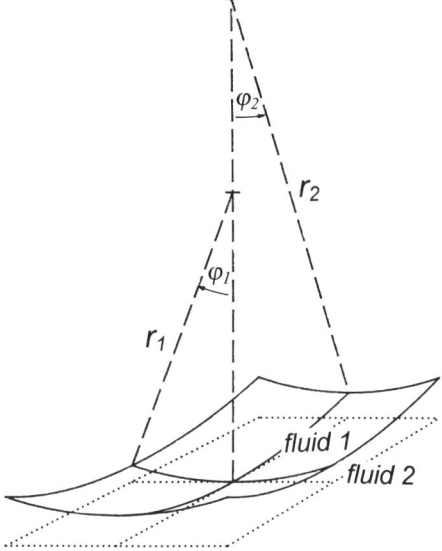

Fig. 3.1. Sketch illustrating the derivation of the capillary equation (balance of forces)

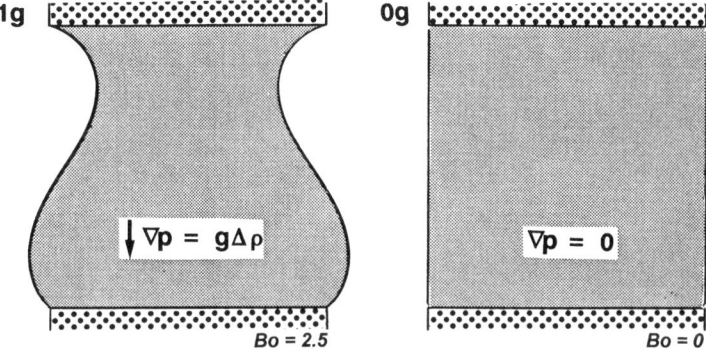

Fig. 3.2. The shape of a liquid column between coaxial circular disks on earth ($1g$) and under microgravity ($0g$). Terrestrially, the increase in fluid-static pressure from *top* to *bottom* gives rise to a corresponding increase in the column's curvature

($1g$), the fluid-static pressure increases from top to bottom, requiring a corresponding increase in curvature. Whereas in the upper region of the column the curvature is convex (= positive) in the axial cross section and concave (= negative) in the meridional cross section, in the lower region of the column the curvature is convex (= positive) in both these directions. Under microgravity ($0g$), owing to the constancy of fluid-static pressure, the mean curvature is constant as well, such that large and perfectly cylindrical columns may be established.

3.2 Minimization of Energy

The alternative to deriving the capillary equation from the balance of forces is to minimize the energy of the fluid [Gauss 1830]. A minimum in the free energy is equivalent to a balance of forces, if no irreversible energy losses such as viscous friction are involved. All forces must be conservative, i.e. they must have a potential. Minimization of the fluid energy is usually subject to several constraints. Most importantly, constancy of the liquid volume has to be required, and in the case of rotation the additional condition of constant angular momentum may arise. Figure 3.3 shows a section of a fluid surface, expressed as

$$z = z(x, y) \tag{3.3}$$

in Cartesian coordinates relative to the x, y plane. The area of the surface element corresponding to $\Delta x \, \Delta y$ equals the absolute value of the vector product

$$\left| \left(\Delta x, 0, \frac{\partial z}{\partial x} \Delta x \right) \times \left(0, \Delta y, \frac{\partial z}{\partial y} \Delta y \right) \right| = \left| \left(-\frac{\partial z}{\partial x}, -\frac{\partial z}{\partial y}, 1 \right) \Delta x \, \Delta y \right| = S \, \Delta x \, \Delta y \,, \tag{3.4}$$

where

$$S = \sqrt{1 + \left(\frac{\partial z}{\partial x} \right)^2 + \left(\frac{\partial z}{\partial y} \right)^2} \,. \tag{3.5}$$

The area A of the fluid surface is given by

$$A = \int dx \int dy \sqrt{1 + \left(\frac{\partial z}{\partial x} \right)^2 + \left(\frac{\partial z}{\partial y} \right)^2} = \int dx \int dy \, S \,. \tag{3.6}$$

For the liquid volume, one obtains

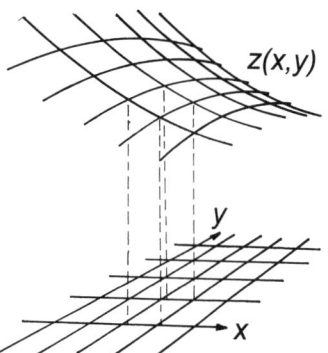

Fig. 3.3. A section of a fluid surface $z = z(x, y)$ relative to the x, y plane

$$V = \int dx \int dy\, z(x,y)\,. \tag{3.7}$$

Contributions to the potential energy E_{pot} arise from gravity and contributions to the kinetic energy E_{kin} arise from rotation:

$$E_{\text{pot}} = \frac{1}{2}g\rho \int dx \int dy\, z^2\,, \tag{3.8}$$

$$E_{\text{kin}} = -\frac{1}{2}\rho\omega^2 \int dx \int dy\, z\left(x^2 + y^2\right)\,. \tag{3.9}$$

The total energy of the fluid is given by

$$E = \sigma A + E_{\text{pot}} + E_{\text{kin}}\,. \tag{3.10}$$

Minimization of the energy under the constraint of constant liquid volume is equivalent to the minimization of $E - pV$, where the capillary pressure p is the Lagrange multiplier.

According to the calculus of variations, minimization of the integral

$$F = \int dx_1 \int dx_2 \ldots \int dx_N\, f\left(z, \frac{\partial z}{\partial x_1}, \frac{\partial z}{\partial x_2}, \ldots, \frac{\partial z}{\partial x_N}\right) \tag{3.11}$$

with respect to the trial function

$$z = z(x_1, x_2, \ldots, x_N) \tag{3.12}$$

of the N variables $x_1, x_2, \ldots, x_N$ is achieved for

$$\sum_{n=1}^{N} \frac{d}{dx_n} \frac{\partial f}{\partial(\partial z/\partial x_n)} = \frac{\partial f}{\partial z}\,. \tag{3.13}$$

Substituting $E - pV$, according to (3.6)–(3.10), into (3.11) for $N = 2$, $x_1 = x$, $x_2 = y$ leads to

$$f\left(z, \frac{\partial z}{\partial x}, \frac{\partial z}{\partial y}\right) = \sigma S + \frac{1}{2}g\rho z^2 - \frac{1}{2}\rho\omega^2 z\left(x^2 + y^2\right) - pz\,. \tag{3.14}$$

The only contribution depending on the derivatives $\partial z/\partial x$, $\partial z/\partial y$ is the surface term σS; this enters the left-hand side of (3.13). The liquid volume and the terms contributing to the potential energy depend on the coordinates x, y, z; these constitute the right-hand side of (3.13). We obtain

$$\sigma\left(\frac{dn_x}{dx} + \frac{dn_y}{dy}\right) = p - g\rho z + \frac{1}{2}\rho\omega^2\left(x^2 + y^2\right)\,, \tag{3.15}$$

where n_x, n_y are the components of the surface normal

$$\boldsymbol{n} = (n_x, n_y, n_z) = \frac{1}{S}\left(-\frac{\partial z}{\partial x}, -\frac{\partial z}{\partial y}, 1\right)\,. \tag{3.16}$$

The curvature of the surface $z(x,y)$ thus is given by the two-dimensional divergence of its normal. The left-hand side of (3.15) is equivalent to the right-hand side of (3.2), since according to (3.1) the sum of the curvatures in any two perpendicular directions φ and $\varphi+\pi/2$ equals $\kappa_1+\kappa_2$. Rotating the coordinate system x,y must not change the surface $z(x,y)$. Any orthogonal coordinate system may be used. In (3.15) the terms making up the potential energy and the kinetic energy have turned into contributions to the pressure p. The pressure includes the capillary pressure and the fluid-static pressure caused by gravity and rotation. The capillary equation is the Euler–Lagrange equation resulting from the variation of the energy, with the pressure p being the Lagrange parameter. More precisely, the differences in pressure Δp and in density $\Delta \rho$ between two adjacent fluids have to be considered.

Most of the analytical work on capillary surfaces has been and still is based on the capillary equation. In particular, the capillary equation offers great advantages if, owing to symmetry arguments, it reduces to an ordinary differential equation. On the other hand, minimization of the energy of the fluid had a renaissance when computer calculations became fast and easy. In the vicinity of its minimum, the energy is a quadratic form in the coordinates, which allows effective mathematical codes. Now, finite-element methods are used much more frequently than finite-difference methods. Considering the energy is also the only way to rigorously check for stability. Stability is not determined locally; an increase in energy in one region of the surface may be overcompensated by a decrease in another region.

Minimization of the energy of the liquid under the constraint of volume conservation actually means that the *free energy* = energy – pressure × volume = $f - pV$ is minimized. From this formalism, it follows that energy and volume are in fact treated equivalently. In consequence, a minimum in the energy E within a branch of solutions also means a minimum in the volume V.

The integrals (3.6)–(3.9) have intentionally been left undetermined. This is adequate for deriving the capillary equation. In reality, a fluid surface is in contact with a solid surface or otherwise is closed. It follows from Young's boundary condition (2.11) that the area A of contact with a solid should be taken into account with weight $-A\sigma \cos\gamma$. Using the notation of Fig. 2.7 and applying (2.11), one has for the total energy of the surfaces of the fluid and the solid

$$\sigma_{12}A_{12}+\sigma_{13}A_{13}+\sigma_{23}A_{23}=\sigma_{12}\left(A_{12}-\cos\gamma_1 A_{13}\right)+\sigma_{23}\left(A_{13}+A_{23}\right) . \tag{3.17}$$

The area A_{13} is the area wetted by fluid 1, it enters the energy with weight $\sigma\cos\gamma$. The area $(A_{13}+A_{23})$ is the total surface area of the solid, which leads to a constant contribution to the energy. It is not affected by the shape of the fluid surface.

For reference, representations of the capillary equation

- in Cartesian coordinates $z(x,y)$
- in cylindrical coordinates $z(r,\varphi)$
- in cylindrical coordinates $r(z,\varphi)$
- in polar coordinates $r(\vartheta,\varphi)$

are presented in Sect. 3.7.

3.3 Analytical Solutions of the Capillary Equation

The capillary equation is a partial differential equation of order two. It reduces to an ordinary differential equation if, owing to symmetry, the capillary surface does not depend on one of the coordinates. The most frequent situations of this kind are plane symmetry and rotational symmetry:

- a single plane in a large liquid reservoir (Fig. 2.9)
- two parallel planes in a large reservoir
- a free liquid drop
- a free liquid jet
- capillary surfaces between coaxial circular disks, i.e. liquid bridges (or zones)
- capillary surfaces between parallel horizontal planes (Fig. 1.6)
- capillary surfaces in cylindrical containers (circular tubes).

In these cases the capillary equation can often be solved analytically or else can be conveniently integrated by a fourth-order Runge–Kutta method. A quite commonly used method is the *shooting method*: a starting point of the solution is chosen and the slope and pressure are iteratively adapted in such a manner that a final point or a final slope at a given distance or height is obtained. A Newtonian method, which simultaneously integrates the surface and its derivatives with respect to slope and pressure, usually guarantees good convergence. For the general three-dimensional case, effective computer codes based on finite elements are available, e.g. SURFACE EVOLVER (static) [Brakke 1995] and FIDAP (dynamic).

3.3.1 Rise of Liquid in a Tube

The best-known solutions of the capillary equation are spherical surfaces, as in free liquid drops. These surfaces are the key to the rise of liquids in capillaries as well. Let us first argue by means of the capillary equation. Assuming that the capillary is sufficiently narrow to enable a spherical surface, we find the radius r of the surface, which forms a contact angle γ with the tube of radius R, to be given by $r = R/\cos\gamma$; see Fig. 3.4. This leads to a curvature of $2\cos\gamma/R$ and a capillary underpressure of $2\sigma\cos\gamma/R$. This underpressure

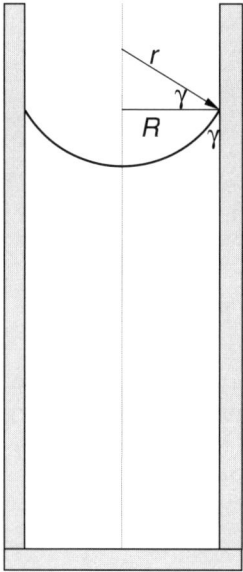

Fig. 3.4. Spherical surface with contact angle γ with a tube of radius R

sucks the liquid into the capillary to height z, where it is balanced by the fluid-static underpressure $g\rho z$. This leads to

$$\frac{2\sigma \cos \gamma}{R} = g\rho z \,. \tag{3.18}$$

An alternative derivation of the same equation may be performed by calculation of the energy of the fluid. The gain in surface energy with an increase in height δz is given by $2\pi R \sigma \cos \gamma \, \delta z$; the corresponding increase in potential energy is equivalent to lifting a layer of thickness δz to the surface at height z, yielding $\pi R^2 g \rho h \, \delta z$. Hence

$$2\pi R \sigma \cos \gamma = \pi R^2 g \rho z \,. \tag{3.19}$$

The latter treatment does not make use of the fact that the capillary surface in the thin capillary actually forms a section of a sphere. Instead, it is assumed that the fluid surface has a nearly constant height z. It is the average height z of a layer of thickness δz placed on top of the surface which enters the potential energy. The right-hand side of (3.19) actually equals the integral of the cross section over the height z or over the pressure p. The extension of this equation to tubes with arbitrary cross section hence reads

$$\int d\Sigma \sigma \cos \gamma = \int d\Omega g\rho z \,, \tag{3.20}$$

where $\int d\Sigma$ is the integral over the periphery and $\int d\Omega$ is the integral over the cross section; see Fig. 3.5. This treatment of the rise of a liquid in a

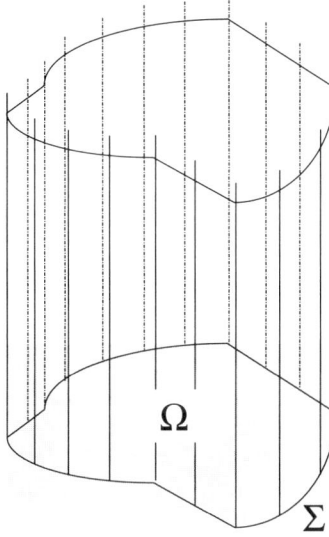

Fig. 3.5. Cylinder with cross section Ω and periphery Σ (domain)

capillary in fact makes the constraint of constant liquid volume redundant. Instead, it is assumed that liquid volume in the reservoir is sufficiently large. And we are not looking for the minimum in energy, but for a balance of the surface energy and the potential energy. Equation (3.20) forms the basis of an effective integral theorem, which allows one to calculate the capillary pressure in a cylindrical tube without explicitly calculating the shape of the surface. This theorem and its consequences are treated explicitly in Sect. 8.2.

3.3.2 Spherical Surfaces

A liquid sphere of radius R forms a contact angle γ with a solid plane, if the distance of the sphere's center from the plane equals $R \cos \gamma$. If several planes $1, 2, \ldots, N$ with contact angles $\gamma_1, \gamma_2, \ldots, \gamma_N$ are present a solution of the capillary equation satisfying the boundary conditions on the contact angles is obtained if the center of the sphere is at distances $R \cos \gamma_1$, $R \cos \gamma_2, \ldots, R \cos \gamma_N$, respectively from the planes (see Fig. 3.6). By choosing appropriate planes, this principle may be used effectively for calculating the volume and surface area of

- liquid drops in a straight wedge
- liquid drops in a wedge formed by touching spherical sections
- liquid drops in a polygonal cylinder
- liquid drops in a tripod
- liquid drops in regular N-pods.

Let us consider the cube shown in Fig. 3.7. If the contact angle of the liquid with the cube material is large, $\gamma > 54.73°$ (or $\gamma > \arccos(1/\sqrt{3})$,

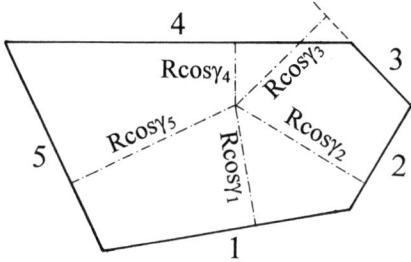

Fig. 3.6. A liquid sphere of radius R forms contact angles $\gamma_1, \gamma_2, \ldots, \gamma_N$ with a set of solid planes $1, 2, \ldots, N$ if the distances of the sphere's center from the planes equal $R \cos \gamma_1, R \cos \gamma_2, \ldots \ldots, R \cos \gamma_N$, respectively

which is the angle between an edge and the space diagonal), the stable liquid configuration under microgravity conditions is a single spherical droplet in one corner. One might consider eight spherical droplets in the eight corners of the cube instead. However, convex droplets are in equilibrium with a vapor pressure that becomes higher as the size of the droplets is reduced. If one of the droplets is only slightly larger than the others, it tends to lower the vapor pressure in the cube, whereas the smaller ones attempt to increase the vapor pressure. The large droplets therefore will grow at the expense of the smaller ones until only the largest droplet is left. That's the way the cookie crumbles!

If the contact angle γ of the liquid with the cube material is between 54.73° and 45° (or $\gamma = \arccos(1/\sqrt{2})$, which is the angle between an edge and a face diagonal), concave spherical droplets in the corners are formed. Concave surfaces cause a capillary underpressure. They are in equilibrium with a vapor pressure that becomes lower as the droplets become smaller. The vapor pressure of the large droplets is therefore higher and they will shrink until all droplets are equal in size. This leads to eight equal concave droplets in the eight corners of the cube; see Fig. 3.7b. Eventually, if the contact angle γ is less than 45°, the liquid penetrates into all edges. We

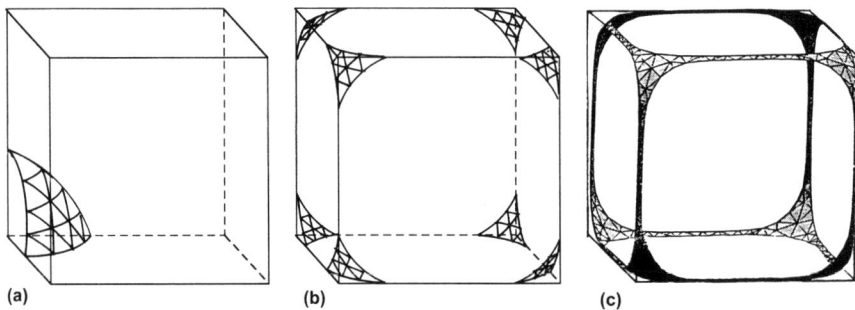

Fig. 3.7. Liquid surfaces in a cube. (**a**) Contact angle $\gamma > 54.7°$, convex spherical surface in one corner; (**b**) $54.7° > \gamma > 45°$, concave spherical surfaces in all corners; (**c**) $45° > \gamma$, concave cylindrical surfaces in wedges connect surplus volumes in corners

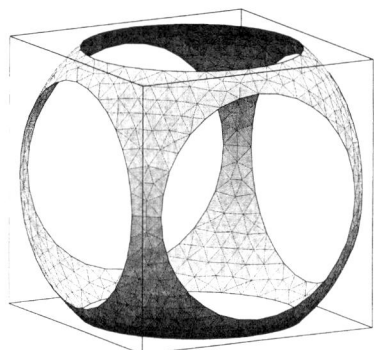

Fig. 3.8. Spherical liquid surface in a cube [Mittelmann 1993]

obtain nearly cylindrical surfaces in the edges, which are connected by nearly spherical surfaces in the corners; see Fig. 3.7c.

With increasing liquid volume, a single spherical solution is obtained, which corresponds to the condition $R \cos \gamma =$ half the edge length (Fig. 3.8). Such single spherical solutions also arise in regular polyhedrons or, more generally, in polyhedrons with equal distances of all faces from the center. The liquid volume has to be chosen accordingly.

3.3.3 Rise of a Liquid in Contact with an Infinite Plane

Another exact analytical solution of the capillary equation is obtained for the one-dimensional problem of a solid plane in an infinite liquid reservoir. A wetting liquid climbs up the plane, as indicated in Fig. 2.9. The capillary underpressure increases linearly with the height z. We obtain the differential equation

$$\frac{\sigma}{\sqrt{1+(\mathrm{d}z/\mathrm{d}x)^2}^3} \frac{\mathrm{d}^2 z}{\mathrm{d}x^2} = g \Delta \rho \, z \,. \tag{3.21}$$

Normalization by twice the capillary length $2(\sigma/g \, \Delta \rho)^{0.5}$ leads to

$$\frac{1}{\sqrt{1+(\mathrm{d}\tilde{z}/\mathrm{d}\tilde{x})^2}^3} \frac{\mathrm{d}^2 \tilde{z}}{\mathrm{d}\tilde{x}^2} = 4\tilde{z}, \quad \tilde{x} = \sqrt{\frac{g \Delta \rho}{4\sigma}} x, \quad \tilde{z} = \sqrt{\frac{g \Delta \rho}{4\sigma}} z \,. \tag{3.22}$$

After multiplication by the derivative $\mathrm{d}\tilde{z}/\mathrm{d}\tilde{x}$, (3.22) may be integrated to give

$$c - \frac{1}{\sqrt{1+(\mathrm{d}\tilde{z}/\mathrm{d}\tilde{x})^2}} = 2\tilde{z}^2 \,, \tag{3.23}$$

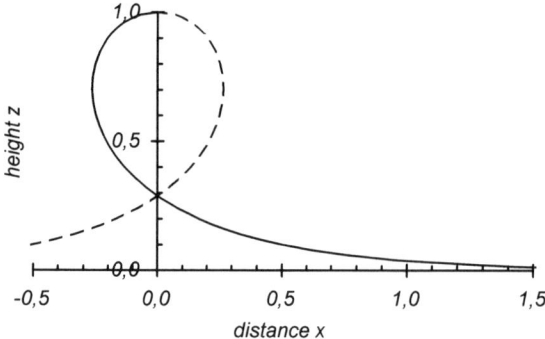

Fig. 3.9. Rise of a liquid in contact with an infinite plane. The surface approaches the level $z = 0$ exponentially at large distances

where c is the first integration constant. The requirement of a horizontal slope $d\tilde{z}/d\tilde{x} = 0$ at infinity for $z = 0$ leads to $c = 1$ and, after a second integration (Fig. 3.9),

$$\frac{1}{2} \log \left(\frac{\tilde{z}}{1 + \sqrt{1 - \tilde{z}^2}} \right) + \sqrt{1 - \tilde{z}^2} = (\tilde{x} - \tilde{x}_0) \ . \tag{3.24}$$

The second integration constant $\tilde{x}_0$ may be equated to zero. The surface approaches the level $\tilde{z} = 0$ exponentially at large distances. For the height z of a surface in contact with an inclined plane at an angle of ψ to the horizontal, one has, from the first integral (3.23),

$$\tilde{z} = \sqrt{\frac{1}{2}[1 - \cos(\psi - \gamma)]} = \sin\left(\frac{1}{2}(\psi - \gamma)\right) \ . \tag{3.25}$$

3.4 Axisymmetric Surfaces

The axisymmetric solutions of the capillary equation in the absence of gravity and rotation were intensively studied by Delaunay [1841]. These solutions are called Delaunay curves. In addition to the well-known sphere, they consist of

- the unduloids; which exhibit a periodic wavy surface (Fig. 3.10). They may be obtained by rolling an ellipse without slip along an axis and rotating the resultant curve around the axis [Finn 1986].
- the periodic nodoids, which intersect themselves (Fig. 3.11), such that only sections of them may be realized.
- the catenoid, which has a mean curvature of zero. Its principal curvatures in the axial and azimuthal directions have opposite sign. It is obtained by rotating a catenary, i.e. a hyperbolic cosine, around the axis (Fig. 3.12).

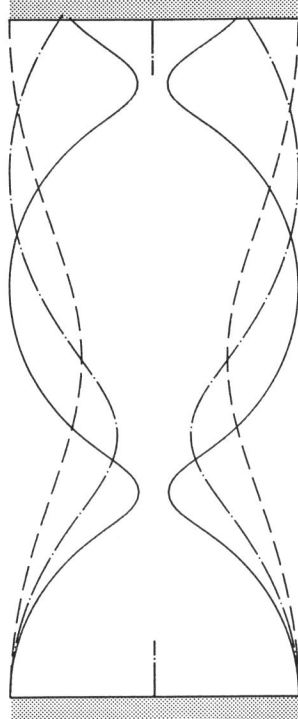

Fig. 3.10. Unduloids with constant maximum radius and varying minimum radius

The link between the families of unduloids and nodoids is a periodic series of coaxial spheres. The sphere and catenoid can be represented by elementary functions, and the unduloid and nodoid by elliptic integrals. For both types of surfaces, it has been proven that they are stable only if their height does not exceed a single period. Otherwise, they break antimetrically in the middle, i.e. one half widens and the other half narrows. This instability is called the Rayleigh instability. Lord Rayleigh proved as early as 1879 [Rayleigh 1879] that a cylindrical liquid jet becomes unstable if its length exceed its circumference; see also Chap. 4. It is distinguished within the family of unduloids by having equal minimum and maximum radii.

Using cylindrical coordinates $z(r, \varphi)$, one finds, for the curvature κ_1 in the radial direction and the curvature κ_2 in the azimuthal direction,

$$\kappa_1 = -\frac{\mathrm{d}}{\mathrm{d}r}\frac{\mathrm{d}z/\mathrm{d}r}{\sqrt{1+(\mathrm{d}z/\mathrm{d}r)^2}}, \quad \kappa_2 = -\frac{1}{r}\frac{\mathrm{d}z/\mathrm{d}r}{\sqrt{1+(\mathrm{d}z/\mathrm{d}r)^2}}. \tag{3.26}$$

Adding up the curvatures leads to the capillary equation,

$$\frac{1}{r}\frac{\mathrm{d}}{\mathrm{d}r}\frac{r\,\mathrm{d}z/\mathrm{d}r}{\sqrt{1+(\mathrm{d}z/\mathrm{d}r)^2}} = \frac{p}{\sigma} - \frac{g\Delta\rho z}{\sigma} + \frac{\Delta\rho\omega^2 r^2}{2\sigma}. \tag{3.27}$$

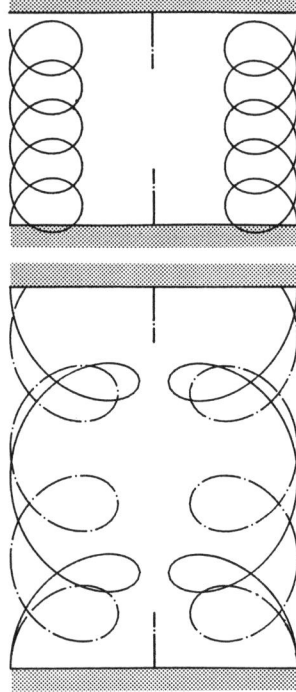

Fig. 3.11. Nodoids with constant maximum radius and varying minimum radius

The expression $(dz/dr)/[1 + (dz/dr)^2]^{0.5}$, the radial component of the surface normal (3.16), equals the sine of the surface's slope. For zero gravity, after multiplication by the radius r, one obtains as a first integral of (3.27)

$$\frac{r\,dz/dr}{\sqrt{1 + (dz/dr)^2}} = r\sin\varphi = r_0 + \frac{pr^2}{2\sigma} + \frac{\Delta\rho\omega^2 r^4}{8\sigma}\ . \tag{3.28}$$

For zero circular frequency of rotation ω, (3.28) is a quadratic equation in the radius r. If the integration constant r_0 and the pressure p/σ have the same sign, the right-hand side of (3.28) has no zero, and $\sin\varphi$ is positive everywhere. We obtain unduloids, which have no radial tangent. Figure 3.10 shows three unduloids with equal maximum radius but differing in pressure.

If the integration constant r_0 in (3.28) differs in sign from p/σ, $\sin\varphi$ may equal zero. We obtain nodoids, which intersect themselves and have a tangent (tangential plane) in the radial direction. Three nodoids differing in pressure are shown in Fig. 3.11.

Finally, for $p/\sigma = 0$, we are left with the equation of the catenoid (Fig. 3.12),

$$\frac{dz/dr}{\sqrt{1 + (dz/dr)^2}} = \frac{r_0}{r}, \quad \frac{r}{r_0} = \cosh\left(\frac{z - z_0}{r_0}\right)\ . \tag{3.29}$$

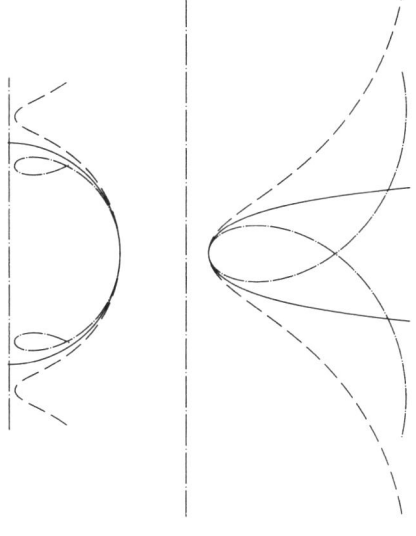

Fig. 3.12. The sphere (*left*) and the catenoid (*right*) represent the link between the family of unduloids and the family of nodoids

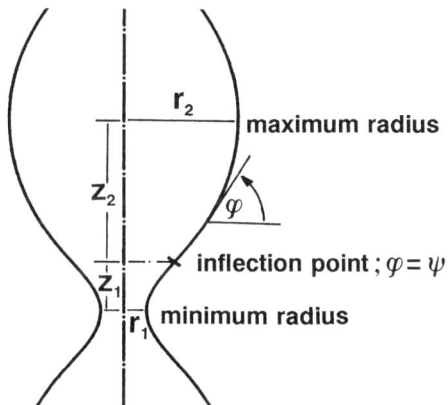

Fig. 3.13. The axial distances z_2 and z_1 of the maximum radius r_2 and of the minimum radius r_1 of the unduloid from the inflection point

A discussion of the unduloids and nodoids can be conveniently given by introducing the minimum and maximum radii r_1 and r_2 of the unduloid, as shown in Fig. 3.13 [Langbein 1992]:

$$r \sin \varphi = \frac{r^2 + r_1 r_2}{r_1 + r_2}, \quad \frac{r_1 r_2}{r_1 + r_2} = r_0, \quad \frac{p}{\sigma} = \frac{2}{r_1 + r_2}. \tag{3.30}$$

In the case of a catenoid r_2 equals infinity, and in the case of a nodoid r_2 equals the negative maximum radius. From (3.30), we obtain for the inflection point of an unduloid

$$r = \sqrt{r_1 r_2}. \tag{3.31}$$

The radius at the inflection point equals the geometric mean of the minimum and the maximum radius. The minimum angle $\psi = \varphi_{\min}$ of the liquid surface at the inflection point of the unduloid is given by

$$\sin\psi = \frac{2\sqrt{r_1 r_2}}{r_2 + r_1}, \quad \cos\psi = \frac{r_2 - r_1}{r_2 + r_1}. \tag{3.32}$$

For the axial distance z_1 of the inflection point from the position of minimum radius r_1, we obtain (see Fig. 3.13)

$$z_1 = \int_{r_1}^{\sqrt{r_1 r_2}} dr \tan\varphi = \int_{r_1}^{\sqrt{r_1 r_2}} dr \frac{r^2 + r_1 r_2}{\sqrt{[r(r_1 + r_2)]^2 - [r^2 + r_1 r_2]^2}}. \tag{3.33}$$

Substituting (3.32) and setting

$$\left(\frac{2r}{r_2 + r_1}\right)^2 = 1 - 2\cos\psi\cos\xi + \cos^2\psi \tag{3.34}$$

yields

$$z_1 = \frac{r_2 + r_1}{2} \int_0^{\psi} d\xi \frac{1 - \cos\psi\cos\varphi}{\sqrt{1 - 2\cos\psi\cos\varphi + \cos^2\psi}}. \tag{3.35}$$

The integral on the right-hand side of (3.35) is an incomplete elliptic integral, which may be transformed into a complete elliptic integral by application of a descending Landen transformation (17.5.1)–(17.5.3) in Abramowitz & Stegun 1965. We obtain

$$z_{1,2} = \frac{r_2 + r_1}{2}\left[E(\cos^2\psi) \mp \cos\psi\right]. \tag{3.36}$$

The negative and the positive sign on the right-hand side of (3.36) apply to the axial distance z_2 from the inflection point of the positions of minimum radius r_1 and maximum radius r_2, respectively. The difference between the axial distances z_2 from the maximum radius to the inflection point and z_1 from the inflection point to the minimum radius thus equals the difference between the maximum and the minimum radii:

$$z_2 - z_1 = r_2 - r_1. \tag{3.37}$$

Similar results are obtained for the family of nodoids. For example, (3.31) with r_2 replaced by $|r_2|$ determines the radius at which there is a radial tangent. Also, (3.36) and (3.37) hold for the axial distances of the positions z_1 and z_2 of minimum and maximum radius from that position;

$$z_2 + z_1 = |r_2| - r_1. \tag{3.38}$$

The unduloid is generated by rolling an ellipse with major axis a and eccentricity e rigidly on an axis without slipping. Let C be the curve swept out by one of the foci of the ellipse. Then the surface generated by rotating C around the axis has a constant curvature $\kappa_1 + \kappa_2 = 1/a$ [Finn 1986].

3.5 Container Shape and Wetting

Further examples of axisymmetric surfaces are depicted in Fig. 3.14. This shows a container filled with a well-wetting liquid (contact angle $\gamma \geq 0$). In Fig. 3.14a, a rod, which is also well wetted by the liquid, is inserted in the center of the container. The resultant surface forms a section of a nodoid. In Fig. 3.14b another rod, with contact angle $\gamma = \pi/2$, is used. The liquid now assumes a nearly spherical shape, the same shape as in the absence of the rod. Finally, in Fig. 3.14c a badly wetting rod is used. The resulting liquid surface forms a section of an unduloid. The situations depicted are maintained if the unduloid, the sphere or the nodoid are slightly deformed due to gravity, see Fig. 3.14.

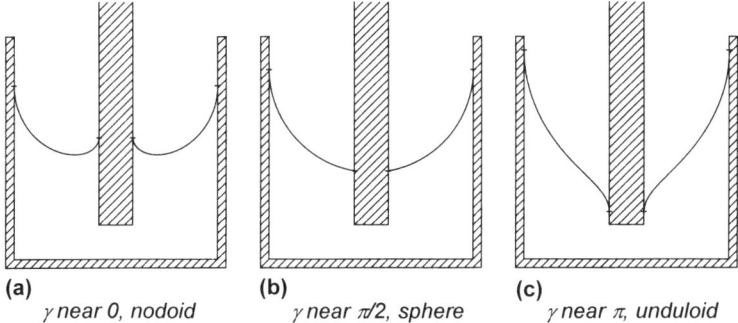

(a) γ near 0, nodoid
(b) γ near $\pi/2$, sphere
(c) γ near π, unduloid

Fig. 3.14. Axisymmetric surfaces arising for different contact angles. The liquid wets the container perfectly: (**a**) γ near 0, nodoid; (**b**) γ near $\pi/2$, sphere; (**c**) γ near π, unduloid. The capillary pressure is obtained from the integral theorem (Sect. 8.2). The rod is driven toward the periphery

From (3.20) we conclude that the capillary pressure is independent of the position of the rod in the container. If R_2 and R_1 are the radii of the container and of the rod, respectively, and γ_2 and γ_1 the corresponding contact angles, we find for the capillary pressure inside the liquid

$$\pi(R_2^2 - R_1^2)p = -\sigma(2\pi R_2 \cos\gamma_2 + 2\pi R_1 \cos\gamma_1) \ . \tag{3.39}$$

If the rod is slightly moved off the center to the right, the negative radial curvature at that side increases and the positive azimuthal curvature increases by the same amount. The liquid surface thus rises on the right-hand side and sinks on the left-hand side. The wetted portions on the right-hand and the left-hand sides of the rod behave differently and a net force on the rod is left. Its sign is determined by the right-hand side of (3.39), i.e. the rod experiences a force toward the periphery if the liquid wets the container wall, i.e. $r_2 \cos\gamma_2 + r_1 \cos\gamma_1 > 0$. It is driven back toward the center if the liquid does not wet the container wall, i.e. $r_2 \cos\gamma_2 + r_1 \cos\gamma_1 < 0$.

58   3. Capillary Shape and Stability

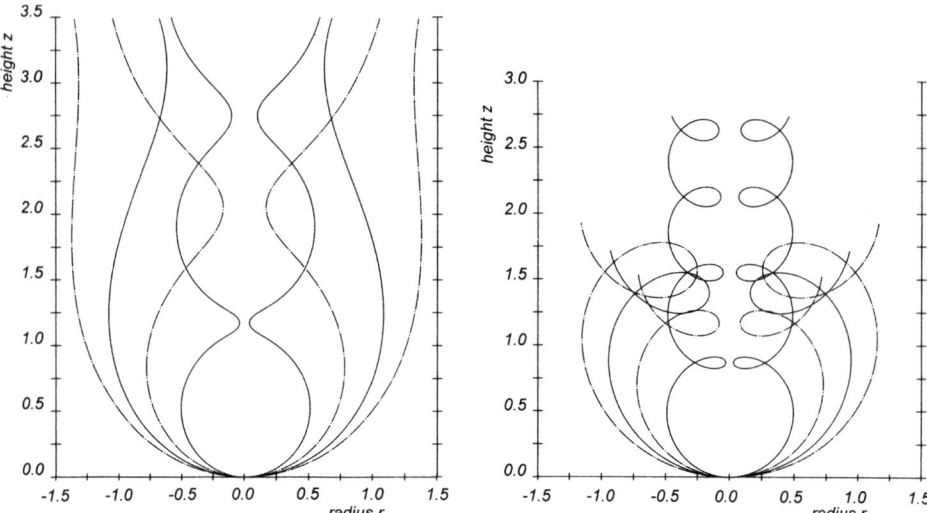

Fig. 3.15. The deformation of the unduloid and nodoid due to gravity

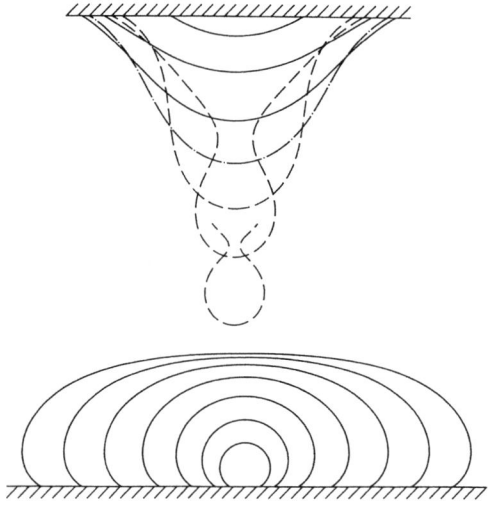

Fig. 3.16. Hanging and sessile drops with increasing liquid volume. *Dash–dot line:* hanging drop with maximum stable volume; *dashed lines:* unstable hanging drops

The same conclusion may be drawn regarding drops and bubbles, although in that case the intermediate states do not necessarily comply with the balance of forces required by the capillary equation and Neumann's boundary condition. It is an advantage of considering a solid rod that it may be fixed at any position until the balance of forces is realized. In the case of drops and bubbles one must require that the adjustment of the surface and pressure is fast compared with the resultant drop motion. Drops and bubbles agglomerate and move to the periphery if the liquid wets the container wall. They move to the center if the liquid does not wet the container wall.

If liquid is added to the sessile drop shown in Fig. 3.16, the drop widens. If liquid is added to the hanging drop shown in Fig. 3.16, the drop lengthens. Later a neck arises, which rapidly becomes thinner and eventually breaks. There exists a maximum for the volume of a hanging drop. One of the methods of determining surface tension is to measure the volume of falling drops. This volume, however, is not really covered by the capillary equation, but rather depends on the dynamics of breakage.

3.6 Drops at Low Bond Numbers

In order to treat the deformation of spherical surfaces at small Bond numbers, in particular the flattening of sessile and the lengthening of hanging drops, we apply the capillary equation in polar coordinates (see (3.60)). Axisymmetry leads to

$$\frac{3-(\mathrm{d}r/r\mathrm{d}\vartheta)\cot\vartheta}{\sqrt{1+(\mathrm{d}r/r\mathrm{d}\vartheta)^2}} - \frac{1+\mathrm{d}^2r/r\mathrm{d}\vartheta^2}{\sqrt{1+(\mathrm{d}r/r\mathrm{d}\vartheta)^2}^3} = \widehat{p}r - Bo\frac{rz}{R^2}, \quad Bo = \frac{g\Delta\rho R^2}{\sigma}. \tag{3.40}$$

Here $\widehat{p}$ is the local curvature determined according to (3.54). Considering nearly spherical surfaces with a basic radius R means that we may expand $r(\vartheta)$ into a Taylor series with respect to the Bond number,

$$r(\vartheta) = R\left[1 + Bo\, r_1(\vartheta) + Bo^2\, r_2(\vartheta) + Bo^3\, r_3(\vartheta) + \ldots\right]. \tag{3.41}$$

Substitution into (3.40) leads to

$$2 - \cot\vartheta \frac{\mathrm{d}r}{r\mathrm{d}\vartheta} - \frac{\mathrm{d}^2r}{r\mathrm{d}\vartheta^2} = \widehat{p}R\left(1 + Bo\, r_1 + Bo^2\, r_2 + \ldots\right)$$
$$+ Bo\left(1 + Bo\, r_1 + \ldots\right)\left[1 - \cos\vartheta(1 + Bo\, r_1 + \ldots)\right] \tag{3.42}$$

and, to zeroth order in the Bond number,

$$\widehat{p}R = 2. \tag{3.43}$$

This is the well-known equation of the curvature of a sphere. To first order in the Bond number, we are left with

$$\frac{\mathrm{d}^2 r_1}{\mathrm{d}\vartheta^2} + \cot\vartheta \frac{\mathrm{d}r_1}{\mathrm{d}\vartheta} + 2r_1 = -(1-\cos\vartheta) = -2\sin^2\left(\frac{\vartheta}{2}\right). \tag{3.44}$$

3. Capillary Shape and Stability

The homogeneous solution of (3.44) reads

$$r_1(\vartheta) = -\frac{1}{3}\left(\sin^2\left(\frac{\vartheta}{2}\right) + \cos\vartheta \log\left[\cos^2\left(\frac{\vartheta}{2}\right)\right]\right)$$

$$= -\sum_{n=2}^{\infty} \frac{n+1}{3(n-1)n} \sin^{2n}\left(\frac{\vartheta}{2}\right). \tag{3.45}$$

The terms quadratic in the Bond number in (3.42) yield the differential equation for $r_2(\vartheta)$

$$\frac{d^2 r_2}{r d\vartheta^2} + \cot\vartheta \frac{dr_2}{d\vartheta} + 2r_2 = -2r_1^2 - r_1(2 - 3\cos\vartheta). \tag{3.46}$$

Solution of the above differential equations for $r_1(\vartheta), r_2(\vartheta)$ and all higher-order contributions $r_n(\vartheta)$ is achieved by means of the general relation

$$\left\{\frac{\partial^2}{\partial \vartheta^2} + \cot\vartheta \frac{\partial}{\partial \vartheta} + 2\right\} \sin^{2n}\left(\frac{\vartheta}{2}\right)$$

$$= n^2 \sin^{2n-2}\left(\frac{\vartheta}{2}\right) - (n-1)(n+2)\sin^{2n}\left(\frac{\vartheta}{2}\right). \tag{3.47}$$

Let us apply the approximate solution found to the flattening of a sessile liquid drop on a solid plate exhibiting a contact angle γ. The liquid volume is given by

$$V(\vartheta) = \pi \int_0^\vartheta d\theta (r\sin\theta)^3. \tag{3.48}$$

At the polar angle ϑ the actual angle γ between the surface normal and the polar axis is given by

$$\gamma = \vartheta - \arctan\left(\frac{dr}{r d\vartheta}\right). \tag{3.49}$$

Up to terms linear in the Bond number, this leads to the following for the volume of a drop with contact angle γ:

$$V(\gamma) \approx \pi R^3 \int_0^\gamma d\theta \sin^3\theta \left\{1 - Bo\left(\sin^2\left(\frac{\theta}{2}\right) + 2\cos\theta \log\left[\cos\left(\frac{\theta}{2}\right)\right]\right)\right\}, \tag{3.50}$$

$$V(\gamma) \approx \frac{\pi R^3}{3}\left\{(2 - 3\cos\gamma + \cos^3\gamma)\right.$$

$$\left. - \frac{Bo}{2}\left(40\sin^6\left(\frac{\gamma}{2}\right) - 34\sin^8\left(\frac{\gamma}{2}\right) - \sin^4\gamma \log\left[\cos\left(\frac{\gamma}{2}\right)\right]\right)\right\}. \tag{3.51}$$

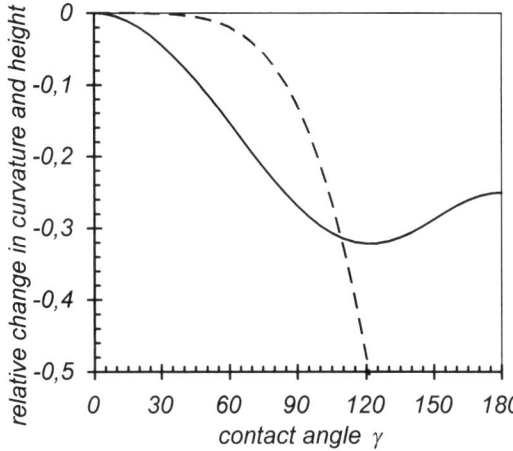

Fig. 3.17. Relative change of pole curvature (*full line*) and height (*dashed line*) of sessile and hanging drops at small Bond numbers. The relative height is defined on the basis of r_0

For a drop with contact angle γ and constant volume, the change of the pole radius R and of the height z with the Bond number (Fig. 3.17) Bo is given by

$$\frac{\delta R}{R} = Bo \frac{20\sin^2(\gamma/2) - 17\sin^4(\gamma/2) - 8\cos^4(\gamma/2)\log[\cos(\gamma/2)]}{12[3 - 2\sin^2(\gamma/2)]}, \quad (3.52)$$

$$\frac{\delta z}{R} = \frac{Bo}{3}\Bigg(\sin^2(\gamma/2) + 2\log[\cos(\gamma/2)] \quad\quad (3.53)$$

$$-\frac{4\sin^4(\gamma/2) + \sin^6(\gamma/2) + 8\sin^2(\gamma/2)\cos^4(\gamma/2)\log[\cos(\gamma/2)]}{2[3 - 2\sin^2(\gamma/2)]}\Bigg).$$

By means of similar reasoning, the pole radius and the height of sessile and hanging drops fixed to a solid disk may be calculated.

3.7 Representations of the Capillary Equation

Representations of the capillary equation in arbitrary orthogonal coordinate systems can all be obtained by differentiation of the components of the surface normal with respect to the independent components according to (3.15) and (3.16). Transformation of the liquid volume and of the different terms of the potential energy and pressure to the required coordinate system usually does not present any difficulties. The essential quantity for the surface normal is the area element S.

In the mathematical literature the mean curvature is sometimes denoted by H, i.e. $p/\sigma = \kappa_1 + \kappa_2 = 2H$. In order to stress that p is the pressure and p/σ is the curvature, we introduce instead

$$\widehat{p} \equiv \frac{p}{\sigma} = \kappa_1 + \kappa_2 . \quad (3.54)$$

3.7.1 Cartesian Coordinates $z(x, y)$

The normal to the surface $z - z(x, y) = 0$ is given by

$$\boldsymbol{n} = \frac{1}{S}\left(-\frac{\partial z}{\partial x}, -\frac{\partial z}{\partial y}, 1\right), \quad S = \sqrt{1 + \left(\frac{\partial z}{\partial x}\right)^2 + \left(\frac{\partial z}{\partial y}\right)^2}. \tag{3.55}$$

Differentiation of the components (= calculation of the two-dimensional divergence) of the normal leads to the capillary equation

$$\widehat{p} = \left(\frac{dn_x}{dx} + \frac{dn_y}{dy}\right)$$

$$= \frac{1}{S^3}\left\{\frac{\partial^2 z}{\partial x^2}\left[1 + \left(\frac{\partial z}{\partial y}\right)^2\right] - 2\frac{\partial z}{\partial x}\frac{\partial z}{\partial y}\frac{\partial^2 z}{\partial x \partial y} + \frac{\partial^2 z}{\partial y^2}\left[1 + \left(\frac{\partial z}{\partial x}\right)^2\right]\right\}. \tag{3.56}$$

For the surface area and the liquid volume one has

$$\text{area} = \int dx \int dy\, S; \quad \text{vol} = \int dx \int dy\, z. \tag{3.57}$$

Contributions to the potential energy arise from gravity E_{grv} and from rotation, E_{rot}:

$$E_{\text{grv}} = \frac{1}{2}g\rho \int dx \int dy\, z^2, \quad E_{\text{rot}} = -\frac{1}{2}\rho\omega^2 \int dx \int dy\, z\left(x^2 + y^2\right). \tag{3.58}$$

Here E_{rot} equals the negative of the kinetic energy of rotation.

3.7.2 Polar Coordinates $r(\vartheta, \varphi)$

The normal to the surface $r - r(\vartheta, \varphi) = 0$ is given by

$$\boldsymbol{n} = \frac{1}{S}\left(1, -\frac{\partial r}{r\,\partial\vartheta}, -\frac{\partial r}{r\sin\vartheta\,\partial\varphi}\right), \quad S = \sqrt{1 + \left(\frac{\partial r}{r\,\partial\vartheta}\right)^2 + \left(\frac{\partial r}{r\sin\vartheta\,\partial\varphi}\right)^2}. \tag{3.59}$$

The other quantities are given by

$$\widehat{p}r = \frac{1}{S}\left(3 - \frac{\partial r}{r\,\partial\vartheta}\cot\vartheta\right) - \frac{1}{S^3}\left\{\frac{\partial^2 r}{r\,\partial\vartheta^2}\left[1 + \left(\frac{\partial r}{r\sin\vartheta\,\partial\varphi}\right)^2\right]\right.$$

$$-\frac{\partial r}{r\,\partial\vartheta}\frac{\partial r}{r\sin\vartheta\,\partial\varphi}\left[\frac{\partial^2 r}{r\,\partial\vartheta(\sin\vartheta\,\partial\varphi)} + \frac{\partial^2 r}{r\sin\vartheta\,\partial\varphi\,\partial\vartheta}\right]$$

$$\left.+ \frac{1}{r\sin^2\vartheta}\frac{\partial^2 r}{\partial\varphi^2}\left[1 + \left(\frac{\partial r}{r\,\partial\vartheta}\right)^2\right]\right\}, \tag{3.60}$$

$$\text{area} = \int r\,d\vartheta \int r\sin\vartheta\,d\varphi\,S, \quad \text{vol} = \int r\,d\vartheta \int r\,d\varphi\,r\sin^2\vartheta\cos\vartheta, \tag{3.61}$$

$$E_{\text{grv}} = \frac{1}{2}g\rho \int r\,d\vartheta \int r\,d\varphi\,r^2\sin^2\vartheta\cos^2\vartheta,$$

$$E_{\text{rot}} = -\frac{1}{2}\rho\omega^2 \int r\,d\vartheta \int r\,d\varphi\,r^3\sin^4\vartheta\cos\vartheta. \tag{3.62}$$

3.7.3 Cylindrical Coordinates $r(\varphi, z)$

The normal to the surface $r - r(\varphi, z) = 0$ is given by

$$\mathbf{n} = \frac{1}{S}\left(1, -\frac{\partial r}{r\,\partial\varphi}, -\frac{\partial r}{\partial z}\right), \quad S = \sqrt{1 + \left(\frac{\partial r}{r\,\partial\varphi}\right)^2 + \left(\frac{\partial r}{\partial z}\right)^2}. \tag{3.63}$$

$$\widehat{p} = \frac{1}{Sr} - \frac{1}{S^3}\left\{\frac{\partial^2 r}{\partial z^2}\left[1 + \left(\frac{\partial r}{r\,\partial\varphi}\right)^2\right]\right.$$
$$- \frac{\partial r}{r\,\partial\varphi}\frac{\partial r}{\partial z}\left[\frac{\partial}{r\,\partial\varphi}\frac{\partial r}{\partial z} + \frac{\partial}{\partial z}\frac{\partial r}{r\,\partial\varphi}\right]$$
$$\left. + \frac{\partial}{r\,\partial\varphi}\left(\frac{\partial r}{r\,\partial\varphi}\right)\left[1 + \left(\frac{\partial r}{\partial z}\right)^2\right]\right\}, \tag{3.64}$$

$$\text{area} = \int r\,d\varphi \int dz\,S, \quad \text{vol} = \frac{1}{2}\int d\varphi \int dz\,r^2, \tag{3.65}$$

$$E_{\text{grv}} = \frac{1}{2}g\rho \int d\varphi \int dz\,r^2 z, \quad E_{\text{rot}} = -\frac{1}{4}\rho\omega^2 \int d\varphi \int dz\,r^4. \tag{3.66}$$

3.7.4 Cylindrical Coordinates $z(r, \varphi)$

The normal to the surface $z - z(r, \varphi) = 0$ is given by

$$\mathbf{n} = \frac{1}{S}\left(-\frac{\partial z}{\partial r}, -\frac{\partial z}{r\,\partial\varphi}, 1\right), \quad S = \sqrt{1 + \left(\frac{\partial z}{\partial r}\right)^2 + \left(\frac{\partial z}{r\,\partial\varphi}\right)^2}. \tag{3.67}$$

The other quantities are given by

$$\widehat{p}S^3 = \frac{1}{r}\frac{\partial}{\partial r}r\frac{\partial z}{\partial r}\left[1 + \left(\frac{\partial z}{r\,\partial\varphi}\right)^2\right]$$
$$+ \frac{1}{r}\frac{\partial z}{\partial r}\left[\left(\frac{\partial z}{\partial r}\right)^2 - 2\frac{\partial z}{r\,\partial\varphi}\frac{\partial^2 z}{\partial\varphi\,\partial r} + \left(\frac{\partial z}{r\,\partial\varphi}\right)^2\right]$$
$$+ \frac{\partial^2 z}{r^2\,\partial\varphi^2}\left[1 + \left(\frac{\partial z}{\partial r}\right)^2\right], \tag{3.68}$$

$$\text{area} = \int dr \int r\, d\varphi\, S, \quad \text{vol} = \int dr \int r\, d\varphi\, z(r,\varphi), \tag{3.69}$$

$$E_{\text{grv}} = \frac{1}{2}g\rho \int d\varphi \int dz\, r^2 z, \quad E_{\text{rot}} = -\frac{1}{4}\rho\omega^2 \int d\varphi \int dz\, r^4. \tag{3.70}$$

3.7.5 Axisymmetry

In axisymmetric situations the capillary equation reduces to an ordinary differential equation. One of the planes of principal curvature κ_1 is the plane through the symmetry axis; the plane of principal curvature κ_2 extends in the azimuthal direction. In cylindrical coordinates $r(z)$, one finds

$$\kappa_1 = -\frac{1}{\sqrt{1+(dr/dz)^2}^3}\frac{d^2 r}{dz^2}, \quad \kappa_2 = \frac{1}{r\sqrt{1+(dz/dr)^2}},$$

$$\widehat{p} = \kappa_1 + \kappa_2, \tag{3.71}$$

whereas in cylindrical coordinates $z(r)$, one obtains

$$\kappa_1 = \frac{d^2 z/dr^2}{\sqrt{1+(dz/dr)^2}^3}, \quad \kappa_2 = \frac{dz/dr}{r\sqrt{1+(dz/dr)^2}},$$

$$\widehat{p} = \kappa_1 + \kappa_2 = \frac{1}{r}\frac{d}{dr}\left(\frac{r\,dz/dr}{\sqrt{1+(dz/dr)^2}}\right). \tag{3.72}$$

The axisymmetric fluid surfaces resulting from equations (3.71), (3.72) may be obtained conveniently by a Runge–Kutta method. Numerical difficulties arising for a vertical slope $dr/dz = 0$ or horizontal slope $dz/dr = 0$ may be effectively avoided by introducing as the independent variable the length ds of arc, according to

$$ds = \sqrt{(dr)^2 + (dz)^2}, \quad dr = ds\cos\varphi, \quad dz = ds\sin\varphi, \tag{3.73}$$

$$\widehat{p} = \frac{d\varphi}{ds} + \frac{\sin\varphi}{r}, \tag{3.74}$$

$$\frac{d^2 r}{ds^2} = -\frac{dz}{ds}\left(\widehat{p} - \frac{1}{r}\frac{dz}{ds}\right), \quad \frac{d^2 z}{ds^2} = \frac{dr}{ds}\left(\widehat{p} - \frac{1}{r}\frac{dz}{ds}\right). \tag{3.75}$$

(Note: in (3.72)–(3.75), φ is not the azimuth but is, rather, the angle corresponding to the slope dz/dr.)

References

1. Abramowitz M, Stegun IA: *Handbook of Mathematical Functions*. Dover, New York (1965)
2. Brakke K: *SURFACE-EVOLVER Manual*, version 1.98. The Geometry Center, Minneapolis (1995)
3. Delaunay CE: Sur la surface de révolution dont la courbure moyenne est constante. J. Math. Pures Appl. **6** (1841) 309–315
4. Finn R: *Equilibrium Capillary Surfaces*. Grundlehren der Mathematischen Wissenschaften, Vol. 284. Springer, Berlin, Heidelberg (1986) p. 82
5. Gauss CF: Principia generalia theoriae figurae fluidorum. Commentarii Societ. Regiae Scientiarum Gottingensis **7** (1830)
6. Hauksbee F: An Experiment touching the Ascent of Water between two Glass planes in an Hyperbolick Figure. Phil. Trans. Roy. Soc. London **27** (1712) 539–540
7. Hauksbee F: Some further Experiments touching the Ascent of Water between two Glass planes in an Hyperbolick Figure. Phil. Trans. Roy. Soc. London **28** (1713a) 153–154
8. Hauksbee F: A further account of the ascent of drops of spirit of wine between two glass planes $20\frac{1}{2}$ inches long; with a table of distances from the touching ends, and the angles of elevation. Phil. Trans. Roy. Soc. London **28** (1713b) 155
9. Langbein D: Stability of liquid bridges between parallel plates. Microgravity Sci. Technol. **5** (1992) 2–11
10. Laplace PS de: *Supplément au livre X du Traitée de Méchanique Céleste*. Couveier, Paris (1805 and 1806)
11. Mittelmann HD: Symmetric capillary surfaces in a cube. Part 2: near the limit angle. Lect. Appl. Math. **29** (1993) 339–361
12. Rayleigh Lord (J.W. Strutt): On the capillary phenomena of jets. Proc. Roy. Soc.London **29** (1879) 71–97
13. Segner JA: De figuris superficierum fluidarum. Commentarii Societ. Regiae Scientiarum Gottingensis **1** 301–372 (1751)
14. Taylor B: *Concerning the Ascent of Water between two Glass planes*. Phil. Trans. Roy. Soc. London **27** (1712) 538
15. Young T: An essay on the cohesion of fluids. Phil. Trans. Roy. Soc. London **95** 65–87 (1805)

4. Stability Criteria

The capillary equation is the Euler–Lagrange equation resulting from minimizing the energy of the liquid under the constraints of constant liquid volume, constant angular momentum, constant frequency of rotation, etc. A solution of the capillary equation, however, is not automatically stable. One may have reached a saddle point or even a maximum of the energy instead. Around an extremum, the energy of the liquid can be represented by a quadratic form in the coordinates, which may be transformed to its principal axes. If all eigenvalues of this transformation are positive, the surface is stable.

A simpler, sufficient but not necessary, stability criterion is the minimum-volume condition. If, within a family of solutions of the capillary equation, the liquid volume exhibits an extremum as a function of the capillary pressure, an instability has been reached. If the volume is plotted versus the pressure or another independent parameter, the stability limits are easily shown up. A further effective method of studying stability limits relies on intuition together with experimental evidence. One may simply guess the most critical deformation and check its relevance by linear stability analysis.

4.1 Stability of Capillary Surfaces

Capillary instability of liquid surfaces is an everyday experience. The best-known example is the free liquid jet. Water flowing from a garden hose or from a tap in the kitchen breaks after a while into droplets, see Fig. 4.1. On the other hand, honey running down from a spoon does not seem to break. This obviously is a matter of viscosity and time. The surface tension attempts to reduce the total surface area of the liquid but has to work against the inertia and viscosity of the liquid. Any gain in energy due to a surface deformation is used up primarily by acceleration of the liquid, i.e. it turns into kinetic energy. The characteristic time of acceleration is given by the ratio of the mass of the liquid to the surface tension, $t = (\rho R^3/\sigma)^{0.5}$. If the liquid surface is stable, this portion of the energy is not lost, but leads to oscillation of the liquid. The circular frequencies ω of such oscillations thus should be scaled by the Weber number [Weber 1931]

68   4. Stability Criteria

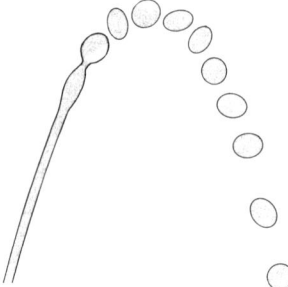

Fig. 4.1. Breakage of a water jet flowing from a garden hose

$$We = \frac{\rho \omega^2 R^3}{\sigma}. \tag{4.1}$$

The Weber number is the ratio of kinetic energy to surface energy. On the other hand, if the surface deformation considered is unstable, the circular frequency ω has an imaginary component, which leads to a neighboring state or, rather, to breakage. Once the liquid has been accelerated, energy is used up by viscous friction. The corresponding characteristic time is determined by the ratio of the dynamic viscosity to the surface tension, $t = \eta R/\sigma$. This time should be scaled by the Ohnesorge number

$$Oh = \sqrt{\frac{\rho \nu^2}{\sigma R}}. \tag{4.2}$$

What is true for a free liquid jet also applies to raindrops running down a window. If a sufficiently large liquid volume has accumulated, it runs down the window along a track that shows a nearly cylindrical surface. The water, however, does not fully drain away; rather, the track breaks into droplets which have nearly equal spacing. When another raindrop hits the window, it revives the former track and follows it, catching its remainder.

Another well-known effect is the breakage of soap bubbles. There is a big gain in energy if the bubble breaks and only a tiny drop is left. On the ground, the breakage of a bubble floating in the air is usually preceded by drainage of the liquid from top to bottom. The liquid film thins at the bubble's upper pole, and a flat, hanging drop arises at the lower pole; see Fig. 4.2. Under microgravity, breakage of the bubble is strongly reduced together with drainage. This also applies to the bubbles on a cup of coffee and to the foam on a glass of beer. Several microgravity experiments on metallic foams have been performed [Sprenger et al. 1987].

Breakage of a free liquid jet and breakage of a soap bubble differ greatly in their quantitative aspects: the former case means a smooth transition to a state with slightly lower energy whereas the latter case involves a big energy jump. A third type of instability is represented by the Kelvin instability, as indicated in Fig. 3.7. Common to all instabilities is that they lead to a state with lower energy. Stability is not a local but an overall effect. A further

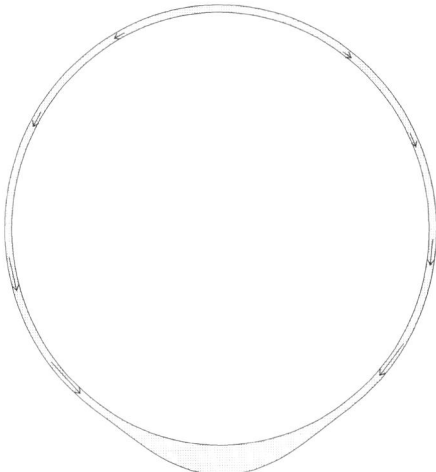

Fig. 4.2. Drainage in a soap bubble under terrestrial conditions

obvious example is that of liquid bridges created between fingers, see Fig. 1.5. The methods of identifying instabilities are:

- Calculation of the normal deformations of the liquid surface and the corresponding eigenvalues. A negative eigenvalue means an unstable deformation.
- Searching for an extremum of the volume within a family of solutions of the capillary equation. The condition of an extremum of the volume strongly simplifies the overall principle that a liquid surface becomes unstable if a surface deformation not requiring energy, i.e. with zero restoring force, exists. This is a sufficient, not a necessary, stability criterion.
- Linear stability analysis. Make a good guess of the critical deformation and test whether it lowers the energy. If it is not the best guess, too weak a stability limit is obtained.

4.2 Breakage of Cylindrical Surfaces

It was proven by Lord Rayleigh [1879, 1945] in the 19th century that a cylindrical liquid jet breaks if its length L exceeds its circumference $2\pi R$. This stability criterion also applies to cylindrical liquid bridges between coaxial circular disks. When lengthened to $L \geq 2\pi R$, one half of the bridge widens, whereas the other half shrinks accordingly (the amphora mode instability). Figure 4.3 demonstrates the breakage of a liquid bridge of silicone oil during the Spacelab D1 experiment "Floating Liquid Zones in Microgravity", described by DaRiva and Martinez [1986]. The picture shows two successive video frames: whereas in the first frame the bridge still exists, in the second frame it has broken into two liquid volumes in contact with the supporting disks. These volumes have not yet assumed their final shape, which clearly

70 4. Stability Criteria

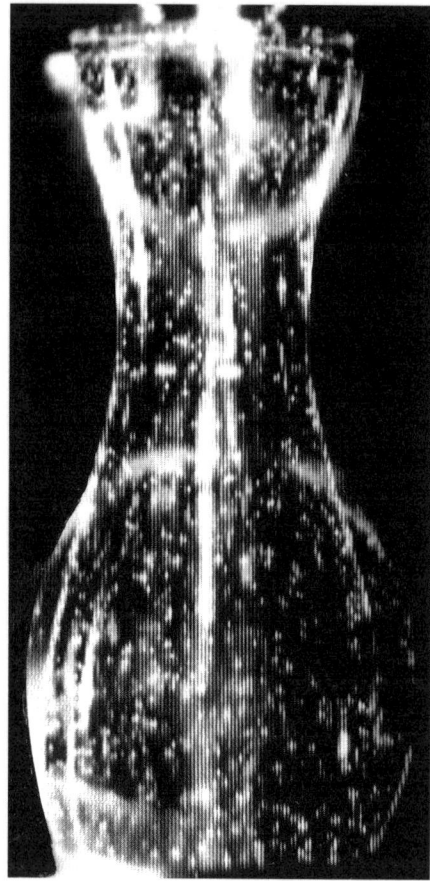

Fig. 4.3. Breakage of a liquid bridge of silicone oil during the Spacelab D1 experiment "Floating Liquid Zones in Microgravity" (DaRiva and Martinez [1986])

are spherical caps. Experimental experience is that the larger of the spherical caps formed encloses 86 percent of the total liquid volume [Martinez and Meseguer 1986, Martinez 1987].

The reason for the effective decrease in surface area (energy) associated with a sinusoidal variation in zone diameter is that the volume of a cylinder depends quadratically on its diameter. The increase in diameter and thus in surface area in those regions which widen is smaller than the decrease in diameter and surface area in those regions which shrink. This effect, however, only produces an energy gain if it is not overcompensated by an increase in surface area due to the corresponding axial variation in diameter. The latter must be smooth enough, which means that the wavelength must be sufficiently long; the wavelength must exceed the circumference.

The first investigations of the influence of gravity, i.e. of the Bond number on the stability of liquid columns were reported by Heywang [1956]. He was interested in the maximum zone height which may be achieved during crystal growth on the ground. Since then, several papers on the stability of axisymmetric liquid columns as a function of the Bond number have been published.

When a cylindrical liquid column is rotated with circular frequency ω, its maximum stable length is reduced to

$$L = \frac{2\pi R}{\sqrt{1 + Rn}}, \qquad (4.3)$$

where Rn is the rotation number

$$Rn = \frac{\Delta\rho \omega^2 R^3}{\sigma}. \qquad (4.4)$$

The rotation number Rn is fully analogous to the Weber number We. Often it is referred to as the Weber number also. Both numbers are the ratio of the kinetic energy to the surface energy. The circular frequency ω_{osc} of oscillation is just replaced by the circular frequency ω_{rot} of rotation. Whereas the Weber number describes the dynamic exchange of kinetic energy and surface energy, the rotation number describes the static surface deformation due to the radial increase in rotational pressure.

During the Skylab mission it was observed that in addition to this axisymmetric breakage, a cylindrical liquid column may undergo a transition to a skipping-rope mode (or C mode). Carruthers and Gibson report from the Skylab IV mission as follows [Carruthers et al. 1975, 1979; Vreeburg 1986]:

Objective: Assessment of the stability of a rotating liquid column for a variety of conditions.

Procedure: From onboard components and material a floating zone apparatus is constructed. The liquid bridge is formed between two end disks of 7/8-inch diameter. The disk surface and edge are treated with contact angle modifier (acetone) and antispread material (Krytox) in order to help establishment of the zone. Rotation is induced manually, by pulling a piece of string that has been coiled around the shafts that support the end disks. Either one or both disks can be rotated.

Performance: 39 different rotation sequences have been recorded. They covered a variety of zone dimensions and liquid material. Rotation rates are about 30 rpm. Vibrations are induced also, to test stability of the zone. Of the 39 zones, 24 proved to be stable.

Results: A "C" mode shape, where the zone moves as a skipping rope, is the preferred one when both end disks rotate and a critical length is exceeded. When only one disk rotates, the zone assumes an axisymmetric bottle shape. The Rayleigh criterion for the maximum stable length of the zone has been exceeded sometimes, albeit with slightly noncylindrical bridges.

The transition of a cylindrical liquid column to the skipping-rope mode arises when

$$L = \frac{\pi R}{\sqrt{Rn}}.\tag{4.5}$$

This implies that at $Rn = 1/3$ there is a competition between two instabilities: the column must decide whether it is going to break axisymmetrically according to (4.3) or to switch to the skipping-rope mode. This bifurcation has been investigated by Martinez [1987] during the D1 mission, and also during the D2 mission.

Another known competition between two instabilities is that between antisymmetric axial breakage with one node and symmetric axial breakage with two nodes. The latter form of breakage is favored in the case of columns showing a strong neck. On the other hand, very thick columns tend to escape sideways, i.e. a nonaxisymmetric instability arises (see Fig. 4.4).

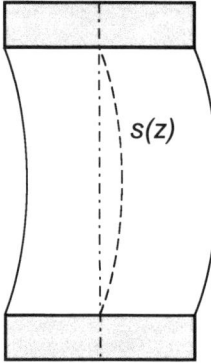

Fig. 4.4. Shape of a liquid column rotating in the skipping-rope mode

The question of whether a rotating cylindrical column with $L = \sqrt{3}\pi R$, $Rn = 1/3$ breaks axisymmetrically or switches into a skipping-rope mode depends on the procedure by which the bifurcation point is reached. An initially stable column may have been lengthened, the liquid volume may have been reduced or the frequency ω of rotation may have been increased. Or, as in the case of crystal growth, the zone length may have increased owing to too strong heating. One may guess that when liquid is extracted on one side axisymmetric breaking is favored; with respect to the other procedures, guesses about the mode of breakage are more difficult. In that case fluid dynamics is required, i.e. the momentum equation has to be considered. And the result will depend on the kinematic viscosity ν of the liquid used.

Let us return to the breakage shown in Fig. 4.3. A cylindrical liquid column breaks if its length equals its circumference. This breakage is antisymmetric; it starts with a sinusoidal deformation (amphora mode). Liquid flows from the neck region to the belly region. But what portion of the liquid volume succeeds in passing through the thinning neck during breakage? If one assumes that the liquid volume passing through the thinning neck decreases with the neck's cross section, one finds that 86 percent and 14 percent of the liquid volume are left on the belly side and on the neck side, respectively. This agrees well with the experimental findings, but is nothing more than a good guess. The fluid flow and thus the volume left on the two sides obviously depend on the viscosity, i.e. on the speed of breakage. This is a question of fluid dynamics.

4.3 Second Variation of Energy

In the capillary equation, the balance of forces is obtained by calculating the minimum of the energy of the liquid with respect to all independent variables. This leads to the shape of the surface $z(x, y)$ as a function of x and y. The procedure used, however, does not guarantee that a real minimum has been achieved. A saddle point or even a maximum in energy may have been found instead. In order to decide on stability, one has to calculate the second variation of the energy, i.e. expand the energy at the extremum into a quadratic form in its variables in terms of the local surface deformation $\zeta(x, y)$, according to $z(x,y) \Rightarrow z(x,y) + \zeta(x,y)$. This expansion may be transformed to its principal axes, i.e. to the normal deformations of the surface. If one of the eigenvalues arising in this transformation turns out negative, the surface found does not represent a minimum but a saddle point or a maximum. Such a surface is unstable.

The energy of the liquid is given by an integral over a function of the surface shape $z(x, y)$ and its partial derivatives $\partial z/\partial x, \partial z/\partial y$ (see (3.11)–(3.13)):

$$F = \int \mathrm{d}x \int \mathrm{d}y\, f\left(z, \frac{\partial z}{\partial x}, \frac{\partial z}{\partial y}\right). \tag{4.6}$$

According to the calculus of variations, minimization of the integral (4.6) with respect to the shape of the liquid surface $z(x, y)$ is achieved for

$$\frac{\mathrm{d}F}{\mathrm{d}z} = \frac{\partial f}{\partial z} - \frac{\mathrm{d}}{\mathrm{d}x}\frac{\partial f}{\partial z_x} - \frac{\mathrm{d}}{\mathrm{d}y}\frac{\partial f}{\partial z_y} = 0, \tag{4.7}$$

where

$$z_x \equiv \frac{\partial z}{\partial x}; \quad z_y \equiv \frac{\partial z}{\partial y}.$$

This leads to the capillary equation presented in Chap. 3. Now, allowing for surface deformations $\zeta(x, y)$ according to

$$z \Rightarrow z(x, y) + \zeta(x, y), \tag{4.8}$$

we obtain the first variation F_1 of the energy of the liquid (4.6):

74   4. Stability Criteria

$$F_1 = \int \mathrm{d}x \int \mathrm{d}y\, \zeta(x,y) \frac{\mathrm{d}F}{\mathrm{d}z(x,y)}$$

$$= \int \mathrm{d}x \int \mathrm{d}y\, \zeta(x,y) \left(\frac{\partial f}{\partial z} - \frac{\mathrm{d}}{\mathrm{d}x} \frac{\partial f}{\partial z_x} - \frac{\mathrm{d}}{\mathrm{d}y} \frac{\partial f}{\partial z_y} \right) \quad (4.9)$$

F_1 vanishes if the minimum in energy, i.e. the balance of forces, is achieved. For the second variation F_2, we obtain

$$F_2 = \int \mathrm{d}x \int \mathrm{d}y\, \zeta(x,y) \int \mathrm{d}\hat{x} \int \mathrm{d}\hat{y}\, \zeta(\hat{x},\hat{y}) \frac{\mathrm{d}^2 F}{\mathrm{d}z(x,y)\,\mathrm{d}z(\hat{x},\hat{y})}, \quad (4.10)$$

$$F_2 = \int \mathrm{d}x \int \mathrm{d}y \left(\frac{\partial^2 f}{\partial z^2} \zeta^2 + 2\zeta\zeta_x \frac{\partial^2 f}{\partial z\,\partial z_x} + 2\zeta\zeta_y \frac{\partial^2 f}{\partial z\,\partial z_y} \right.$$
$$\left. + \zeta_x^2 \frac{\partial^2 f}{\partial z_x^2} + 2\zeta_x \zeta_y \frac{\partial^2 f}{\partial z_x\,\partial z_y} + \zeta_y^2 \frac{\partial^2 f}{\partial z_y^2} \right). \quad (4.11)$$

After partial integration with respect to x and y according to

$$\int \mathrm{d}x \int \mathrm{d}y \left[2\zeta\zeta_x \frac{\partial^2 f}{\partial z\,\partial z_x} + 2\zeta\zeta_y \frac{\partial^2 f}{\partial z\,\partial z_y} \right]$$
$$= -\int \mathrm{d}x \int \mathrm{d}y\, \zeta^2 \left[\frac{\mathrm{d}}{\mathrm{d}x} \frac{\partial^2 f}{\partial z\,\partial z_x} + \frac{\mathrm{d}}{\mathrm{d}y} \frac{\partial^2 f}{\partial z\,\partial z_y} \right], \quad (4.12)$$

one is left with

$$F_2 = \int \mathrm{d}x \int \mathrm{d}y \left[\zeta^2 \left(\frac{\partial^2 f}{\partial z^2} - \frac{\mathrm{d}}{\mathrm{d}x} \frac{\partial^2 f}{\partial z\,\partial z_x} - \frac{\mathrm{d}}{\mathrm{d}y} \frac{\partial^2 f}{\partial z\,\partial z_y} \right) \right.$$
$$\left. + \zeta_x^2 \frac{\partial^2 f}{\partial z_x^2} + 2\zeta_x \zeta_y \frac{\partial^2 f}{\partial z_x\,\partial z_y} + \zeta_y^2 \frac{\partial^2 f}{\partial z_y^2} \right]. \quad (4.13)$$

F_2 is a quadratic form in the local deformation $\zeta(x,y)$ and its partial derivatives ζ_x, ζ_y with respect to x and y. In order for F_2 to be positive for all allowed deformations $\zeta(x,y)$, its transformation to its principal axes must not lead to a negative eigenvalue. Those deformations which conserve the liquid volume are allowed. We therefore consider the extremes of F_2 under the constraints of constant liquid volume and normalization of $\zeta(x,y)$:

$$\int \mathrm{d}x \int \mathrm{d}y\, \zeta(x,y) = 0\,; \quad \int \mathrm{d}x \int \mathrm{d}y\, \zeta^2(x,y) = \text{const}\,. \quad (4.14)$$

By multiplication of the volume constraint with the Lagrange multiplier $2\lambda_1$, and the normalization integral with the Lagrange multiplier λ_2, we obtain

$$\frac{\mathrm{d}}{\mathrm{d}x} \left(\frac{\partial \zeta}{\partial x} \frac{\partial^2 f}{\partial z_x^2} + \frac{\partial \zeta}{\partial y} \frac{\partial^2 f}{\partial z_x\,\partial \zeta_y} \partial y \frac{\partial^2 f}{\partial z_x\,\partial z_y} \right) + \frac{\mathrm{d}}{\mathrm{d}y} \left(\frac{\partial \zeta}{\partial x} \frac{\partial^2 f}{\partial z_x\,\partial z_y} + \frac{\partial \zeta}{\partial y} \frac{\partial^2 f}{\partial z_y^2} \right)$$
$$= \left(\frac{\partial^2 f}{\partial z^2} - \frac{\mathrm{d}}{\mathrm{d}x} \frac{\partial^2 f}{\partial z\,\partial z_x} - \frac{\mathrm{d}}{\mathrm{d}y} \frac{\partial^2 f}{\partial z\,\partial z_y} \right) \zeta(x,y) - \lambda_1 - \lambda_2 \zeta(x,y)\,. \quad (4.15)$$

Equation (4.15) is the differential equation for the normal surface deformations for constant liquid volume. If all the eigenvalues λ_2 are positive, the liquid surface is stable. A stability limit is reached if $\lambda_2 = 0$.

4.4 Normal Deformations of Liquid Zones

4.4.1 Instabilities of Periodic Surfaces

For axisymmetric surfaces, the capillary equation reads, in cylindrical coordinates $r(\varphi, z)$,

$$\widehat{p} + \frac{\rho \omega^2 r^2}{2\sigma} = \frac{1}{r\sqrt{1+(dr/dz)^2}} - \frac{1}{\sqrt{1+(dr/dz)^2}^3} \frac{d^2 r}{dz^2} \tag{4.16}$$

(see (3.71)). If the deformation is expressed as $r(z) \Rightarrow r(z) + s(z)$, the differential equation (4.15) for the normal surface deformations may be rewritten as

$$\frac{d}{dz}\left(\frac{r}{\sqrt{1+(dr/dz)^2}^3}\frac{ds}{dz}\right) + \left(\frac{1}{r\sqrt{1+(dr/dz)^2}} + \frac{\rho\omega^2 r^2}{\sigma} + \lambda_2\right) s(z)$$
$$+ \lambda_1 r(z) = 0 \,. \tag{4.17}$$

These equations can be conveniently solved numerically by a Runge–Kutta integration method. There are also a few special cases where analytical solutions may be found.

In particular, we are now able to prove that an axisymmetric liquid column $r(z)$ between coaxial circular disks with equal radii R is unstable if it is symmetric with respect to the middle plane and has a vertical slope $dr/dz = 0$ at the supporting disks. From that result it follows that an unduloid between solid disks becomes unstable if the separation of the disks is equal to the period of the unduloid. If the unduloid degenerates to a cylinder, the period is given by $2\pi R$, i.e. the cylinder becomes unstable if its length exceeds its circumference. (A cylinder is obtained by rolling an ellipse with eccentricity zero, i.e. a circle, on the axis.) This applies to free capillary surfaces such as the free liquid jet (= Rayleigh instability) in the same way as it does to capillary bridges contacting solid disks. The Rayleigh instability is thus a special case of the above principle. Likewise, a nodoid supported between solid disks becomes unstable if it has an axial tangent, i.e. if the disk separation equals the period of the nodoid. The nodoid, however, is more likely to undergo an earlier, nonaxisymmetric instability.

By differentiating the capillary equation (4.16) with respect to z, we learn that the equation for the normal deformations (4.17) is solved by

$$s(z) = \frac{dr}{dz} \,; \quad \lambda_1 = \lambda_2 = 0 \,. \tag{4.18}$$

The constraint of constant liquid volume is automatically satisfied by the symmetry of the surface $r(z)$ with respect to the middle plane and the corresponding antimetry of the perturbation $s(z)$. This not only proves the instability of the unduloid and of the nodoid, but also of all the modified surfaces of revolution which arise owing to rotation.

4.4.2 Normal Deformations of a Circular Cylinder

In the case of a cylindrical surface $r(z) \Rightarrow r(z)+s(z)$, the differential equation for the normal deformations (4.17) becomes

$$R^2 \frac{d^2 s}{dz^2} + \left(1 + \frac{\rho\omega^2 R^3}{\sigma} + \lambda_2 R\right) s(z) + \lambda_1 R^2 = 0 . \tag{4.19}$$

Equation (4.19) represents the inhomogeneous equation of harmonic oscillations, which is quite generally solved by

$$s(z) = a_1 \sin(qz) + a_2 \cos(qz) + a_3 . \tag{4.20}$$

Substitution of (4.20) into (4.19) leads to

$$(qR)^2 = 1 + \frac{\rho\omega^2 R^3}{\sigma} + \lambda_2 R , \quad a_3 q^2 + \lambda_1 = 0 . \tag{4.21}$$

According to (4.20), $s(z)$ must vanish at the disks (for $r = R$, $z = L/2$) and must not change the liquid volume. This yields the conditions

$$\pm a_1 \sin\left(\frac{qL}{2}\right) + a_2 \cos\left(\frac{qL}{2}\right) + a_3 = 0 ,$$

$$a_2 \sin\left(\frac{qL}{2}\right) + a_3 \left(\frac{qL}{2}\right) = 0 . \tag{4.22}$$

Equation (4.22) has solutions

$$a_1 \neq 0, \quad a_2 = a_3 = 0, \quad qL = 2\pi n, \quad n = 1, 2, \ldots , \tag{4.23}$$

and

$$a_1 = 0, \quad a_2, a_3 \neq 0, \quad \tan\left(\frac{qL}{2}\right) = \left(\frac{qL}{2}\right) . \tag{4.24}$$

The solutions $s(z)$ corresponding to (4.23) are antimetric with respect to the middle plane $z = 0$. The nth solution has $2n - 1$ nodes in addition to the two nodes at the solid disks. The surfaces become unstable if the eigenvalue λ_2 vanishes. From (4.17) and (4.19) we obtain [Gillis 1961, Vega & Perales 1983]

$$\frac{L}{2R} = \frac{n\pi}{\sqrt{1+Rn}} . \tag{4.25}$$

The solutions $s(z)$ corresponding to (4.24) are symmetric with respect to the middle plane $z = 0$. The equation is satisfied for

$$qL = (2.860593, 4.918048, 6.941779, 8.954817, \ldots)\,\pi \ . \tag{4.26}$$

The cylindrical surface hence becomes unstable with respect to a deformation with $2n$ nodes for

$$\frac{L}{2R} = \frac{(n+1/2)\pi}{\sqrt{1+Rn}} \tag{4.27}$$

$$\times \left\{ 1 - \frac{1}{[(n+1/2)\pi]^2} - \frac{2}{3}\frac{1}{[(n+1/2)\pi]^4} - \frac{13}{15}\frac{1}{[(n+1/2)\pi]^6} - \ldots \right\}.$$

The symmetric solution $qL = 2.8606$ shows a higher decrement than the antimetric solution $n = 1$. The free liquid jet is therefore more likely to break according to the symmetric mode.

4.4.3 The Symmetric Instability of the Catenoid

Another special case in which the differential equation for the normal deformations can be solved analytically is the symmetric catenoid between circular disks [Erle et al. 1970]. The two principal curvatures of the catenoid are equal with opposite sign. If the radius and height are nondimensionalized by use of the minimum radius $r_0 = R$ between the disks, the equation (3.2a) for the catenoid reads

$$r = \cosh z \ . \tag{4.28}$$

Substitution of this expression into (4.17) for zero angular frequency ω leads to

$$\frac{d}{dz}\left(\frac{1}{\cosh^2 z}\frac{ds}{dz}\right) + \left(\frac{1}{\cosh^2 z} + \lambda_2 R\right) s(z) + \lambda_1 R \cosh z = 0 \ . \tag{4.29}$$

A particular analytical solution to (4.29) for $\lambda_2 = 0$ has been reported by Erle et al. [1970]:

$$s(z) = a_1 \sinh z + a_2 \left(z \sinh z - \cosh z\right) + a_3 \cosh^3 z \tag{4.30}$$

where

$$4a_3 + \lambda_1 R = 0 \ .$$

The three terms on the right-hand side of (4.30) represent a homogeneous antimetric solution, a homogeneous symmetric solution and the inhomogeneous solution, respectively. The boundary condition $s(z) = 0$ at the disks $z = \pm L/2R$ leads to $a_1 = 0$. The antimetric contribution $\sinh z$ vanishes. For the symmetric contribution, we obtain at the disks $z = \pm L/2R$

4. Stability Criteria

$$a_2 \left(z \sinh z - \cosh z \right) + a_3 \cosh^3 z = 0 . \tag{4.31}$$

The constraint of constant liquid volume yields

$$a_2 \left(4z \cosh^2 z - 6z - 6 \sinh z \cosh z \right)$$
$$+ a_3 \left(3z + 3 \sinh z \cosh z + 2 \sinh z \cosh^3 z \right) = 0 . \tag{4.32}$$

Elimination of a_2 and a_3 leads to

$$3z^2 - z \coth z \left(2 \cosh^4 z - 7 \cosh^2 z + 6 \right) + \left(4 \cosh^4 z - 3 \cosh^2 z \right) = 0 . \tag{4.33}$$

Equation (4.33) has a zero at

$$z = L/2R = 2.239180 . \tag{4.34}$$

The catenoid becomes unstable if its length exceeds the diameter of its neck by a factor 2.2392. The unstable deformation obtained is symmetric with respect to the center plane and has two nodes between the solid disks.

Initial estimates of the eigenvalues λ_2 of the normal deformations $n \geq 1$ may be obtained by using $\xi = \cosh z$ rather than z as the independent variable. For the antimetric and the symmetric solutions, respectively, we put

$$s(z) = \begin{cases} \sinh z\, s_1(\xi) & n = 1, 3, 5, \ldots \\ \cosh z\, s_2(\xi) & n = 2, 4, 6, \ldots \end{cases} \tag{4.35}$$

The antimetric solutions $n = 1, 3, 5, \ldots$ require $\lambda_1 = 0$, yielding

$$(\xi^2 - 1) \frac{d^2 s_1}{d\xi^2} + \left(\xi + \frac{2}{\xi} \right) \frac{ds_1}{d\xi} + \lambda_2 R \xi^2 s_1 = 0 , \tag{4.36}$$

whereas for the symmetric solutions $n = 2, 4, 6, \ldots$, we have

$$(\xi^2 - 1) \frac{d^2 s_2}{d\xi^2} + \left(\xi + \frac{2}{\xi} \right) \frac{ds_2}{d\xi} + \left(2 + \lambda_2 R \xi^2 \right) s_2 + \lambda_1 R \xi^2 = 0 . \tag{4.37}$$

For large ξ (i.e. large distance from the middle plane) the second term in each of the parentheses in (4.36), (4.37) may be neglected. From (4.36) we obtain the differential equation for the Bessel functions of order zero. This equation has the solution

$$s_1(z) \propto \sinh z\, J_0 \left(\sqrt{\lambda_2 R} \cosh z \right) . \tag{4.38}$$

The boundary condition of zero deformation at the disks together with the equation for the zeros of the Bessel function, leads to

$$\sqrt{\lambda_2 R} \cosh \left(\frac{L}{2R} \right) \approx \left(n - \frac{1}{4} \right) \pi \quad n = 1, 3, 5, \ldots . \tag{4.39}$$

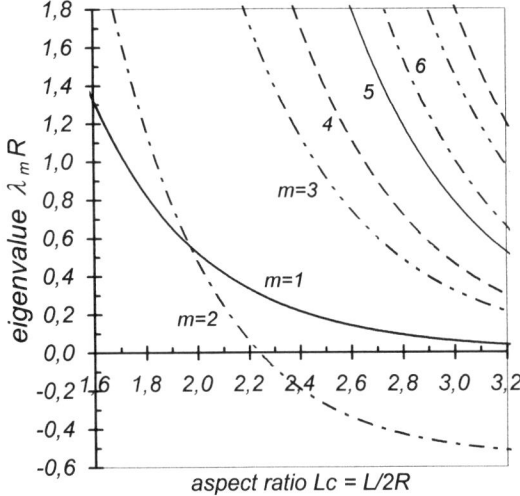

Fig. 4.5. The eigenvalues $\lambda_2(m)R$ of the normal deformations for $m = 1$ to 8 of the symmetric catenoid, versus the aspect ratio $L_c = L/2R$

The antimetric deformations described by (4.38), (4.39) with increasing numbers $n = 1, 3, 5, \ldots$ of nodes do not give rise to additional instabilities. The solution of (4.37) for the symmetric deformations is more intricate owing to the additional parameter λ_1 resulting from the volume constraint. Figure 4.5 shows the eigenvalues λ_2 of the normal deformations versus the aspect ratio $L_c = L/2R$.

4.5 Nonaxisymmetric Instabilities

4.5.1 Lateral Deformations of the Center Line

Up to now we have restricted ourselves to axisymmetric deformations $s(z)$ of axisymmetric surfaces $r(z)$. Such deformations satisfy the differential equation (4.17). Further possibly unstable deformations are:

- deformations $s(z)$ of the center line with an unaltered circular cross section, as observed during the Skylab IV mission; see Sect. 4.2
- variations of the shape of the cross section, in particular oscillations of free drops or columns.

In the following, we study infinitesimal lateral deformations $s(z)$ of the centre line at constant circular cross section (Langbein et al. 1986). This allows us to apply a Taylor expansion with respect to $s(z)$. The area between two circular cross sections at a distance dz apart, with radii $r(z)$ and $r(z) + (dr/dz)dz$ and with centers shifted by $s(z) + (ds/dz)dz$, is given by

$$d^2 A = dz\, r\, d\varphi \sqrt{1 + \left(\frac{dr}{dz} + \frac{ds}{dz} \cos\varphi\right)^2}. \tag{4.40}$$

The area of the surface shown in Fig. 4.4 hence equals

$$A = \int_0^L dz \int_0^{2\pi} r\,d\varphi \sqrt{1 + \left(\frac{dr}{dz} + \frac{ds}{dz}\cos\varphi\right)^2}$$

$$\approx \int_0^L dz\, 2\pi r \left[\sqrt{1 + \left(\frac{dr}{dz}\right)^2} + \frac{1}{4}\frac{(ds/dz)^2}{\sqrt{1+(dr/dz)^2}^3}\right]. \quad (4.41)$$

The volume of the liquid column is not changed by the lateral shift considered here. For the kinetic energy of rotation, we find

$$E_{\text{kin}} = \int_0^L dz \int_0^{r(z)} dr \int_0^{2\pi} r\,d\varphi \frac{1}{2}\left(r^2 + s^2 + 2rs\cos\varphi\right)$$

$$= \int_0^L dz\,\pi \left(\frac{r^4}{4} + \frac{r^2 s^2}{2}\right). \quad (4.42)$$

By substitution of (4.41) and (4.42) into the expression for the total energy of the liquid ((3.6)–(3.10)) and by variation of $A - pV$ with respect to $r(z)$ and $s(z)$, we obtain a coupled system of differential equations of order two for $r(z)$ and $s(z)$. The lateral deformation $s(z)$ enters the equilibrium condition (4.16) for $r(z)$ quadratically, such that this condition is still valid.

In order to check whether the extremum of the total energy found represents a minimum, the second variation of the total energy $A - pV$ has to be transformed to the principal axes with respect to the deformation $s(z)$:

$$\frac{d}{dz}\left(\frac{r}{\sqrt{1+(dr/dz)^2}^3}\frac{ds}{dz}\right) + \left(\frac{\rho\omega^2}{\sigma}r^2 + \lambda_2\right)s(z) = 0. \quad (4.43)$$

Equation (4.43) differs from the corresponding differential equation (4.17) for the axial deformations by the term $r/[1+(dr/dz)^2]^{0.5}$ and the Lagrange parameter λ_1, which accounts for the volume constraint. In the case of a cylindrical surface $r(z) = R$ we are left with

$$R\frac{d^2 s}{dz^2} + \left(\frac{\rho\omega^2 R^2}{\sigma} + \lambda_2\right)s(z) = 0. \quad (4.44)$$

Equation (4.44) is solved by

$$s(z) = \cos(qz). \quad (4.45)$$

A vanishing deformation at the disks, of spacing L, leads to

$$q = \frac{(2n+1)\pi}{L}, \quad (qR)^2 = \left(\frac{\rho\omega^2 R^3}{\sigma} + \lambda_2 R\right). \quad (4.46)$$

The condition for λ_2 to vanish, i.e. for the rotating cylindrical column to become unstable, is

$$\frac{L}{2R} = \frac{(n+1/2)\pi}{\sqrt{Rn}}, \quad Rn = \frac{\rho\omega^2 R^3}{\sigma}. \tag{4.47}$$

The assumption that the cross section of the liquid surface stays circular clearly is a first approximation. The pressure rises on the side to which the column is displaced, and falls on the opposite side. The cross section thus assumes an oval shape. The liquid volume in the most strongly displaced region of the column, i.e. in the middle, most likely increases, while it decreases near the disks. Also, the kinematic viscosity ν of the liquid has to be taken into account.

4.5.2 Liquid Rings

Let us consider a free liquid drop into which a thin rod has been introduced; see Fig. 4.6a. Thin in this context means that the rod diameter is much less than the drop diameter. The drop will stay basically spherical; however, depending on its contact angle, its surface will become a section of an unduloid or of a nodoid; see Fig. 4.6b. If the liquid volume is slowly reduced through holes in the rod, the near-spheroid shrinks to a ring (with the capillary surface still being an unduloid or a nodoid). For the same reason as for axisymmetric surfaces, we may expect that this ring must eventually become unstable. This will happen not in the axial but in the azimuthal direction: if the ring has become sufficiently slim, the increase in surface area in those regions to which liquid is flowing will be smaller than the decrease in surface area in

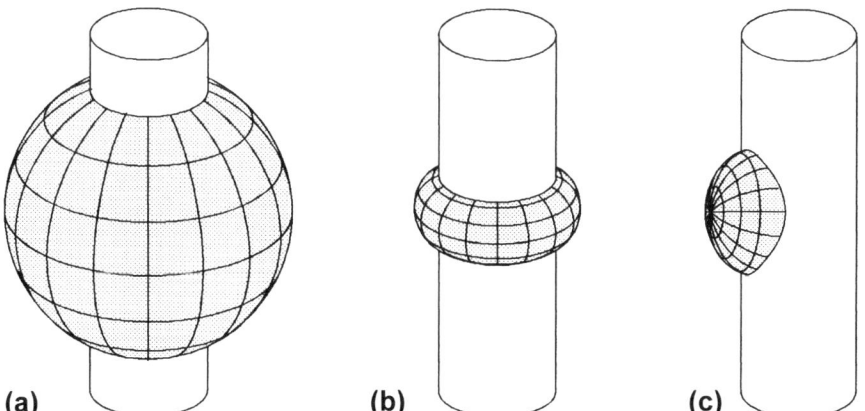

Fig. 4.6. Breakage of a liquid ring around a rod with decreasing liquid volume

those regions from which the liquid is flowing. A net decrease in surface area will be left, if it is not compensated by the increase of the surface area in the azimuthal direction, i.e. the liquid ring must be sufficiently long and thin, which is just the same condition as for the axial instabilities discussed in Sect. 4.4.2.

Breakage of the liquid ring leads to a single drop on one side of the rod; see Fig. 4.6c. If, now, liquid is slowly injected again, the drop will wind back around the rod, until eventually its flanks meet on the rear side and the former liquid ring is reobtained. This, however, requires a much higher liquid volume than that observed at breakage of the ring. This is a typical case of hysteresis (Fig. 4.7).

Another obvious example is a liquid bridge between coaxial circular disks. Up to now we have concentrated on their axisymmetric instabilities. However, if the liquid volume becomes large relative to the corresponding cylindrical volume, the liquid is increasingly squeezed out of the space and changes from an axisymmetric to a nonaxisymmetric configuration. For the same reasons as discussed for the liquid ring above, the liquid is pressed out more strongly on one side than on the other. This nonaxisymmetric deformation has to compete with the axisymmetric breakage of the liquid. Once again, let us mention the C-mode or skipping-rope instabilities discussed in the preceding section.

A transition from an axisymmetric configuration to a nonaxisymmetric configuration and back to an axisymmetric configuration is found if the drop volume is maintained constant and, instead, the curvature of the solid surface in contact with it is continuously varied from convex (forming a rod) to plane

Fig. 4.7. Formation of three symmetrically arranged drops rather than a torus during ejection of colored water from a cone during a parabolic flight

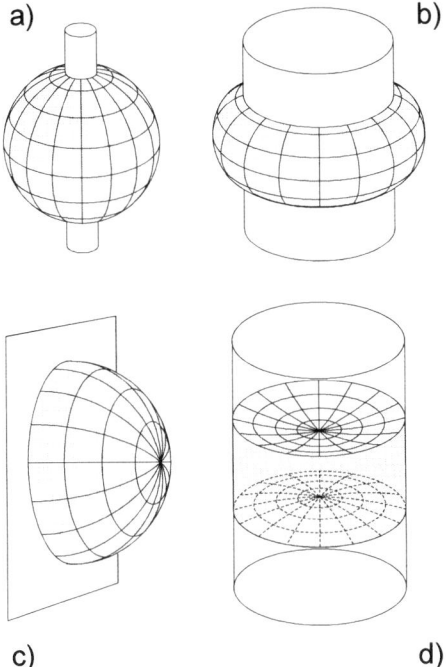

Fig. 4.8. Shapes of drops that arise when the curvature of the supporting solid surface is varied from strongly convex (forming a rod) to plane and, finally, to strongly concave (forming a tube)

and then to concave (thus forming a tube); see Fig. 4.8. For a thin rod, we obtain the approximate sphere described above. With decreasing curvature of the rod, i.e. increasing rod radius, the sphere contracts to a ring and subsequently breaks. For zero curvature, in the case of a plane, the drop forms a spherical section, as depicted earlier in Fig. 2.8. If the curvature of the solid becomes concave, i.e. if the liquid drop is located inside a tube, the drop becomes more and more compressed until, eventually, it meets itself on the opposite side of the tube. It will then close the hole in the center and form an axisymmetric plug.

4.6 The Minimum-Volume Condition

A very common method of investigating the stability limits of liquid surfaces is based on the minimum-volume condition. If, within a family of solutions of the capillary equation, the liquid volume exhibits a minimum, maximum or saddle point as a function of the capillary pressure, an instability is reached. Thus, if the volume is plotted versus the pressure or any other independent parameter, the stability limits show up. Needless to say, one has to find and include all possible solutions of the capillary equation.

Let us recall that the capillary equation is obtained by minimizing the energy of the liquid with respect to all surface deformations that conserve the liquid volume. The pressure is the corresponding Lagrange multiplier. Whenever a family of solutions of the capillary equation contains two neighboring solutions with equal liquid volume, these two solutions cannot truly represent a minimum in energy. A saddle point of the energy has been obtained instead. The question in which direction the liquid surface is going to deform and eventually to break depends on the third derivative of the energy of the liquid with respect to the corresponding surface deformation. The stable branch is usually distinguished by $dV/dp < 0$, i.e. only those families of capillary surfaces may be stable whose curvature decreases with increasing volume. One might object that a minimum in volume is not compatible with the minimization of the energy at constant liquid volume. The minimization of the energy, however, does not exclude the possibility that two solutions of the capillary equation may have equal liquid volumes.

This leads to the minimum-volume condition for the stability of a liquid surface, which should actually be termed the extremum-volume condition, since it applies to all points with zero slope of the liquid volume as a function of pressure or any equivalent parameter. This means that the liquid surface may be deformed without the need for energy or may even lead to a gain in energy, i.e. to an instability. If a minimum in the liquid volume occurs, only the branch with the lower energy is stable.

The extremum-volume condition strongly simplifies the overall principle that a liquid surface becomes unstable if a surface deformation not requiring energy, i.e. having zero restoring force, exists. For the identification of a stability limit, it is no longer necessary to consider all possible surface deformations. Rather, it is sufficient to take into account those deformations leading to neighboring solutions of the capillary equation. An important extension of the extremum-volume condition is the statement that a bifurcation into two families of solutions of the capillary equation means an instability point if the liquid surfaces in the two families coincide at the bifurcation point. Any infinitesimal violation of the symmetry causes the (V, p) curves to split up, as shown in Fig. 4.9, such that extrema of the volume appear, i.e. a stability limit is reached.

For the formal proof of the extremum-volume condition, let us assume that $z_1(x, y)$ and $z_2(x, y)$ both satisfy the equilibrium condition (4.7) for equal capillary pressure and equal liquid volume. By total differentiation of (4.7) with respect to z according to

$$\frac{d}{dz} = \frac{\partial}{\partial z} - \frac{d}{dx}\frac{\partial}{\partial(\partial z/\partial x)} - \frac{d}{dy}\frac{\partial}{\partial(\partial z/\partial y)} + \frac{d^2}{dx^2}\frac{\partial}{\partial(\partial^2 z/\partial x^2)} + \frac{d^2}{dx\,dy}\frac{\partial}{\partial(\partial^2 z/\partial x \partial y)} + \frac{d^2}{dy^2}\frac{\partial}{\partial(\partial^2 z/\partial y^2)} \tag{4.48}$$

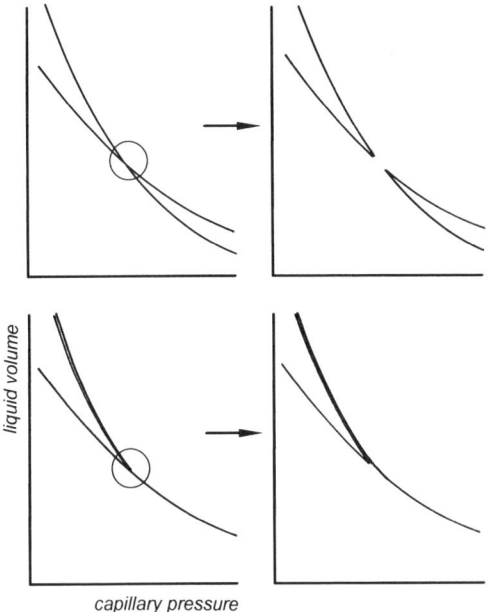

Fig. 4.9. Any infinitesimal violation of the symmetry causes the (V, p) curve to split, such that extrema of the volume appear, i.e. a stability limit is reached

and application of the commutation relations

$$\frac{\partial}{\partial z}\frac{\mathrm{d}}{\mathrm{d}x} = \frac{\mathrm{d}}{\mathrm{d}x}\frac{\partial}{\partial z}, \quad \frac{\partial}{\partial(\partial z/\partial x)}\frac{\mathrm{d}}{\mathrm{d}x} = \frac{\mathrm{d}}{\mathrm{d}x}\frac{\partial}{\partial(\partial z/\partial x)} + \frac{\partial}{\partial z},$$

$$\frac{\partial}{\partial(\partial^2 z/\partial x^2)}\frac{\mathrm{d}}{\mathrm{d}x} = \frac{\partial}{\partial(\partial z/\partial x)}, \quad (4.49)$$

we reobtain (4.15) with $\lambda_1 = 0$ and $\lambda_2 = 0$. This proves the equivalence of the extremum-volume condition to calculating the eigenvalues of the normal deformations: a liquid surface which according to (4.3) is an equilibrium surface becomes unstable if an infinitesimally different surface with infinitesimally different pressure exists and encloses the same liquid volume.

The extremum-volume condition is a sufficient, but not necessary condition for an instability.

4.7 Linear Stability Analysis

There is yet another most effective method of analysing stability limits: intuition together with experimental evidence. Experimental evidence may indicate that a certain deformation of a fluid surface becomes very sensitive, i.e. it oscillates with a very low frequency (= low restoring force) or even causes breakage. Most scientists involved in microgravity experiments have experienced several such situations and strongly hoped that breakage might be avoided. Several times, payload specialists have reacted successfully to prevent breakage (with the exception of those experiments where breakage was intended). In many of these practical cases, linear stability analysis has been used effectively. Its main features are as follows:

- The theoretical task is to find an analytical solution for the critical deformation in question and to test whether it lowers the energy of the liquid. Since finding an exact solution is usually rather difficult, the alternative is to make a good guess, which should be made flexible enough by inclusion of several adjustable parameters.
- The suggested deformation must be allowed, i.e. it must not affect the constraints and boundary conditions of the problem; usually, it should maintain the liquid volume and the contact angle.
- The deformation is assumed to be sufficiently small that linear expansions of the relevant equations, be it the capillary equation or the constraints, are possible.
- An exception concerning linearity is the total energy. An exact capillary surface that minimizes the total energy is a precondition for a linear stability analysis. The deformations therefore cause quadratic contributions to the energy.
- If the infinitesimal deformation considered is an exact solution of the capillary equation itself, the exact stability criterion is reobtained.
- A linear stability analysis is not suited to describe the dynamic behaviour. A strong change in energy nevertheless hints at an oscillation with high frequency or at a fast breakage. Acceleration of the fluid and viscous friction have to be taken into account.
- During a linear stability analysis, it is often very useful to determine the capillary pressure in advance by means of the integral theorem, see Sect. 8.2.
- Often, a sinusoidal test function $\sin(qx)$ in one of the independent variables x helps a lot. The wavenumber q, which initially is arbitrary, must be adjusted such that the volume and the contact angle are not altered. The volume constraint may often be satisfied by the symmetry properties, whereas constancy of the contact angle along the contact line decides on the wavenumber.
- A negative value of q^2 means an imaginary wavenumber q and thus a deformation increasing exponentially in the direction x considered. Such deformations are clearly not infinitesimal and thus contradict the assumptions of linear stability analysis. They are not allowed at all.
- If an allowed infinitesimal deformation lowering the energy has been found, this does not necessarily mean that the most critical instability has been obtained. There may be concurrent deformations which become effective earlier. One finds only those deformations and instabilities which one is looking for.
- A deformation lowering the energy is a sufficient, but not necessary condition for an instability.

Let us look at a short conversation from the second run of the STACO (Stability of Liquid Columns) experiment during the Spacelab D1 mission. A liquid column of silicone fluid had been created between coaxial circular disks 35 mm in diameter. The principal investigator was I. Martinez, Madrid.

R. Furrer, payload specialist: *This are 95 millimeters, I try now to establish a perfectly cylindrical shape.*
U. Merbold, crew interface coordinator: *We all are excited about the column, it is just gorgeous.*
R. Furrer: *There are oscillations within the column with a period of about 4 seconds or so. These were the oscillations which made the column to break in the first run, because I was not aware how long the time delay is, which feeding of liquid needs in order to stabilize.*

Linear stability analysis was used in Sect. 4.5.1 and will be applied several times more in this book.

References

1. Carruthers JR, Gibson EG, Klett MG, Facemire BR: Studies of rotating liquid floating zones on Skylab IV. AIAA Paper 75-692, Denver, CO (1975)
2. Carruthers J, Gibson EG: Liquid floating zone. NASA TM-78234 (1979) pp. 6–9 to 6–15
3. Vreeburg JPB: Summary review of microgravity fluid science experiments. ESA, Amsterdam (1986)
4. DaRiva I, Martinez I: Floating liquid zones. Naturwissenschaften **73** (1986) 345–347
5. Erle MA, Gilette RD, Dyson DC: Stability of interfaces of revolution with constant surface tension. The case of the catenoid. Chem. Eng. J. **1** (1970) 97
6. Gilette RD, Dyson DC: Smallest volume. Chem. Eng. **2** (1971) 44
7. Gillis J: Stability of a column of rotating viscous liquid. Proc. Camb. Phil. Soc. **57** (1961) 152
8. Heywang W: 1956 Zur Stabilität senkrechter Schmelzzonen. Z. Naturforsch. **11a** (1956) 238–243
9. Langbein D, Rischbieter F: Form, Schwingungen und Stabilität von Flüssigkeitsgrenzflächen. Forschungsbericht W 86-29 des BMFT (1986) 1–130
10. Martinez I: Stability of long liquid columns in Spacelab-D1. ESA-SP 256 (1987) 235–240
11. Martinez I, Meseguer J: Floating liquid zones in microgravity. In: *Scientific Results of the German Spacelab Mission D1*. P.R. Sahm, R. Jansen, M. Keller (eds.) (1986) 105–112
12. Sprenger HJ, Pötschke J, Potard C, Rogge V: Composites. In: *Fluid Sciences and Materials Science in Space*. H.U. Walter (ed.) Springer Berlin Heidelberg (1987), Chap. 16, 567–597
13. Rayleigh, Lord (J.W. Strutt): On the capillary phenomena of jets. Proc. Roy. Soc. London **29** (1879) 71–97

14. Rayleigh, Lord (J.W. Strutt): *The Theory of Sound.* Dover, New York (1945) Sect. 364
15. Vega JM, Perales JM: Almost cylindrical isorotating liquid bridges for small Bond numbers. In: *Materials Sciences in Space.* ESA SP-191 (1983) 247–252
16. Weber C: Zum Zerfall eines Flüssigkeitsstrahles. Z. angew. Math. Mech. **11** (1931) 136–155
17. Vreeburg JPB: Summary review of microgravity fluid science experiments. ESA, Amsterdam (1986)

5. Axisymmetric Liquid Columns at Rest and Under Rotation

An axisymmetric liquid surface has the advantage that the capillary equation reduces to an ordinary differential equation of order two. In the absence of gravity and rotation its solutions are unduloids or nodoids. The solutions in the presence of gravity and rotation are easily obtained by a Runge–Kutta integration. Liquid columns between coaxial circular disks have repeatedly served as model systems for studying convection during crystal growth. Their axial deformations can be classified by the number of nodes in the axial direction, and their lateral deformations by the number of nodes in the azimuthal direction.

Rotating free liquid drops have repeatedly served as cosmological models, for example for the formation of the planets. Such drops may float freely or may be positioned by acoustic or electromagnetic levitators. In theory, they may also be supported by coaxial circular disks. By requiring a radial slope at the disks and by increasing the aspect ratio of the disk separation to the disk diameter towards infinity, a free drop can be simulated. The instabilities of liquid columns and of free liquid drops are investigated in this chapter.

5.1 Introduction

Since the advent of research under microgravity conditions, there has been strong interest in the stability of liquid bridges between coaxial circular disks. This has been motivated by many fundamental problems in fluid physics, on the one hand, and by numerous applications, on the other hand. Oscillations of liquid columns may be excited accurately by vibration of the supporting disks. Various sensors may be integrated into the disks thus allowing quantitative measurements. During the D2 Spacelab mission the resonance frequencies of liquid bridges were analyzed in situ by means of pressure sensors located in the center of the supporting disks [Langbein 1991]. Liquid bridges also are the primary subject of investigations of steady and oscillatory Marangoni convection.

A particular stimulation for studies of the bridge geometry is its application to crystal growth from a floating zone. The first calculation of the maximum stable height of an axisymmetric bridge was reported by Heywang [1956]. Numerous papers on the stable height as a function of gravity and of the frequency of rotation have been published since then [Padday 1973, Coriell et al. 1977, Da-Riva et al. 1979, Meseguer 1983a,b, Martinez 1984, 1987, Martinez et al. 1987]

An additional advantage of investigating liquid bridges between coaxial circular disks is the independence of the findings of the contact angle γ of the liquid with the solid material of the disks used. The contact angle is well known to show hysteresis and to be particularly vulnerable to surface contamination; see Sect. 2.4. However, along a solid edge it may vary freely within the angular interval determined by contact of the liquid with the two solid faces. This effect is known as canthotaxis. Sharp solid edges increase the interval of possible contact angles. A solid edge may be understood as a surface showing infinite curvature. The consequences of this principle are discussed in more detail in Chap. 7. Canthotaxis is regularly used for preventing liquids from spreading along solid faces and spilling out under microgravity conditions.

Liquid bridges may oscillate with $m = 1, 2, 3, \ldots$ nodes longitudinally and with $n = 1, 2, 3, \ldots$ nodes laterally. We denote the corresponding normal deformations generally by $D\{m,n\}$. The symmetry of the lateral oscillations is described by $\cos(n\varphi)$.

5.2 The Normal Deformations

The stability of a liquid column is governed by three strongly competing normal modes:

- The best-known unstable deformation is the *antimetric longitudinal mode* $D\{1,0\}$, also called the Rayleigh instability, amphora mode or longitudinal instability. One half of the column widens, and the other half shrinks accordingly (see Fig. 4.3). This mode usually requires a larger liquid volume and a longer column than does the symmetric mode $D\{2,0\}$ described next.
- The *symmetric longitudinal mode* $D\{2,0\}$. This arises at the minimum liquid volume that allows a solution of the capillary equation at a given aspect ratio $L_c = L/2R$ of the disks. The resulting deformation is symmetric with respect to the middle plane: the column shrinks in the middle and widens close to the two supporting disks. The deformation is axisymmetric, it has an azimuthal wavenumber $n = 0$. Breakage according to the symmetric mode $D\{2,0\}$ usually has a high decrement, i.e. breakage is faster than that according to the antimetric mode $D\{1,0\}$. The volume shift due to the deformation $D\{2,0\}$ is smaller than that due to the antimetric mode $D\{1,0\}$. In the case of a cylindrical column the mode $D\{2,0\}$ requires a long column and therefore is difficult to observe. On the other hand, for free liquid jets it is the main instability.

5.2 The Normal Deformations

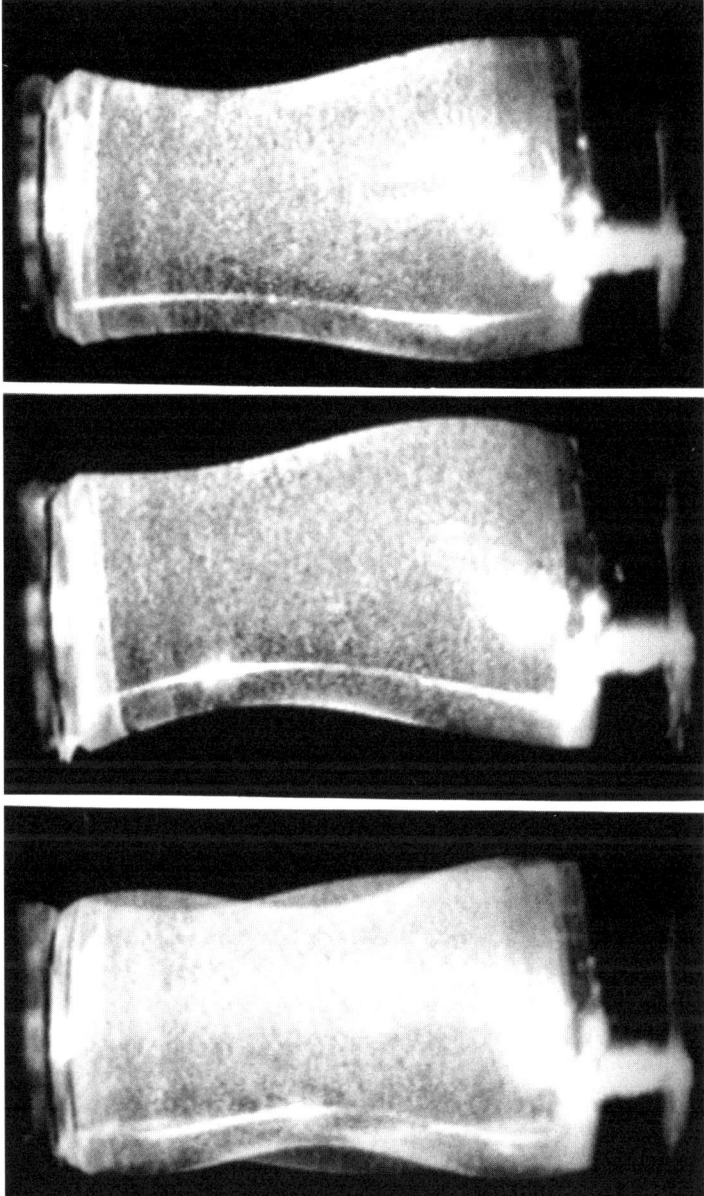

Fig. 5.1. A skipping-rope mode of a column of silicone oil 60 mm in diameter and 96 mm in length [Martinez 1984]

- The *lateral mode* $D\{0,1\}$, also called the skipping-rope mode, jumping-rope mode or C mode. The liquid is squeezed out of the space between the disks sideways. This instability usually does not involve breakage of the column. It requires a large liquid volume or a high rotation number Rn. It corresponds to the azimuthal wavenumber $n = 1$. The lateral mode was first observed during the Skylab mission by Carruthers et al. [1975]; the bridge rotated around its axis like a skipping-rope. Figure 5.1 shows a skipping-rope mode of a column of silicone oil 60 mm in diameter and 96 mm in length [Martinez 1984].

- In addition to these instabilities, there exist many further instabilities with azimuthal wavenumbers $n > 1$. However, realization of these modes is difficult. The column will break because of one of the other modes described above before the critical configuration is achieved. An instability may only become effective if another one does not arise earlier or is faster. (Those who come too late will be punished by life.)

5.2.1 The Symmetric Mode $D\{2,0\}$

In the axisymmetric case, the capillary equation reduces to an ordinary differential equation of order two in the radius r and the axial coordinate z. The solutions thus contain two adjustable parameters, for instance the first and second derivatives of the radius at a given position. For a liquid column between coaxial circular disks, these parameters may be chosen as the pressure (second derivative) and the slope (first derivative) at one of the disks. The parameters have to be mutually adjusted such that the surface correctly meets the other disk. A one-parameter family of possible surfaces results. In numerical calculations, it is convenient to calculate the shape of the surface together with its derivatives with respect to the pressure and slope at the starting position and to improve an initial good guess by a Newtonian method (= shooting method).

Figure 5.2a shows the result of this adaptation process for a liquid column with aspect ratio $L_c = L/2R = 1$. The capillary pressure P_c, the angle of inclination Ψ_c, the surface area A_c and the neck or belly radius R_c are plotted versus the liquid volume V_c. The smallest neck radius $R_c = 0$ is obtained for two half-spheres touching each other, with pressure $P_c = 2$, inclination $\Psi_c = 1$, area $A_c = 1$, and volume $V_c = 2/3$. In this branch of solutions the pressure, inclination and area decrease, whereas the neck radius increases with decreasing volume V_c. The minimum volume $V_c = 0.3869$ is obtained for a pressure of 1.2002 and neck radius of 0.2447. The corresponding surface area A_c equals 0.7467. This is an unstable branch of surfaces, such surfaces break according to the symmetric mode $D\{2,0\}$ into two spherical sections not touching each other (except for the limiting case $V_c = 2/3$).

At $V_c = 0.3869$ the unstable branch smoothly joins the stable branch. If now the liquid volume is increased, the neck radius increases further, whereas the pressure and inclination first continue to decrease, exhibit a minimum and then increase as well. For a cylindrical surface all the reduced parameters assume a value of 1. If the volume is increased further, the neck radius turns into a belly radius, another spherical solution shows up and, eventually, at

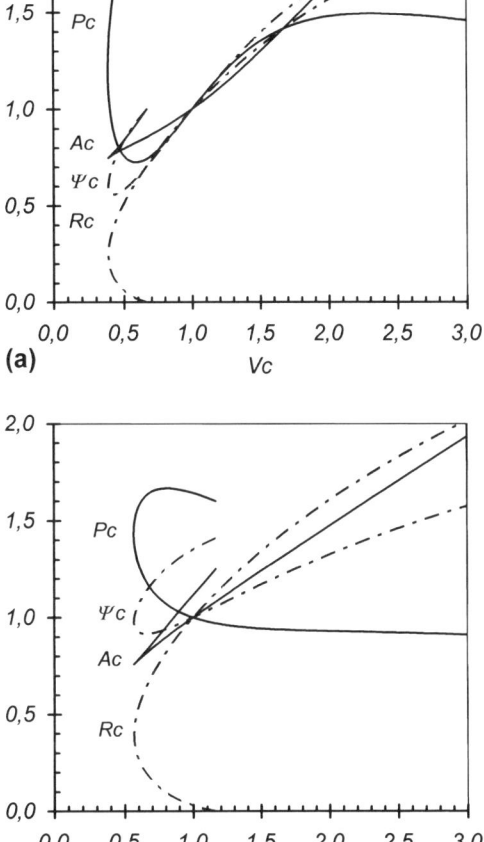

Fig. 5.2. (a) Nondimensional pressure P_c, slope I_c, area A_c and extremum radius R_c versus the volume V_c of a liquid column with aspect ratio $L_c = L/2R = 1$. The area A_c shows a singularity at the minimum volume. Only the branch of surfaces with lower area (= lower energy) is stable (b) Equivalent plot for $L_c = 2$

$\Psi_c = 2, V_c = 2.9210$, one obtains a surface that has a radial tangent at the disks. The minimum volume represents the lower stability limit and the surface with the radial tangent represents the upper stability limit.

The dependence of the area (surface energy) A_c on the volume V_c deserves particular attention. During the derivation of the capillary equation, energy and volume are treated as fully equivalent. Minimization of the liquid energy under the constraint of volume conservation actually means that the *free energy* = energy − pressure × volume is minimized. Thus, a minimum in energy within a family of solutions also means a minimum in volume. In the vicinity of a minimum both quantities depend quadratically on any independent parameter representing the family of solutions, e.g. the capillary pressure. Therefore, in the vicinity of the minimum volume, the surface area and volume follow the relation

$$A_c - A_{c\,\min} = c_1 (V_c - V_{c\,\min}) \pm c_2 (V_c - V_{c\,\min})^{3/2} + \ldots, \tag{5.1}$$

where c_1 and c_2 are adjustable parameters. The unstable branch $dV/dp > 0$ clearly has a higher energy for equal liquid volume.

Figure 5.2b depicts the same relations between V_c, P_c, Ψ_c, A_c and R_c as does Fig. 5.2a, but applies to the aspect ratio $L_c = L/2R = 2$. Figure 5.3 shows the disk separation L_c versus the calculated minimum volume V_c required for solutions of the capillary equation to exist. This figure represents the stability diagram with respect to the symmetric deformation $D\{2,0\}$. The parameter of the various curves is the rotation number $Rn = -2, -1, 0, 1, 2, 3, 4$. The minimum volume required clearly increases with increasing aspect ratio L_c and with increasing rotation number Rn. Independently of Rn, the highest minimum volume which may arise is the cylindrical volume $V_c = 1$. This situation is obtained for the aspect ratio

$$L_c = \frac{\xi}{\sqrt{1+Rn}}, \tag{5.2}$$

where $\xi = 4.493409$ is the leading positive zero of $\xi = \tan \xi$. (See also Sect. 4.4.2 on the normal deformations $D\{2,0\}$ of cylindrical surfaces.) From (5.2) it follows that the volume V_c required remains smaller than 1 for all negative rotation numbers $Rn < -1$.

5.2.2 The Antimetric Mode $D\{1,0\}$

Usually, the symmetric mode is the principal instability. The other normal modes become effective only if they belong to the stable branch $dV/dp < 0$. In

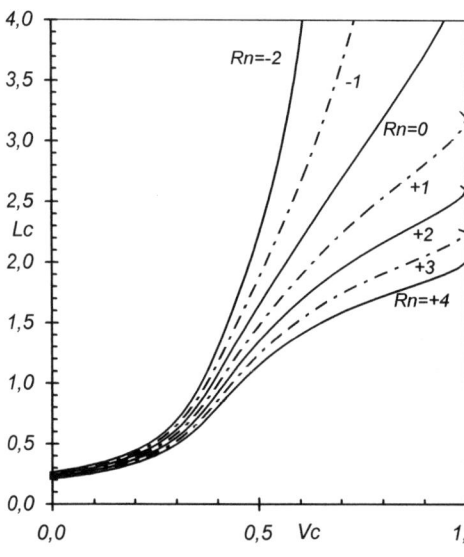

Fig. 5.3. Stability diagram with respect to the symmetric deformation $D\{2,0\}$. The aspect ratio L_c is plotted versus the minimum volume V_c required. The parameter of the various curves is the rotation number $Rn = -2, -1, 0, 1, 2, 3, 4$. Higher rotation numbers clearly require lower aspect ratios

Sect. 4.4.1 it was shown that an axisymmetric liquid column becomes unstable if it is symmetric with respect to the middle plane and has an axial slope at the supporting disks, which means that a full period fits between the disks. Figure 5.4 shows the disk separation L_c versus the volume V_c required by these periodic surfaces. This figure represents the stability diagram with respect to the antimetric deformation $D\{1,0\}$. The parameter of the curves is once more the rotation number $Rn = -2, -1, 0, 1, 2, 3, 4$.

The symmetric stability curve for $D\{2,0\}$ and the antimetric stability curve for $D\{1,0\}$ contact each other smoothly with a parallel tangent. Up to this bifurcation point the $D\{2,0\}$ mode governs the stability, whereas for larger aspect ratios the Rayleigh instability $D\{1,0\}$ is the critical mode. In particular, for a cylindrical column, $D\{1,0\}$ arises for

$$L_c = \frac{\pi}{\sqrt{1+Rn}} \tag{5.3}$$

(see also Sect. 4.4.2, (4.25)). The critical length L_c of a cylindrical column with respect to the antimetric mode $D\{1,0\}$ according to (5.3) is clearly shorter than that for the symmetric mode $D\{2,0\}$ according to (5.2), i.e. cylindrical columns will generally break owing to the antimetric mode $D\{1,0\}$. On the other hand, (5.2) and (5.3) are not applicable to rotation numbers $Rn < -1$. As a function of L_c, one finds two bifurcations of $D\{2,0\}$ and $D\{1,0\}$ for $-1.1722 < Rn < -1$. No bifurcation exists for $Rn < -1.1722$: in that case the periodic surfaces all belong to the unstable branch $dV/dp > 0$. For $Rn < -1.1722$, arbitrarily long columns are stable, provided sufficient liquid volume, according to Fig. 5.3, is available. These columns do not extend over a full period.

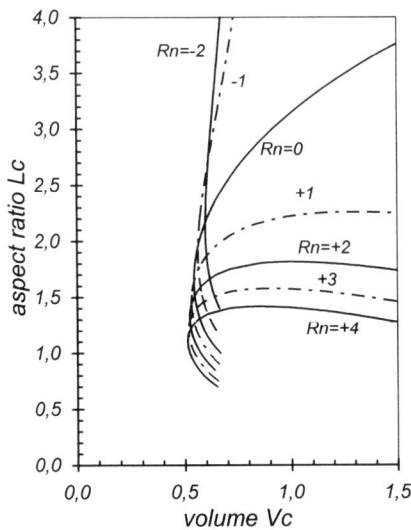

Fig. 5.4. Stability diagram with respect to the antimetric deformation $D\{1,0\}$. The aspect ratio L_c is plotted versus the volume V_c required by the periodic surfaces. The parameter of the various curves is the rotation number $Rn = -2, -1, 0, 1, 2, 3, 4$

5.2.3 The Lateral Instability $D\{0,1\}$

Lateral deformations of the fluid surface are conveniently treated by linear stability analysis. Assuming that the deformation has an angular wavenumber n, we put

$$r(\varphi, z) = r_0(z) + r_1(z)\cos(n\varphi) + r_2(z)\cos(2n\varphi) + \ldots \quad (5.4)$$

($n = 1$ means a lateral shift). Expansion of the capillary equation (3.64) inclusive of the rotational pressure term $\Delta\rho\omega^2 r^2/2\sigma$, up to terms linear in $r_1(z)$, yields

$$\hat{p} = \frac{1}{r_0(z)\sqrt{1+(dr_0/dz)^2}} - \frac{d^2 r_0}{dz^2}\frac{1}{\sqrt{1+(dr_0/dz)^2}^3}$$

$$+ \left(\frac{(n^2-1)r_1(z)}{r_0^2(z)\sqrt{1+(dr_0/dz)^2}} - \frac{d^2 r_1/dz^2}{\sqrt{1+(dr_0/dz)^2}^3}\right.$$

$$\left. - \frac{(dr_0/dz)(dr_1/dz)}{r_0(z)\sqrt{1+(dr_0/dz)^2}^3} + 3\frac{d^2 r_0}{dz^2}\frac{(dr_0/dz)(dr_1/dz)}{\sqrt{1+(dr_0/dz)^2}^5}\right)\cos(n\varphi), (5.5)$$

where

$$\hat{p} = \hat{p}_0 + \frac{1}{2}Rn\,r_0^2(z) + Rn\,r_0(z)r_1(z)\cos(n\varphi). \quad (5.6)$$

By comparison of the coefficients of 1 and $\cos(n\varphi)$, we obtain

$$\hat{p}_0 + \frac{1}{2}Rn\,r_0^2(z) = \frac{1}{r_0(z)}\frac{1}{\sqrt{1+(dr_0/dz)^2}} - \frac{d^2 r_0}{dz^2}\frac{1}{\sqrt{1+(dr_0/dz)^2}^3}, \quad (5.7)$$

$$Rn\,r_0^2(z)r_1(z) = -\frac{dr}{dz}\left(\frac{r_0(z)}{\sqrt{1+(dr_0/dz)^2}^3}\frac{dr_1}{dz}\right)$$

$$+ \frac{(n^2-1)r_1(z)}{r_0(z)\sqrt{1+(dr_0/dz)^2}}. \quad (5.8)$$

Alternatively, if the radius r is chosen as the independent variable and the axial coordinate z as the dependent variable, one obtains from (3.68)

$$z(r,\varphi) = z_0(r) + z_1(r)\cos(n\varphi) + z_2(r)\cos(2n\varphi) + \ldots, \quad (5.9)$$

$$\hat{p} = \frac{1}{rS}\frac{\partial z}{\partial r} + \frac{1}{S^3}\left\{\frac{\partial^2 z}{\partial r^2}\left[1+\left(\frac{\partial z}{r\,\partial\varphi}\right)^2\right]\right.$$

$$\left. - \frac{\partial z}{\partial r}\frac{\partial z}{r\,\partial\varphi}\left(\frac{\partial}{\partial r}\frac{\partial z}{r\,\partial\varphi} + \frac{\partial z}{r\,\partial\varphi}\frac{\partial z}{\partial r}\right) + \frac{\partial^2 z}{r^2\,\partial\varphi^2}\left[1+\left(\frac{\partial z}{\partial r}\right)^2\right]\right\} \quad (5.10)$$

and

$$\widehat{p}_0 + \frac{1}{2} Rn r^2 = \frac{1}{r} \frac{\mathrm{d}}{\mathrm{d}r} r \frac{\mathrm{d}z_0/\mathrm{d}r}{\sqrt{1 + (\mathrm{d}z_0/\mathrm{d}r)^2}}, \tag{5.11}$$

$$0 = \frac{\mathrm{d}^2 z_1}{\mathrm{d}r^2} + \left(\frac{1}{r} - 3 \frac{(\mathrm{d}z_0/\mathrm{d}r)(\mathrm{d}^2 z_0/\mathrm{d}r^2)}{1 + (\mathrm{d}z_0/\mathrm{d}r)^2} \right) \frac{\mathrm{d}z_1}{\mathrm{d}r} - \frac{n^2 z_1}{r^2} \left[1 + \left(\frac{\partial z_0}{\partial r} \right)^2 \right]. \tag{5.12}$$

Figure 5.5 shows the disk separation L_c versus the volume V_c that gives rise to the lateral mode $D\{0,1\}$. This figure represents the stability diagram with respect to that deformation $D\{0,1\}$. As before, the parameter of the various curves is the rotation number $Rn = -2, -1, 0, 1, 2, 3, 4$. Unlike Figs. 5.3 and 5.4, where the stability curves give the smallest stable volumes, the stability curves in Fig. 5.5 represent the upper limits of the stable volumes. The stability curves for $D\{0,1\}$ smoothly touch the stability curves for $D\{2,0\}$ and cross over with the stability curves for $D\{1,0\}$. (A liquid ring anchored at the edges of the disks is more stable than a liquid ring with constant contact angle around a rod.)

From (5.8), we conclude in particular that the stability limit for the lateral mode $D\{0,1\}$ and $Rn = 0$ is given by the condition that the radial slope $\mathrm{d}z/\mathrm{d}r$ of the surface at the disks is zero. The unstable mode means a constant lateral shift $r_1(z) = \mathrm{const}$.

The various stability limits are collated in Fig. 5.6. For rotation numbers $Rn > 2.5$ the lateral stability curve for $D\{0,1\}$ touches the symmetric curve for $D\{2,0\}$ at lower volumes than does the antimetric curve for $D\{1,0\}$. In that case the antimetric longitudinal mode $D\{1,0\}$ may no longer show up;

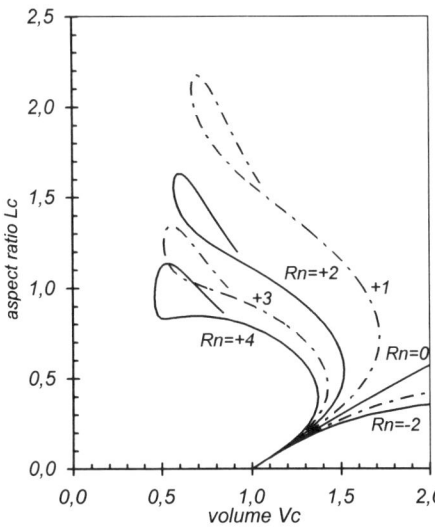

Fig. 5.5. Stability diagram with respect to the lateral mode $D\{0,1\}$. As before, the parameter of the various curves is the rotation number $Rn = -2, -1, 0, 1, 2, 3, 4$. Whereas Figs. 5.3 and 5.4 show the smallest stable volumes, the stability curves here show the largest stable volumes

98   5. Axisymmetric Liquid Columns at Rest and Under Rotation

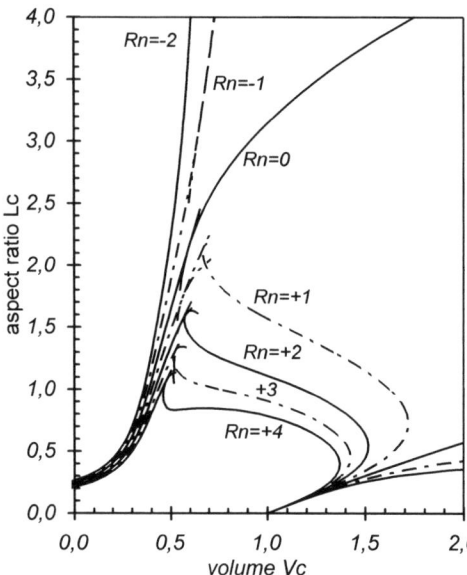

Fig. 5.6. Stability diagram of rotating liquid columns under weightlessness. For rotation numbers $Rn > 2.5$ the lateral stability curve for $D\{0,1\}$ touches the symmetric curve for $D\{2,0\}$ at lower volumes than does the antimetric curve for $D\{1,0\}$. In that case $D\{1,0\}$ may no longer show up; it is fully blocked by the other instabilities

the other instabilities fully block it. This has been observed with relatively short columns in Skylab IV. For $0 < Rn < 2.5$, one finds small intervals where lowering the liquid volume leads to the instability $D\{1,0\}$, whereas increasing the volume leads to the instability $D\{0,1\}$. For $-1.7222 < Rn \leq 0$, long columns generally break on reduction of the liquid volume according to $D\{1,0\}$. And for $Rn < -1.7222$ the columns break symmetrically according to $D\{2,0\}$.

Figure 5.7 shows the cross over points of the various instabilities. The aspect ratio L_c is plotted versus the rotation number Rn. For $Rn < 0$ only the bifurcation of $D\{2,0\}$ vs $D\{1,0\}$ is relevant, for $2.5 < Rn$ only the bifurcation of $D\{2,0\}$ vs $D\{0,1\}$ is relevant.

5.2.4 Stability of a Liquid Ring

The stability limit with respect to the lateral mode $D\{0,1\}$ for $Rn = 0$ is determined by the condition that the radial slope dz/dr of the surface at the disks is zero. The corresponding surfaces are the nodoids discussed in Sect. 3.4. In the case of nodoids, the primary integral (3.30) of the capillary equation (5.7) reads

$$r \sin \varphi = \frac{r^2 - r_1 r_2}{r_2 - r_1}, \quad \frac{p}{\sigma} \equiv \widehat{p} = \frac{2}{r_2 - r_1}, \tag{5.13}$$

where r_2 and r_1 are the maximum and minimum radii, respectively. Integration of the length from slope zero, $\varphi = 0$ (radial slope), at the disks to slope $\varphi = \pi/2$ (axial slope) at the belly leads to

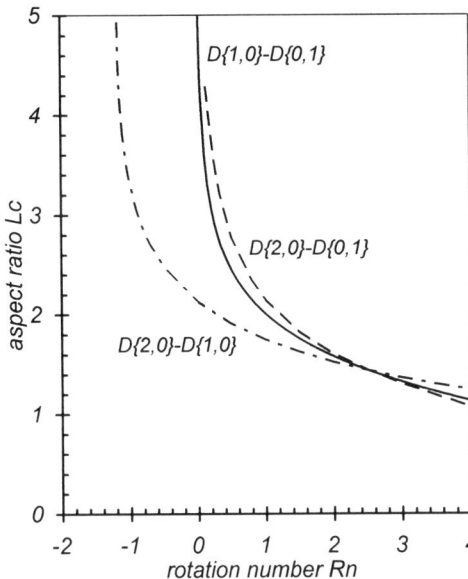

Fig. 5.7. The cross over points of the various instabilities. The aspect ratio L_c is plotted versus the rotation number Rn. For $Rn < 0$ only the bifurcation of $D\{2,0\}$ vs $D\{1,0\}$ is relevant, and for $2.5 < Rn$ only the bifurcation of $D\{2,0\}$ vs $D\{0,1\}$ is relevant

$$\frac{L}{2} = \int_{\sqrt{r_1 r_2}}^{r_2} dr \tan\varphi = \frac{r_2 - r_1}{2} \int_0^{\pi/2} d\varphi \sin\varphi \left(1 + \frac{c \sin\varphi}{\sqrt{1 - c^2 \cos^2\varphi}}\right), \quad (5.14)$$

$$L_c \equiv \frac{L}{2\sqrt{r_2 r_1}} = \frac{c}{\sqrt{1-c^2}} \left(1 + \frac{1}{c} E(c^2) - \frac{1-c^2}{c} K(c^2)\right)$$

$$\approx \frac{c}{\sqrt{1-c^2}} \left(1 + \frac{\pi}{4}c + \frac{\pi}{32}c^3 + \frac{3\pi}{256}c^5 + \dots\right), \quad (5.15)$$

where

$$c = \frac{r_2 - r_1}{r_2 + r_1}, \quad (5.16)$$

and $K(c^2)$ and $E(c^2)$ are the complete elliptic integrals of the first and second kind, respectively. The volume and surface area may be integrated and expressed in terms of these complete elliptic integrals in a similar way. One finds

$$\frac{V}{2\pi} = \int_{\sqrt{r_1 r_2}}^{r_2} dr \, r^2 \tan\varphi \quad (5.17)$$

$$= \left(\frac{r_2 + r_1}{2}\right)^3 \int_0^{\pi/2} d\varphi \sin\varphi \frac{c \left(c \sin\varphi + \sqrt{1 - c^2 \cos^2\varphi}\right)^3}{\sqrt{1 - c^2 \cos^2\varphi}},$$

$$V_c \approx \frac{1}{1-c^2}\left[1 + c\frac{\pi}{2} - c^2\left(\frac{\pi^2}{8} - \frac{5}{3}\right)\right.$$
$$\left. + c^3\left(\frac{\pi^3}{32} - \frac{17\pi}{48}\right) - c^4\left(\frac{\pi^4}{128} - \frac{7\pi^2}{96}\right) + \cdots\right]. \tag{5.18}$$

Equations (5.14) to (5.18) give the stability limit for the lateral mode $D\{0,1\}$ and zero rotation $Rn = 0$. Turning to rotating columns $Rn \neq 0$, one obtains, by expanding the surface shape and the lateral modes according to (5.7) and (5.8) around that exact solution, the following generalized relations between the inclination, volume, surface area, belly radius and length:

$$P_c L_c = 1 + \left(\frac{\pi}{4} - \frac{1}{2}Rn\right) L_c - \left(\frac{\pi^2}{16} + \frac{\pi}{4}Rn\right) L_c^2$$
$$+ \left[\left(\frac{\pi^3}{32} - \frac{3\pi}{32}\right) + \left(\frac{3\pi^2}{16} - 2\right) Rn\right] L_c^3 \pm \cdots, \tag{5.19}$$

$$\left[\frac{dz}{dr}\right]_{\text{disks}} = -Rn\, L_c^2 \left[1 + \frac{\pi}{4} L_c + (0.3266 - 0.8644 Rn)\, L_c^2 + \cdots\right], \tag{5.20}$$

$$V_c = 1 + \frac{\pi}{2} L c - \left(\frac{\pi^2}{4} - \frac{8}{3}\right) L_c^2$$
$$+ \left[\left(\frac{5\pi^3}{32} - \frac{23\pi}{16}\right) + \left(\frac{\pi^2}{4} - 4\right) Rn\right] L_c^3$$
$$- \left[\left(\frac{7\pi^4}{64} - \frac{33\pi^2}{32}\right) + 1.4196 Rn\right] L_c^4 \pm \cdots \tag{5.21}$$

$$A_c = \frac{\pi}{2} - \left(\frac{\pi^2}{8} - 2\right) L_c$$
$$+ \left[\left(\frac{\pi^3}{16} - \frac{5\pi}{8}\right) + \left(\frac{\pi^2}{8} - 2\right) Rn\right] L_c^2 \pm \cdots, \tag{5.22}$$

$$R_c = 1 + L_c - \left(\frac{\pi}{4} - \frac{1}{2}\right) L_c^2$$
$$+ \left[\left(\frac{\pi^2}{8} - \frac{\pi}{4}\right) + \left(\frac{\pi}{4} - \frac{3}{2}\right) Rn\right] L_c^3 \mp \cdots. \tag{5.23}$$

The coefficients of L_c^n in (5.19)–(5.23) are polynomials of order $n-1$ in Rn. In particular, at the lateral stability limit the slope of the liquid surface at the disks is proportional to Rn and to the square of the aspect ratio; $dz/dr \approx -Rn\, L_c^2[1 + (\pi/4)L_c]$. The coefficients of the terms in L_c^4 in (5.20) and (5.21) have been calculated numerically.

Both the lateral mode $D\{0,1\}$ and the antimetric mode $D\{1,0\}$ bifurcate with the symmetric mode $D\{2,0\}$ for higher volumes. With liquid bridges exhibiting small neck radii, the symmetric instability is most likely. For nearly cylindrical bridges and rotation numbers $Rn < 1/3$, the amphora-mode instability is favored, whereas for $Rn > 1/3$ the skipping-rope instability $D\{0,1\}$ will arise. One should keep in mind, however, that breakage due to the former two instabilities is favored once the skipping-rope mode has arisen.

Figure 4.7 depicts the lateral instability of a nodoidal ring surrounding a solid cone (from parabolic flights in a KC-135 in 1988). Figure 2.5 shows the spreading and subsequent rupture of a liquid bridge due to heating of one of the supporting disks above the consolute temperature of the paraffin oil/benzyl benzoate system (from the D1-mission, 1985).

5.3 Nearly Cylindrical Surfaces

5.3.1 Fourier Expansion of an Axisymmetric Surface

An analytical treatment of the normal modes is also possible for nearly cylindrical surfaces with $V_c \approx 1$, a situation of basic interest in crystal growth under weightlessness. The solutions of the capillary equation at rest or under rotation are periodic in the height z, which enables a convergent Fourier expansion according to

$$r_0(z) = 1 + \sum_{n=0}^{\infty} a_n \cos(nq_0 z) . \qquad (5.24)$$

The coefficient a_0 in the sum in (5.24) represents the deviation of the mean radius of the surface from the disk radius $R = 1$. For small deviations a_0 has the same order as a_1, and a_n has order a_1^n. Substitution of (5.24) into the capillary equation (5.7) and comparison of coefficients of $\cos(nq_0 z)$ for $n = 0$ to 3 leads to

$$\widehat{p}_0 = \left(1 - \frac{Rn}{2}\right) - (1 + Rn)a_0$$
$$+ \left(1 - \frac{Rn}{2}\right)\left(a_0^2 + \frac{a_1^2}{2}\right) - \frac{1}{4}a_1^2 q_0^2 , \qquad (5.25)$$

$$(1 + Rn) - q_0^2 = \left(1 - \frac{Rn}{2}\right)(2a_0 + a_2)$$
$$- 3a_0^2 - \left(\frac{3}{4} - \frac{q_0^2}{8} + \frac{3q_0^4}{8}\right)a_1^2 - q_0^2 a_2 , \qquad (5.26)$$

$$(1 + Rn)a_2 - 4q_0^2 a_2 = \left(\frac{1}{2} - \frac{Rn}{4} + \frac{q_0^2}{4}\right)a_1^2$$
$$+ \left(1 - \frac{Rn}{2}\right)2a_0 a_2 - \left(\frac{3}{2} + \frac{q_0^2}{4}\right)a_0 a_1^2 , \qquad (5.27)$$

$$(1 + Rn)a_3 - 9q_0^2 a_3 = \left(1 - \frac{Rn}{2} + q_0^2\right) a_1 a_2 - \left(\frac{1}{4} + \frac{q_0^2}{8} - \frac{3q_0^4}{8}\right) a_1^3. \quad (5.28)$$

From (5.26), we find for the wavenumber q_0, to lowest order,

$$q_0 = \sqrt{1 + Rn}. \quad (5.29)$$

The boundary conditions $r = R$ at the disks at $z = \pm L/2$ and conservation of the liquid volume yield

$$a_0 + a_1 \cos\left(\frac{qL}{2}\right) + a_2 \cos(qL) + a_3 \cos\left(\frac{3qL}{2}\right) = 0, \quad (5.30)$$

$$V = 2\pi \int_0^{L/2} dz$$
$$\times \left(1 + 2[a_0 + a_1 \cos(qz) + a_2 \cos(2qz)] + [a_0 + a_1 \cos(qz)]^2\right), \quad (5.31)$$

$$V_c = 1 + \left(2a_0 + a_0^2 + \frac{1}{2}a_1^2\right)$$
$$+ 2(a_1 + a_0 a_1) \frac{\sin(qL/2)}{qL/2} + \left(2a_2 + \frac{1}{2}a_1^2\right) \frac{\sin(qL)}{qL}. \quad (5.32)$$

5.3.2 The Symmetric Instability $D\{2,0\}$

Equations (5.25)–(5.28) and (5.30) represent five equations between the pressure $\widehat{p}_0$, the wavenumber q_0 and the coefficients a_0, a_1, a_2, a_3, which means that just one of these quantities may be chosen independently. One additional condition is required. The symmetric instability $D\{2,0\}$ corresponds to the minimum-volume condition $dV/da_0 = 0$. By calculating the total derivative of the volume (5.32) under the constraints of (5.25)–(5.30), we obtain

$$V_c = 1 - \frac{(1 + Rn)^2}{2 - Rn/2} \left(\frac{\xi}{\sqrt{1 + Rn}} - L_c\right)^2, \quad (5.33)$$

where $\xi = 4.493409$ as in (5.2).

5.3.3 The Antimetric Instability $D\{1,0\}$

An axisymmetric liquid column becomes unstable if it is symmetric with respect to the middle plane and has an axial slope at the supporting disks, i.e. if its period corresponds to the disk separation. By requiring

$$\frac{q_0 L}{2} = \pi, \quad \cos\left(\frac{q_0 L}{2}\right) = -1, \quad (5.34)$$

we obtain

$$L_c = \frac{\pi}{\sqrt{1+Rn}} \left(1 + \frac{2-Rn}{4(1+Rn)}(V_c - 1)\right.$$
$$\left. - \frac{5 + 16Rn - Rn^2 + 3Rn^3/4}{16(1+Rn)^2}(V_c - 1)^2\right). \quad (5.35)$$

(see Fig. 5.4). Equation (5.35) extends (5.3) to terms in $(V_c - 1)^2$.

5.3.4 The Lateral Mode $D\{0,1\}$

The lateral mode $D\{0,1\}$, in contrast to $D\{2,0\}$ and $D\{1,0\}$, requires specific modeling. For a perfectly cylindrical surface with $a_0 = a_1 = 0$ in (5.24), we obtain from (5.8)

$$\frac{d^2 r_1}{dz^2} = -Rn\, r_1(z), \quad r_1(z) = b_1 \cos(q_1 z), \quad (5.36)$$
$$q_1^2 = Rn. \quad (5.37)$$

Now taking into account the terms linear in a_0 and a_1, we obtain from (5.8)

$$\frac{d^2 r_1}{dz^2} = -Rn\, r_1(z) - a_1 b_1 [Rn \cos(q_0 z)\cos(q_1 z) + q_0 q_1 \sin(q_0 z)\sin(q_1 z)]. \quad (5.38)$$

Application of the addition theorem for the cosine and integration leads to

$$\frac{r_1(z)}{b_1} = \cos(q_1 z) + a_1 \frac{q_1}{q_0}\left(\frac{3\sqrt{Rn(1+Rn)}}{1-3Rn}\cos(q_0 z)\cos(q_1 z)\right.$$
$$\left. + \frac{1+3Rn}{1-3Rn}\sin(q_0 z)\sin(q_1 z)\right), \quad (5.39)$$

$$V_c = 1 + 2a_0\left(1 - \frac{\tan X}{X}\right) \quad \text{where } X = \frac{\pi}{2}\frac{\sqrt{1+Rn}}{\sqrt{Rn}} \quad (5.40)$$

and

$$\left(1 - \frac{2L_c\sqrt{Rn}}{\pi}\right)\left(1 - \frac{\tan X}{X}\right) = \frac{V_c - 1}{2}\left[\frac{1}{2} + \frac{1+3Rn}{1-3Rn}\frac{\tan X}{X}\right]. \quad (5.41)$$

Equation (5.41) describes the behavior of the lateral instability curves at $V_c \approx 1$.

5.3.5 Nonzero Bond Number

In view of their numerous applications under microgravity conditions and on the ground, extensive stability diagrams of liquid bridges as a function of gravity have been calculated [Martinez 1984, Martinez et al. 1987, Langbein et al. 1986, Langbein 1993]. The stability is governed by the following arguments:

- Short columns with aspect ratio $L_c < 2.1323$ break symmetrically according to the normal deformation $D\{2,0\}$ if the minimum liquid volume for the existence of a capillary surface is present.
- There is also an unstable branch of solutions enclosing larger liquid volumes up to the full period of the unduloids. If, owing to additional measures, one succeeds in establishing those surfaces they break antimetrically according to the normal deformation $D\{1,0\}$.
- For columns with aspect ratio $L_c > 2.1323$, the antimetric breakage according to $D\{1,0\}$ becomes the principal instability. There exist solutions of the capillary equation enclosing smaller volumes, which, however, have negative eigenvalues with respect to $D\{1,0\}$.
- Gravity acting in the axial direction breaks the symmetry between top and bottom. The normal deformations $D\{1,0\}$ and $D\{2,0\}$ are no longer antimetric and symmetric. Since with increasing gravity long columns become less likely, the minimum-volume condition and the normal deformation $D\{2,0\}$ are the main reasons for breakage.
- For thick columns, lateral deformations have to be taken into account. At zero gravity and zero rotation the surface becomes unstable for a radial slope at the supporting disks (see (5.20)). Since gravity bends the slope downward at both disks, the lateral instability becomes effective at lower liquid volumes than under zero-gravity conditions.

If lateral instabilities and thus a radial slope of the surface may be excluded, it is convenient to use cylindrical coordinates $r(\varphi, z)$. Substitution of (5.4) into the capillary equation (3.64) and comparison of the coefficients of 1 and $\cos(n\varphi)$ leads to

$$\widehat{p}_0 + \frac{1}{2} Rn\, r_0^2(z) - Bo\, z$$
$$= \frac{1}{r_0(z)} \frac{1}{\sqrt{1+(\mathrm{d}r_0/\mathrm{d}z)^2}} - \frac{\mathrm{d}^2 r_0}{\mathrm{d}z^2} \frac{1}{\sqrt{1+(\mathrm{d}r_0/\mathrm{d}z)^2}^3}\,, \qquad (5.42)$$

together with (5.8), which is not changed by gravity. The Bond number is defined and r and z are nondimensionalized by means of the disk radius R; $Bo = g\Delta\rho R^2/\sigma$.

For a large liquid volume, when a nearly radial slope and thus lateral instability become likely, it is convenient to switch from cylindrical coordinates $r(z,\varphi)$ to cylindrical coordinates $z(r,\varphi)$. In that case the relevant equations read

$$\widehat{p}_0 + \frac{1}{2} Rn\, r^2 - Bo\, z_0(r) = \frac{1}{r} \frac{d}{dr} r \frac{dz_0/dr}{\sqrt{1 + (dz_0/dr)^2}}, \quad (5.43)$$

$$-Bo\, z_1(r) = \frac{d^2 z_1}{dr^2} + \left(\frac{1}{r} - 3\frac{(dz_0/dr)(d^2 z_0/dr^2)}{1 + (dz_0/dr)^2} \right) \frac{dz_1}{dr}$$
$$- \frac{n^2 z_1}{r^2} \left[1 + \left(\frac{\partial z_0}{\partial r} \right)^2 \right]. \quad (5.44)$$

Using (5.8) and (5.44) for the lateral deformations means applying linear stability analysis with respect to these modes. On the basis of (5.42), (5.8), (5.43) and (5.44), the stability diagram shown in Fig. 5.8 has been obtained using a fourth-order Runge–Kutta method. For any given aspect ratio L_c and Bond number Bo there exists a minimum volume V_c that allows a solution of the capillary equation inclusive of the boundary conditions. And there is a corresponding maximum liquid volume. Realization of these surfaces,

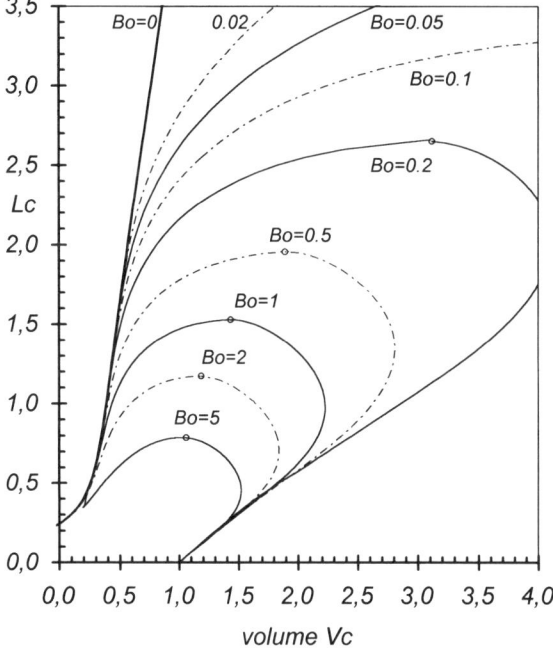

Fig. 5.8. Stability diagram of liquid bridges as a function of gravity (Bond number). The aspect ratio L_c has been plotted versus the minimum stable liquid volume (*left-hand branches*) and versus the lateral instability (*right-hand branches*). The two stability limits meet with parallel positive slope. The maximum possible aspect ratio is limited by the lateral instability. The parameter of the various curves is the Bond number $Bo = \Delta \rho\, g R^2 / \sigma$

however, is difficult, as they require unrealistic slopes at the disks, and the disks must be concave.

The difference between the minimum and the maximum liquid volume decreases with increasing aspect ratio L_c, i.e. for each Bond number there exists a maximum aspect ratio that allows for axisymmetric capillary surfaces. The maximum aspect ratio decreases with increasing Bond number Bo in proportion to $Bo^{-1/3}$ for $Bo < 1$ and in proportion to $Bo^{-1/2}$ for $Bo > 1$. However, in no case may this maximum aspect ratio be realized; it is affected in all cases by the lateral instability. The curves describing the lateral instability according to (5.8) and (5.44) meet the curves corresponding to the minimum volume at smaller volumes than those required for the maximum aspect ratio.

At large liquid volumes the lateral instability always prevails.

5.4 Rotating Free Drops

5.4.1 Motivation

Free liquid drops have repeatedly served as cosmological models, e.g. for the formation of the planets. These liquid drops may float freely or may be positioned by acoustic or electromagnetic levitators. In theory, they may also be supported by coaxial circular disks. If a radial slope at the disks is required and the aspect ratio is increased towards infinity, a free drop can be simulated. Experimentally, however, this does not work at all. We have seen in Sect. 5.2.4 that a radial slope represents the stability limit with respect to the lateral mode $D\{0,1\}$ even for zero rotation number Rn. A drop with radial slope will easily escape sideways.

Free liquid drops have been studied in the Drop Dynamics Module, an acoustic levitator. This system has been successfully operated by Principal Investigator T. Wang during the Spacelab 3 mission, and a second time by E. Trinh during Spacelab 2. Also, towards metallurgical objectives, the electromagnetic levitator TEMPUS (*T*iegelfreies *e*lektro*m*agnetisches *P*rozessieren *u*nter *S*chwerelosigkeit) has flown successfully in the Spacelab missions IML-2 and MSL-1. Undercooling, viscosity, surface tension and other parameters of metallic melts were investigated.

For the historical background and the scientific motivation, let us cite some earlier work:

If a freely floating drop of oil with a spherical shape is spun up by means of a disk, the increasing values of the angular frequency ω cause different shapes: first the drop appears ellipsoidal, then it dimples inward at both poles and, finally, a ring, which participates in the rotation, becomes detached at the equator (Minkowski [1921] on Plateau's work).

In another celebrated experiment conducted by Plateau, a spherical oil drop was rotated in a mixture of water and alcohol with the same density. The drop first looked ellipsoidal, then it appeared oblate (as is true for our rotating planet earth) and, finally, a ring separated at the equator. Subsequently this ring disintegrated into separate parts. Since this could be associated with the formation of planets and their satellites, it served as an experimental verification of Laplace's cosmological hypothesis. This analogy, however, is only superficial [Myshkis et al. 1976].

In many ways it is remarkable that even though Plateau's original investigations were explicitly motivated by "cosmological" considerations, the almost complete similarity of Plateau's problem to one under active theoretical consideration at that time by Jacobi, Riemann, Poincaré, Bryan and Darwin [cf. Jeans 1919] seems to have been overlooked. Apparently, the emergence of a sequence of toroidal figures of equilibrium and their striking similarity (in appearance!) with systems such as Saturn's rings succeeded in drawing attention away from the real significance of Plateau's problem and toward the astronomical problem of the time [Chandrasekar 1965].

5.4.2 Shape of Rotating Drops

As mentioned above, a free liquid drop may be simulated by a drop between disks with infinite aspect ratio $L_c = L/2R$ and a radial slope at the disks. Axisymmetric rotating drops may thus be treated by means of (5.7) and (5.11) close to the equator and close to the poles, respectively. The corresponding equations for the lateral deformation are (5.8) and (5.12). Rewriting (5.7) as

$$\frac{1}{r}\frac{d}{dr}\left(r\frac{dz/dr}{\sqrt{1+(dz/dr)^2}}\right) = -\widehat{p}_0 - \frac{\Delta\rho\omega^2}{2\sigma}r^2 , \qquad (5.45)$$

one obtains the primary integral

$$\frac{r\,dz/dr}{\sqrt{1+(dz/dr)^2}} = -\left(b_0 + b_1 r^2 + b_2 r^4\right) , \qquad (5.46)$$

where

$$b_1 = \frac{\widehat{p}_0}{2}, \quad b_2 = \frac{\Delta\rho\omega^2}{8\sigma} .$$

The quantity b_2 is determined by the experimental variables, and b_0 and b_1 are adjustable parameters; b_n has dimensions (length)$^{-(2n-1)}$.

Several options for introducing dimensionless parameters, depending on the choice of a characteristic length l, have been used. The common methods are [Chandrasekar 1965, Myshkis et al. 1976, Brown & Scriven 1980]:

108   5. Axisymmetric Liquid Columns at Rest and Under Rotation

- equate l to the actual equatorial radius r_2 during rotation: $l = r(0) = r_2$ \hfill (5.46a)
- introduce a capillary length l by requiring $\rho \omega^2 l^3 / (8\sigma) = 1$ \hfill (5.46b)
- deduce l from the spherical shape of the liquid volume V at rest: $V = (4\pi/3)l^3$. \hfill (5.46c)

The radius r, height z, volume V and moment I of inertia transform according to

$$\tilde{r} = rl^{-1}, \quad \tilde{z} = zl^{-1}, \quad \tilde{V} = Vl^{-3}, \quad \tilde{I} = Il^{-5}. \tag{5.47}$$

The choice of l as defined by (5.46b) leads to $b_2 = 1$. The choice (5.46a) leads to the rotation number $\Sigma \equiv Rn$ introduced by Chandrasekar:

$$Rn = \frac{\Delta \rho \omega^2 \left(\tfrac{1}{2} r_2\right)^3}{\sigma}. \tag{5.48}$$

In Runge–Kutta codes for the integration of closed rotating surfaces it is convenient to start with $b_2 = 1$ (i.e. option (5.46b)) and to adapt the parameters to constant volume or maximum radius afterwards. The shape adopted by the surface at rest, the sphere, does not fit into this scheme. For bubbles or for less dense liquids in denser liquids, one has to use $b_2 = -1$.

Figure 5.9 shows the contours of rotating free liquid drops as a function of the angular frequency ω of rotation (see also Table 5.1). Equal drop volumes have been assumed. The resting drop is a sphere. With increasing frequency ω of rotation, the drops flatten and develop a dimple at the poles. For $\rho \omega^2 r_2^3 / 8\sigma = 2.3291$, the poles touch in the center. The stability limit with respect to the lateral deformation $D\{0,2\}$ is given by $\rho \omega^2 r_2^3 / 8\sigma = 0.4488$. Here r_2 is the equatorial radius. In addition, the capillary equation allows several unstable toroidal shapes of rotating drops.

For the position r_1, z_1 of maximum height of the surface and for the equatorial radius r_2, one finds

maximum height: $\quad dz/dr = 0, \quad b_0 r^{-1} + b_1 r + b_2 r^3 = 0,$ \hfill (5.49)
at the equator(s): $\quad dz/dr = \pm\infty, \quad b_0 r^{-1} + b_1 b_1 r + b_2 r^3 = \pm 1.$ \hfill (5.50)

Singly connected surfaces require zero slope, $dz/dr = 0$, at $r = 0$, which means $b_0 = 0$. This leads to the relations

$$b_1 r_2 + b_2 r_2^3 = \pm 1, \tag{5.51}$$

$$b_1 z + 3b_2 \frac{V}{2\pi} = -\int_{\text{equator}}^{\text{pole}} dz \frac{d}{dr} \frac{dz/dr}{\sqrt{1+(dz/dr)^2}}$$

$$= \left[\sqrt{1+(dz/dr)^2}^{-1}\right]_{\text{equator}}^{\text{pole}} = 1. \tag{5.52}$$

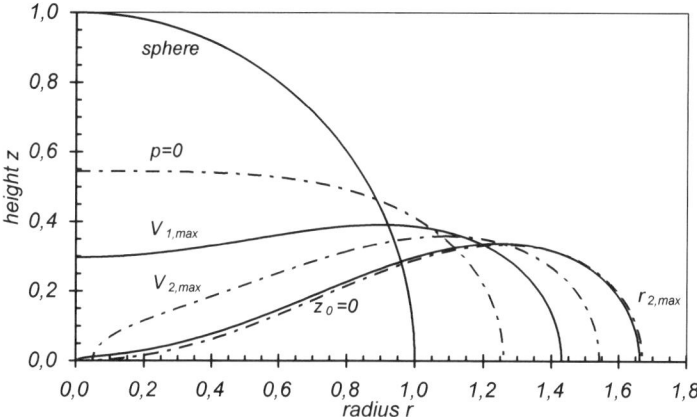

Fig. 5.9. Contours of freely rotating liquid drops as a function of the frequency of rotation. Equal drop volumes have been assumed. The resting drop is a sphere. With increasing angular frequency ω of rotation, the drops flatten and develop a dimple at the poles. For $(\rho\omega^2/8)r_1^3 = 2.3291$, the poles touch each other in the center

From (5.52), it follows that drops with zero polar pressure, $b_1 = 0$, and zero polar height, $z(0) = 0$, enclose equal liquid volumes. In between, the liquid volume reaches a maximum.

Also shown in Fig. 5.9 are a toroid with very small hole diameter and another toroid which has its maximum thickness closer to the rotation axis. With increasing frequency of rotation, the radius r_1 at the maximum thickness z_1 of the toroid initially decreases and, together with the angular momentum, reaches a minimum. The toroid then takes a nearly circular axial cross section (as observed, for example, in acoustic levitators). These toroids are long columns bent to a ring. By expanding at large radii r_1, where r_0 is the radius of the inner equator, r_1 is the radius at maximum height z and r_2 is the radius of the outer equator, one obtains

Table 5.1. Parameters of the contours shown in Fig. 5.9. Here r_2 is the equatorial radius, and z_1 is the maximum height (the polar height or the thickness of the toroid)

b_0	b_1	b_2	r_2	z_1	$A/(4\pi)$	$V/(2\pi)$	$\Delta\rho\omega^2 r_2^3/8\sigma$
0.00	−1.00	0.00	1.00	1	1	2/3	0.00
0.00	0.00	1.00	1.00	0.431185	0.701091	1/3	1.00
0.00	−0.557981	1.00	1.184234	0.245602	0.897515	0.379014	1.660780
0.00	−1.002687	1.00	1.325552	0.00	1.085922	1/3	2.329114
−0.038435	−0.728517	1.00	1.247047	0.290107	0.975269	0.352530	1.939315
−0.280703	0.186708	1.00	1.026497	0.308838	0.586436	0.240461	1.081615

$$b_0 r_1^{-1} = -\frac{3}{4}\left(8b_2 r_1^3\right) + \frac{9}{16}\left(8b_2 r_1^3\right)^{-1} + O\left[\left(8b_2 r_1^3\right)^{-3}\right],$$

$$b_1 r_1 = \frac{1}{4}\left(8b_2 r_1^3\right) - \frac{9}{16}\left(8b_2 r_1^3\right)^{-1} + O\left[\left(8b_2 r_1^3\right)^{-3}\right], \quad (5.53)$$

$$r_2 = r_1\left[1 + \left(8b_2 r_1^3\right)^{-1} + \frac{5}{8}\left(8b_2 r_1^3\right)^{-3}\right],$$

$$r_0 = r_1\left[1 - \left(8b_2 r_1^3\right)^{-1} - \frac{5}{8}\left(8b_2 r_1^3\right)^{-3}\right], \quad (5.54)$$

$$z_1 = r_1\left\{\left(8b_2 r_1^3\right)^{-1} + \frac{1}{8}\left(8b_2 r_1^3\right)^{-3} + O\left[\left(8b_2 r_1^3\right)^{-5}\right]\right\}. \quad (5.55)$$

The toroid, with radius r_1 around the rotation axis, has a radius $(8b_2 r_1^2)^{-1}$ in the perpendicular cross section. All relevant parameters turn out to be power series in $(8b_2 r_1^3)^{-1}$. The ratio of length to circumference increases in proportion to $2\pi r_1/(2\pi/8r_1^2) = (2r_1)^3$. The toroid must break azimuthally.

For bubbles or for less dense liquids in denser liquids, one obtains negative rotation numbers Rn and cigar-like shapes. The cigars reach a length of infinity for

$$b_1 r_2 + b_2 r_2^3 = 1, \quad b_1 + 3b_2 r_2^2 = 0, \quad (5.56)$$

$$b_2 = -1, \quad b_1 = 3/\sqrt[3]{4}, \quad r_2 = 1/\sqrt[3]{2}, \quad (5.57)$$

where r_2 is the equatorial radius. The height diverges logarithmically towards infinity as r approaches that maximum radius.

Figure 5.10 shows the volume of the possible surfaces versus the rotation number Rn according to (5.52). Negative rotation numbers mean bubbles. Singly connected surfaces exist for $0 \leq Rn \leq 2.3291$. At the latter value of Rn the toroids branch off. The branch of singly connected surfaces is joined also to a branch of self-intersecting surfaces (with negative polar radius), as shown in Fig. 5.11. Sections of these surfaces come into existence in the presence of rotating rods or rotating wedges. The stability limit with respect to the lateral mode $D\{0,2\}$ is given by $(\rho\omega^2/8)r_1^3 = 0.4488$. The stability limits of the lateral modes $D\{0,n\}$ with $n = 3$ and $n = 4$ are also shown. In addition, there exist several unstable toroidal shapes of rotating drops.

5.4.3 Stability

As stated already in Sect. 5.2.4, for zero rotation number Rn a radial slope represents the stability limit with respect to the lateral mode $D\{0,1\}$. The drop is just able to float sideways. For positive values of Rn, even a negative slope at the disks leads to lateral instability.

The critical drop deformation is therefore the lateral mode $D\{0,n\}$ with azimuthal wavenumber $n = 2$. Like all lateral modes with $n > 0$, within the linear stability analysis applied here, it conserves angular momentum. Figure 5.12 shows the stability diagram with respect to $D\{0,2\}$. This diagram

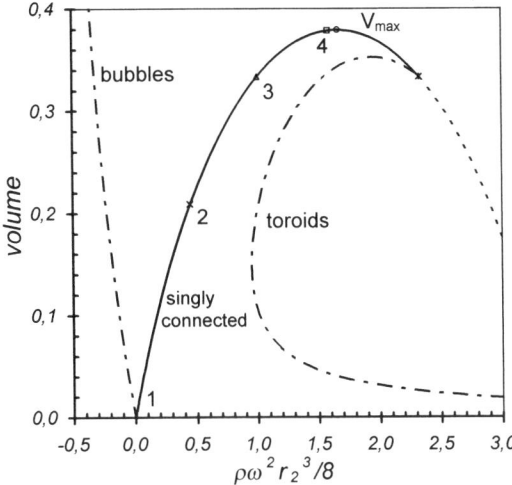

Fig. 5.10. The volumes of liquid drops versus the rotation number Rn. Negative rotation numbers mean less dense drops or bubbles. Singly connected surfaces exist for $0 \leq Rn \leq 2.3291$. At that value the toroids branch off. The branch of singly connected surfaces is also joined to a branch of self-intersecting surfaces (with negative polar radius). Sections of these surfaces come into existence in the presence of rotating rods or rotating wedges

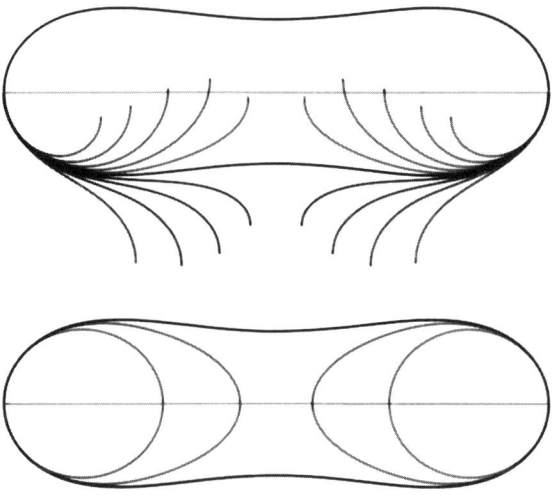

Fig. 5.11. Self-intersecting rotating surfaces

is based on (5.6) and (5.8) with $r_1(z)$ satisfying the boundary conditions $dr_1/dz = 0$ at the equator and $r_1(z) = 0$ at the poles. It corresponds to Fig. 5.5, which relates to the first lateral mode $D\{0,1\}$. In addition to the stability curves for $Rn = 0, 0.5, 1, 2, 3, 4$, the stability limits for a radial and an axial slope at the supporting disks are shown. The curve for an axial slope represents the stability limit with respect to the antimetric mode $D\{1,0\}$ (which is equivalent to a full period of the solution), and the curve for a radial slope corresponds to a freely rotating drop. In the limit of the free drop (infinite aspect ratio) one finds the ratio of the equatorial radius to the polar radius to be 1.477737.

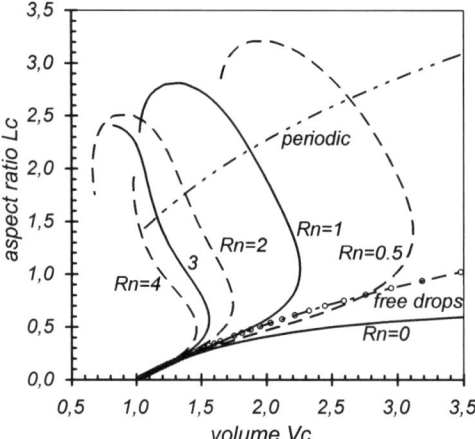

Fig. 5.12. Stability diagram with respect to the lateral mode $D\{0,2\}$. The parameter of the various curves is the rotation number $Rn = 0, 1, 2, 3, 4$. The curve relevant to free liquid drops is that for a slope of 0. The curve for a slope of $\pi/2$ (= periodic) represents the instability with respect to the antimetric mode $D\{1,0\}$

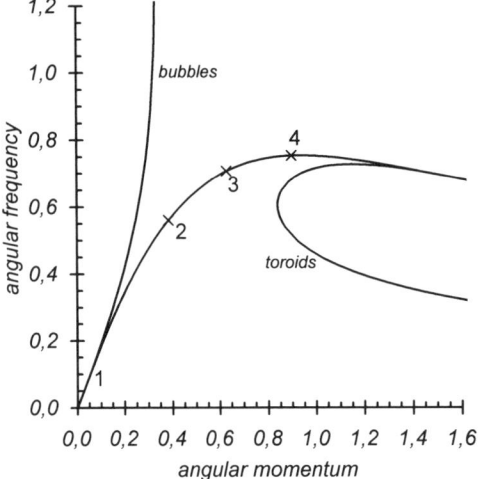

Fig. 5.13. Angular frequency ω versus angular momentum ωI of freely rotating drops. Instabilities arise owing to the lateral modes $D\{0,n\}$ with $n \leq 2$. With free drops, only the mode $D\{0,2\}$ may be realized. The lateral mode $D\{0,3\}$ becomes unstable for a drop with a polar pressure of zero and a ratio of equatorial radius to polar radius of 2.319191

Further instabilities of a freely rotating drop arise owing to the lateral modes $D\{0,n\}$ with $n \leq 2$. The lateral mode $D\{0,3\}$ becomes unstable for a drop with a polar pressure of zero and a ratio of equatorial radius to polar radius of 2.319191. These stability limits, as well as that for wavenumber $n = 4$, are indicated in Fig. 5.13.

In Table 5.2, the circular frequency ω has been nondimensionalized by means of the liquid volume according to (5.46c) the nondimensionalized frequency is given by

$$\tilde{\omega}^2 = \frac{\Delta \rho \, \omega^2}{8\sigma} \frac{3V}{4\pi} . \tag{5.58}$$

Table 5.2. The stability limits of freely rotating drops arising from the lateral modes $D\{0,n\}$ with wavenumbers $n = 1$ to 4

n	$\tilde{\omega}$	Equatorial radius r_2	Equator/pole	Rn
1	0.00	1.00	1.00	0.00
2	0.559928	1.127066	1.477737	0.448863
3	$2^{-1/2} = 0.707107$	$2^{1/3} = 1.259921$	2.319191	1.00
4	0.753560	1.410701	4.378508	1.594191

5.4.4 Conservation of Angular Momentum

In contrast to liquid bridges between coaxial circular disks, whose circular frequency ω is determined by the disks driving it, during an oscillation or deformation a free liquid drop will maintain its angular momentum ωI rather than its circular frequency ω. This, however, does not strongly affect the stability, since the critical lateral deformations $D\{0,n\}$ with wavenumbers $n > 1$ maintain the angular momentum anyway. The change in the moment of inertia I owing to a deformation according to (5.4) is given by

$$I = \rho \int_{-z_0}^{+z_0} dz \int_0^{2\pi} d\varphi \frac{r^4}{2} = \rho \int_{-z_0}^{+z_0} dz \left(\pi r_0^4 + \int_0^{2\pi} d\varphi\, 2 r_0^2 r_1^2 \cos^2(n\varphi) \right)$$

$$= \pi \rho \int_{-z_0}^{+z_0} dz \left(r_0^4 + 2 r_0^2 r_1^2 \right). \tag{5.59}$$

The moment of inertia I increases quadratically with $r_1(z)$. Within the linear approach used, conservation of the angular momentum ωI means a constant circular frequency ω. The circular frequency ω decreases quadratically with the amplitude of the deformation $r_1(z)$. The deformations $D\{n,0\}$ are possible solutions of the linear stability analysis, independent of whether constancy of the angular frequency ω or of the angular momentum ωI is required.

On the other hand, the reduction of the angular frequency ω means a reduction of the potential energy of rotation according to $E_{\text{rot}} = -\omega^2 I/2$. The balance between the surface energy σA and potential energy E_{rot} is thus violated in favor of the surface energy. If conservation of the angular momentum is required, the total free energy of the drop increases quadratically with the deformation $r_1(z)$. A drop is more stable, or at least no less stable, if it rotates at constant angular momentum ωI than if it rotates at constant circular frequency ω.

In order to analyze the stability, the total energy of the drop

$$E = \sigma A + E_{\text{rot}},$$

where

$$A = \int_{-z_0}^{+z_0} dz \, 2\pi r \sqrt{1 + (dr/dz)^2} \,, \quad E_{\rm rot} = -\frac{1}{2}\omega^2 I \,, \tag{5.60}$$

has to be minimized under the constraints of constant volume V and constant angular momentum ωI:

$$V = \int_{-z_0}^{+z_0} dz \pi r^2 = \text{const} \,, \quad \omega I = \text{const} \,. \tag{5.61}$$

The constraint on the angular momentum ωI gives rise to a second Lagrange parameter in the Euler–Lagrange equation in addition to the pressure (which is the Lagrange parameter with respect to the volume). Since the angular momentum ωI shows the same functional dependence as the potential energy $-\omega^2 I/2$, these two contributions add together without changing the character of the resulting capillary equation.

For quantitative investigations, one has to ask to what extent the angular velocity may be balanced radially during one period of the oscillation. If the angular velocity undergoes a rapid modulation, conservation of the angular momentum is affected only marginally. However, at the stability limit the surface deformations have zero frequency. The angular velocity has plenty of time to adapt. The period of the corresponding normal deformation has to be compared with the time required for the radial adaptation of the angular frequency. The former is given by $(\rho R^3/\sigma)^{1/2}$, and the latter by R^2/ν, where ν is the kinematic viscosity. If the latter falls short of the former, there is an irreversible loss in energy.

It may be anticipated that during breakage of a rotating drop, the fragments adopt differing circular frequencies. Radial breakage, as described by Plateau, see Sect. 5.4.1, must lead to an inner, singly connected drop with a higher circular frequency ω and an outer toroid with a lower circular frequency ω.

5.4.5 Finite-Element Analysis

A detailed finite-element analysis of the shape and stability of free liquid drops has been performed by Brown and Scriven [1980]; see Figs. 5.14 and 5.15 here. We read the following in Brown and Scriven [1980].

- The stability inequalities for driven and isolated rotating drops are identical except for the term $8L^2(\delta I)^2/I^3$, which is always positive and appears when the perturbation must conserve angular momentum as well as drop volume. Hence an equilibrium shape is more stable, or at least no less stable, if it rotates freely in isolation from its surroundings. (L is the angular momentum.)

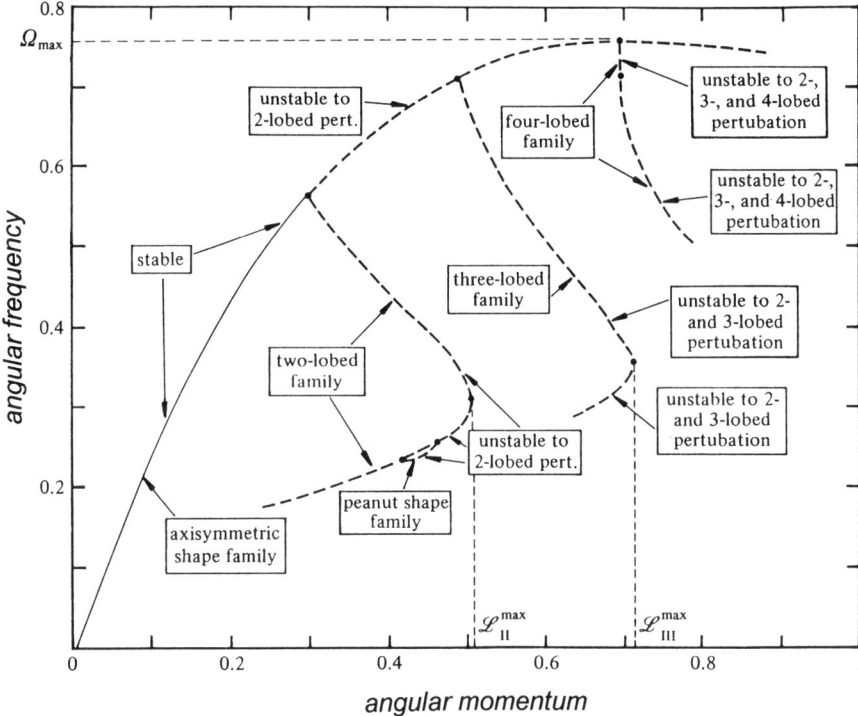

Fig. 5.14. Shape families, their stability and bifurcations, for drops driven to rotate at fixed angular velocity. The shapes are stable along the *solid curves*, unstable along the *dashed curves* [Brown and Scriven 1980]. The n-lobed perturbations are the deformations $D\{0,n\}$

- A real drop is subject to asymmetric disturbances, not just axisymmetric ones. The stability calculations show that the drop becomes unstable at angular velocities well below $\Omega_{\max} = 0.7539$. ($\Omega = (\rho\omega^2 r^3/\sigma)^{0.5}$). At $\Omega_2 \approx 0.5599$ the axisymmetric shape driven at constant angular velocity, is neutrally stable to a two-lobed perturbation. This is found to mark the bifurcation of a family of two-lobed shapes.
- All axisymmetric shapes that spin more rapidly than Ω_2 prove to be unstable to that two-lobed perturbation. However, there are additional points of neutral stability along the axisymmetric family. At $\Omega_3 = 0.7072$ the axisymmetric shape is neutrally stable to a three-lobed perturbation, and at higher angular momentum the axisymmetric shapes are unstable to such perturbations as well as to two-lobed ones. (The n-lobed perturbations are the deformations $D\{0,n\}$ considered here. We find $\Omega_3 = 1/\sqrt{2}$)
- All these perturbations change neither the angular momentum nor the moment of inertia of the drop, and therefore leave its angular velocity unaltered. Indeed, for axisymmetric drops the eigenvalues and eigenfunctions

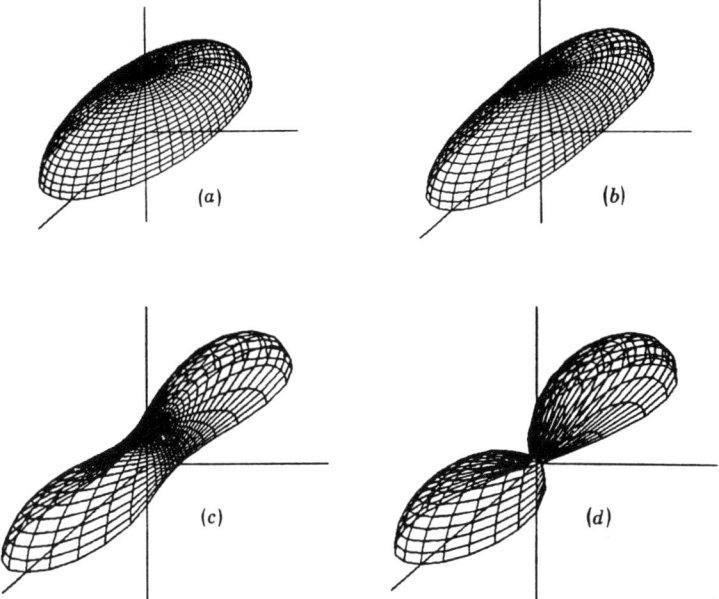

Fig. 5.15. Sample two-lobed shapes (upper half only) calculated with a finite-element algorithm. The stability with respect to angular-momentum-conserving perturbations is indicated. (**a**) $\Omega = 0.548$, $L = 0.311$, stable; (**b**) $\Omega = 0.528$, $L = 0.349$, stable; (**c**) $\Omega = 0.316$, $L = 0.506$, stable; (**d**) $\Omega = 0.224$, $L = 0.387$, unstable

for two-, three- and four-lobed perturbations are the same when the drop is isolated as when it is driven.
- Isolated drops are either more stable or no less stable than drops driven at fixed angular velocity. In other words, removing the isolation constraint and allowing momentum exchange between the drop and its surroundings may destabilize but never stabilize the shape.
- Axisymmetric rotating drops are stable up to a (dimensionless) angular velocity $\Omega_2 \approx 0.5599$, where the two-lobed family branches off. This result of the finite-element analysis appears at first to contradict Chandrasekar's analysis of the stability of axisymmetric drops to inviscid perturbations of infinitesimal amplitude. Chandrasekar found, by a moment method, that though the drop is neutrally stable to a two-lobed perturbation at Ω_2, this type of perturbation does not cause instability at higher angular velocities. The reason is that the type of perturbation considered by Chandrasekar is more highly constrained than the type that we admit. His hypothesis of negligible viscosity implies an energy conservation constraint. Such shapes are traditionally said to be ordinarily stable but secularly unstable.
- The bifurcations from the axisymmetric family are all, with respect to angular velocity, subcritical – the lobed families branch off to lower angular

frequencies. The stability or relative stability of drops driven at constant angular frequency is lost at those bifurcations. However, the bifurcations to the two-lobed and three-lobed families are supercritical with respect to angular momentum, and at these there is no loss, or further loss, of stability of isolated drops. In contrast, the bifurcation to the four-lobed family is subcritical with respect to angular momentum, and, indeed, isolated drops lose relative stability at this bifurcation. The turning points on the two-lobed and three-lobed families are subcritical with respect to angular momentum, and at these points isolated drops lose stability or relative stability. The four-lobed family again provides a contrast: the turning point there is supercritical and, sure enough, isolated drops gain relative stability at this point.
- The predictions made here by computer-aided finite-element analysis may be tested with experiments that are planned for the low-effective-gravity environment of Spacelab [Wang et al. 1976]. In these experiments the drops are to rotate freely, apart from the hopefully slight drag exerted on them by the gas in which they are suspended, etc.

Several experiments of this kind have been successfully performed in Spacelab 3, Spacelab 2, USML-1 and USML-2 [Lee et al. 1988, 1998; Trinh et al. 1982, 1990; Wang et al. 1976, 1986, 1994, 1996].

References

1. Brown RA, Scriven LE: The shape and stability of rotating liquid drops. Proc. Roy. Soc. London **A371** (1980) 331–357
2. Carruthers JR, Gibson EG, Klett MG, Facemire BR: Studies of rotating liquid floating zones on Skylab IV. AIAA Paper 75-692, Denver, CO (1975)
3. Chandrasekar S: The stability of a rotating liquid drop. Proc. Roy. Soc. London **A286** (1965) 1–26
4. Coriell SR, Hardy SC, Cordes MR: Stability of liquid zones. J. Colloid Interf. Sci. **60** (1977) 126–135
5. Da-Riva I, Martinez I: Floating zone stability. Proceedings of the Third European Symposium on Materials Sciences in Space, ESA SP-142 (1979) 67–73
6. Heywang W: Zur Stabilität senkrechter Schmelzzonen. Z. Naturforsch. **11a** (1956) 238–243
7. Jeans JH: Problems of cosmology and stellar dynamics. Cambridge University Press, Cambridge (1919) 40
8. Langbein D, Rischbieter F: Form, Schwingungen und Stabilität von Flüssigkeitsgrenzflächen. Forschungsbericht W 86-29 des BMFT (1986) 1–130
9. Langbein D: Fluid physics. In: *Research in Space – The German Spacelab Missions.* P.R. Sahm, M.H. Keller, B. Schiewe (eds.) WPF Cologne (1993) 91–114
10. Langbein D, Falk F, Großbach R, Heide W, Bauer H, Meseguer J, Perales S, Sanz A: LICOR – liquid columns' resonances. *Scientific Results of the German Spacelab Mission D2.* P.R. Sahm, M.H. Keller, B. Schiewe (eds.) WPF Cologne (1995) 209–219

11. Lee CP, Wang TG: The centering dynamics of a thin liquid shell in capillary oscillations. J. Fluid Mech. **188**, 411–435 (1988)
12. Lee CP, Anilkumar AV, Hmelo AB, Wang TG: Equilibrium of liquid drops under the effects of rotation and acoustic flattening: results from USML-2 experiments in space. J. Fluid Mech. **354** (1998) 43–67
13. Martinez I: Liquid column stability – experiment 1 ES-331. ESA SP-222 (1984) 31–36
14. Martinez I.: Stability of long liquid columns in Spacelab-D1. ESA SP-256 (1987) 235–240
15. Martinez I, Haynes JM, Langbein D: Fluid statics and capillarity. In: *Fluid Science and Materials Science in Space*, HU Walter (ed.), Springer, Berlin, Heidelberg (1987) 53–81
16. Meseguer J: The breaking of axisymmetric slender liquid bridges. J. Fluid Mech. **130** (1983a) 123–151
17. Meseguer J: The influence of axial microgravity on the breakage of axisymmetric slender liquid bridges. J. Cryst. Growth **62** (1983b) 577–586
18. Minkowski H: Kapillarität. In: *Encyklopädie der Mathematischen Wissenschaften*, Vol. 5. A. Sommerfeld (ed.) Teubner, Leipzig (1921)
19. Myshkis AD, Babskii VG, Kopachevskii ND, Slobozhanin LA, Tyuptsov AD: *Low-Gravity Fluid Mechanics* [translated by R.S. Wadhwa]. Springer, Berlin, Heidelberg (1976)
20. Padday JF, Pitt AR: The stability of axisymmetric menisci. Phil. Trans. Roy. Soc. London **275** (1973) 489–528
21. Trinh E, Wang TG: Large-amplitude free and driven drop-shape oscillations: experimental observations, J. Fluid Mech. **122** (1982) 315–338
22. Trinh EH, Leung E: Ground-based studies of the vibrational and rotational dynamics of acoustically levitated drops and shells. AIAA-90-0315 (1990)
23. Wang TG, Saffren MM, Elleman DD: Drop dynamics in space. ESA-SP 114 (1976) 405–419
24. Wang TG, Trinh E, Croonquist AP, Elleman DD: The shapes of rotating free drops: Spacelab experimental results. Phys. Rev. Lett. **56** (1986) 452–455
25. Wang TG, Anilkumar AV, Lee CP, Lin KC: Bifurcation of rotating liquid drops: results of USML-1 experiments in space. J. Fluid Mech. **276** (1994) 389–403
26. Wang TG, Anilkumar AV, Lee CP: Oscillations of liquid drops: results from USML-1 experiments in space. J. Fluid Mech. **308** (1996) 1–14
27. Vega JM, Perales JM: Almost cylindrical isorotating liquid bridges for small Bond numbers. ESA SP-191 (1983) 247–252

6. Liquid Zones

A more common situation than a liquid bridge between coaxial circular disks is that of a liquid bridge between spheres or between parallel plates. Such bridges give rise to attraction and thus may lead to an rearrangement of arrays of spheres. The stability of liquid bridges between parallel plates is considered in this chapter. In the absence of gravity, these bridges are sections of unduloids, catenoids or nodoids. If the contact angles with the two plates are equal, the unduloids become unstable if their inflection point coincides with the contact line. If the contact angles differ, the stability is limited by the minimum-volume condition.

A double float zone is composed of two coaxial liquid bridges coupled in the middle by a freely floating disk. The condition of equilibrium in a double float zone is that the two zones exert equal forces on the separating disk. A single cylindrical zone becomes unstable if the length equals the circumference. If the same zone is separated into two zones by an additional disk, it tends to switch into the antimetric mode in a similar way. This, however, does not cause breakage, but leads to two unequal zones. A small hole in the separating disk would lead to a condition of equal pressure rather than of equal force. Liquid would flow from the slim zone to the fat zone and the double zone would break.

6.1 Liquid Bridges Between Parallel Plates

6.1.1 Introduction

The stability characteristics of liquid zones are strongly changed if liquid bridges between parallel plates are considered. Whereas a cylindrical zone between coaxial disks becomes unstable for an aspect ratio $L_c = \pi$, a cylindrical zone with a contact of angle $\pi/2$ between parallel plates becomes unstable for an aspect ratio $L_c = \pi/2$. The various problems caused by advancing and receding contact angles and hysteresis have hitherto excluded quantitative investigations of such liquid bridges under microgravity conditions. Only a few experiments on liquid surfaces between spheres and on metallic melts between parallel plates have been performed. On the other hand, there is much interest in the behavior of ensembles of spheres, which serve as model

120   6. Liquid Zones

systems for porous media. Parallel plates may be understood as two coaxial spheres with infinite radius.

A Spacelab-1 experiment on liquid surfaces between three equal spheres touching each other was performed by Haynes [1978,1984]. The behavior of bridges of metallic melts between parallel plates was studied in the MASER-1 sounding-rocket mission [Rositto et al. 1987, Passerone et al. 1988].

A mathematical treatment of the stability of liquid bridges between parallel plates has been given by Vogel [1987, 1989, 1991a,b]. Vogel considers liquid surfaces with constant mean curvature, i.e. the case of zero gravity, and identifies two stability criteria, namely

- a stable liquid surface must not have an inflection point between the plates
- within a family of stable liquid surfaces, the enclosed liquid volume must increase with increasing capillary pressure.

6.1.2 Branches of Solutions of the Capillary Equation

The prerequisite for analyzing stability is to obtain a complete overview of all possible surfaces. The possible surfaces once again are unduloids, catenoids and nodoids, as discussed in Sect. 3.4.

6.1.2.1 Contact Angle Less than a Right Angle. A catenoid and the inner section of a nodoid fit between two parallel plates if the contact angle γ of the liquid with the plates is smaller than a right angle; see Figs. 6.1a,b. An unduloid may fit also. This, however, requires that its slope at the inflection point is smaller than γ. In that case the unduloid fits even in four different ways, i.e. the lower inflection point and the upper inflection point may each lie either inside or outside the plates; see Figs. 6.1c–e [Langbein 1992a, 1992b].

Two of the families of unduloids, one with both inflection points inside and the other with both inflection points outside, actually belong to the same symmetric family of solutions. They are smoothly linked to each other via an unduloid with slope γ at the inflection point. The asymmetric families of unduloids with one inflection point lying inside and the other one lying outside the parallel plates are degenerate. They join the family of symmetric unduloids at the unduloid with slope γ at the inflection point, such that the latter unduloid actually belongs to all families mentioned.

A qualitative sketch of the volume of the liquid bridges versus the capillary pressure is given in Fig. 6.2. There are nodoids with negative pressure, there is a catenoid with pressure zero and there is a symmetric branch of unduloids, which, at the unduloid with minimum slope γ, is touched by an asymmetric, degenerate branch of unduloids. The symmetric family of unduloids is completed by two spherical sections touching each other. There are, however, further branches of solutions, which extend over more than one wavelength. These solutions are always unstable.

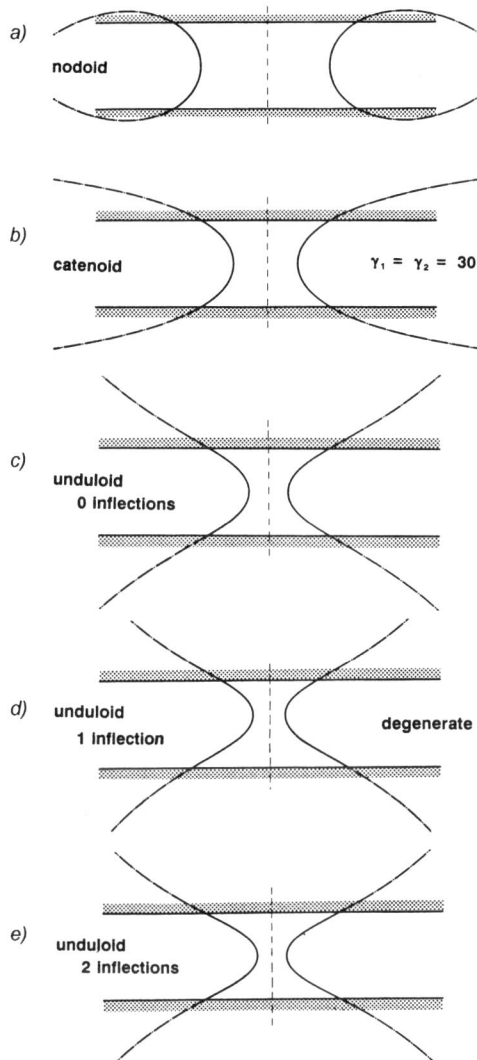

Fig. 6.1. Liquid surfaces with equal contact angles $\gamma_1 = \gamma_2 < \pi/2$ between parallel plates: (**a**) nodoid; (**b**) catenoid; (**c**) unduloid with both inflection points outside; (**d**) unduloid with one inflection point outside and the other one inside; (**e**) unduloid with both inflection points inside

6.1.2.2 Contact Angle Larger than a Right Angle. If the contact angle γ of the liquid with the solid plates exceeds a right angle, the family of possible surfaces includes the outer sections of nodoids (Fig. 6.3a), spherical sections (Fig. 6.3b) and, once more, unduloids. The slope of the unduloids at the inflection point must exceed the contact angle γ. The inflection points may again lie inside or outside the parallel plates; see Figs. 6.3c–f. The undu-

122   6. Liquid Zones

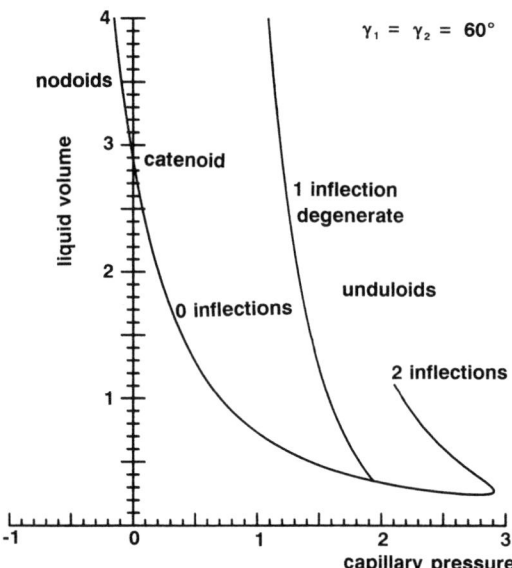

Fig. 6.2. Volume of liquid bridges with contact angle $\gamma < \pi/2$ between parallel plates versus capillary pressure

loid with slope γ at the inflection point belongs to the family of symmetric solutions and to the degenerate family of asymmetric solutions. The family of symmetric unduloids is completed by a sphere between the parallel plates (Fig. 6.3f). Figure 6.4 shows the volume of liquid bridges with contact angle $\gamma > \pi/2$ versus the capillary pressure. Again, there are further symmetric and asymmetric branches of solutions extending over more than one wavelength of the unduloids.

6.1.2.3 Contact Angle Equal to a Right Angle. The various branches of solutions all join if the contact angle γ of the liquid with the plates equals $\pi/2$. The possible solutions are cylinders and unduloids, which have their minimum or maximum radius at the plates.

There is a family of unduloids with one inflection point between the plates. These unduloids have their minimum radius at one plate and their maximum radius at the other plate. They are asymmetric and are degenerate. This family of unduloids joins the family of cylinders when the minimum and the maximum radius become equal, i.e. when their length equals half their circumference.

There is also a family of unduloids with two inflection points between the plates. These unduloids may have either their minimum radius or their maximum radius at the plates. This makes no difference to their volume and therefore brings about a degeneracy, too. This family of unduloids also joins the family of cylinders when they have equal minimum and maximum radii, that is, if their length equals their circumference.

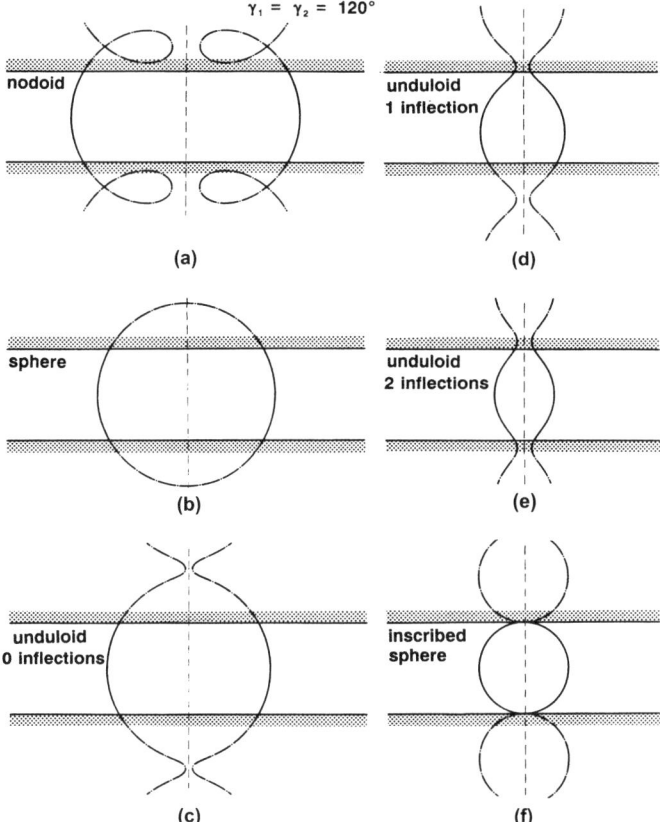

Fig. 6.3a–f. Liquid surfaces with equal contact angles $\gamma_1 = \gamma_2 > \pi/2$ between parallel plates

Generalizing, we find families of unduloids with N inflection points between the plates. Each of these families is degenerate, starts with zero minimum radius, i.e. with $N/2$ half-spheres fitting between the plates, and joins the family of cylinders for equal minimum and maximum radii, i.e. when their length equals $N/2$ times their circumference. Figure 6.5 depicts the resulting plot of liquid volume versus capillary pressure.

6.1.3 Properties of the Inflection Point

An important extension of the minimum volume condition (see Sect. 4.6) is the statement that the bifurcation point of two families of solutions of the capillary equation means an instability if the liquid surfaces at that bifurcation coincide. Any infinitesimal violation of the symmetry causes the V, p curves to split as shown in Fig. 4.9, such that extrema of the volume appear, i.e. a stability limit is reached. Since the families of unduloids discussed

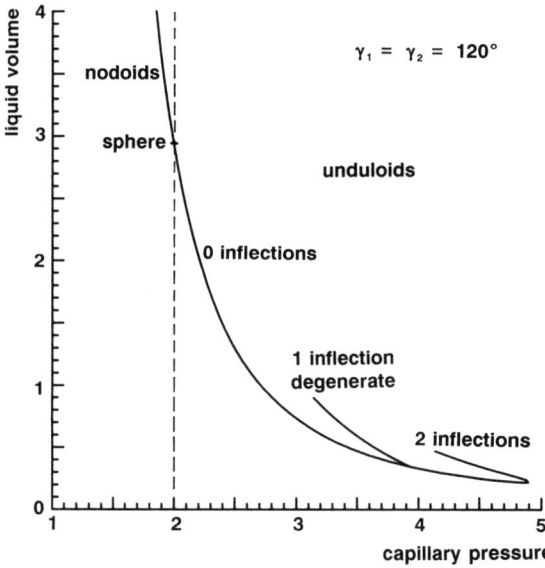

Fig. 6.4. Volume of liquid bridges with contact angle $\gamma > \pi/2$ between parallel plates versus capillary pressure

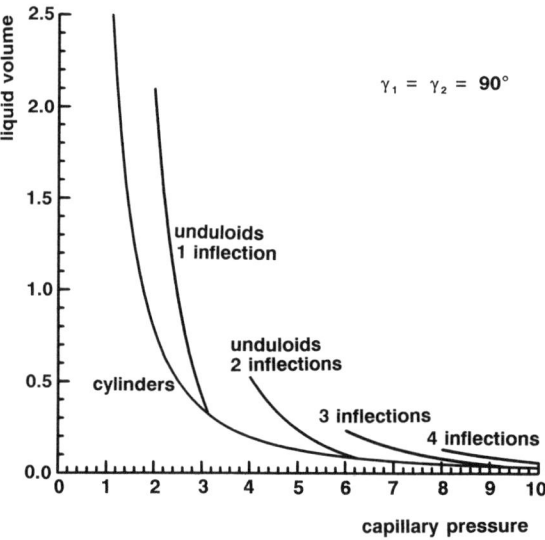

Fig. 6.5. Volume of liquid bridges with contact angle $\gamma = \pi/2$ between parallel plates versus capillary pressure. The degenerate families of unduloids with N inflection points within the plates join the nondegenerate family of cylinders

bifurcate when their slope ψ at the inflection point equals the contact angle γ, we find this condition to cause a stability limit. Only unduloids that do not have an inflection point between the plates may be stable. Unduloids with one or more inflection points between the plates are always unstable. The mathematical proof of this principle has been given by Vogel [1989].

Whether the unduloid whose slope at the inflection point equals the contact angle γ represents the leading stability limit depends on wether the V, p curves have already reached a minimum volume prior to the bifurcation point, i.e. at lower capillary pressure. The task of finding the primary stability limit becomes one of calculating the minimum volumes within the families of symmetric unduloids and comparing the capillary pressure at the minimum volume with that at the bifurcation point.

The most important properties of the inflection points of unduloids are presented in Sect. 3.4; see (3.30)–(3.38). In particular, the difference between the axial distances from the position of maximum radius to the inflection point ($= z_2$) and from the inflection point to the position of minimum radius ($= z_1$) equals the difference between the maximum and the minimum radii:

$$z_2 - z_1 = r_2 - r_1 \tag{6.1}$$

(3.37). In order to judge the stability, one additionally needs the surface areas and the volumes enclosed. The surface areas A_1 and A_2 of the sections from the position of minimum radius to the inflection point and from the inflection point to the position of maximum radius, respectively, are given by

$$A_{1,2} = 2\pi \left(\frac{r_2 + r_1}{2}\right)^2 \left[2E(\cos^2 \psi) - \sin^2 \psi K(\cos^2 \psi) \mp 2\cos\psi\right], \tag{6.2}$$

where $\psi = \varphi_{\min}$ is the minimum angle of the liquid surface at the inflection point. For the corresponding volumes, one obtains

$$V_{1,2} = \frac{\pi}{3}\left(\frac{r_2 + r_1}{2}\right)^3 \tag{6.3}$$
$$\times \left[(7+\cos^2\psi)E(\cos^2\psi) - 4\sin^2\psi K(\cos^2\psi) \mp 9\cos\psi \pm \cos^3\psi\right].$$

Equations (6.2)–(6.4) leads to the further valuable relations

$$3V_1 - (r_2 + r_1)A_1 = -\pi r_1 r_2 z_2,$$
$$3V_2 - (r_2 + r_1)A_2 = -\pi r_1 r_2 z_1. \tag{6.4}$$

6.1.4 The Instability Due to the Bifurcation (Due to $D\{1,0\}$)

We are now well prepared to determine the stability limits of the family of symmetric unduloids due to the bifurcation and to the minimum-volume condition. Let the separation of the parallel plates be L and the contact angle of the liquid be γ. For $\gamma < \pi/2$, an unduloid with slope $\psi = \gamma$ at the inflection point fits between the plates if the distance z_1 from the position of minimum radius to the inflection point equals $L/2$, yielding the dimensionless pressure

$$\tilde{p} \equiv \hat{p}L = 2\left[E(\cos^2 \gamma) - \cos \gamma\right]. \tag{6.5}$$

The dimensionless volume of this unduloid is given by

$$\tilde{V} \equiv \frac{V}{L^3} \tag{6.6}$$

$$= \frac{2\pi}{3\tilde{p}^3}\left[(7 + \cos^2 \gamma)E(\cos^2 \gamma) - 4\sin^2 \gamma K(\cos^2 \gamma) - 9\cos \gamma + \cos^3 \gamma\right].$$

For a contact angle $\gamma > \pi/2$, an unduloid with slope $\psi = \pi - \gamma$ at the inflection point fits between the plates if the distance z_2 from the inflection point to the position of maximum radius equals $L/2$. Since in that case $\cos \gamma = -\cos \psi$, (6.5) and (6.6) are reobtained. These equations give the capillary pressure $\tilde{p}$ and the volume $\tilde{V}$ at the stability limit of the unduloids due to the bifurcation. Breakage is due to the antimetric axisymmetric mode $D\{1,0\}$. Figure 6.6 shows $\tilde{V}$ versus $\tilde{p}$. The asterisks and circles indicate the contact angle γ in steps of $\pi/12$.

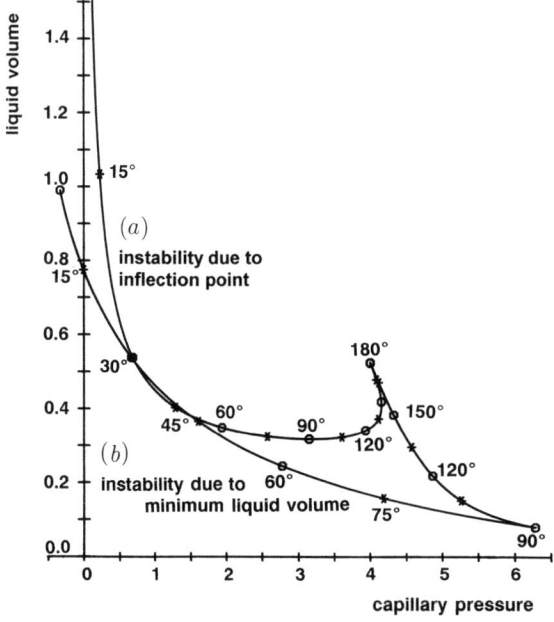

Fig. 6.6. Volume of liquid bridges versus capillary pressure at the stability limit (a) due to the bifurcation (inflection point) and (b) due to the minimum liquid volume

The smallest stable volume arises for $\gamma = \pi/2$. This corresponds to the cylinder whose length equals half its circumference. Its capillary pressure is $\tilde{p} = \pi$, and its volume is $\tilde{V} = 1/\pi$.

For $\sin^2 \gamma \ll 1$, we find from (6.5) and (6.7)

$$\tilde{V} = \frac{8\pi}{3\tilde{p}^2} \, . \tag{6.7}$$

At small contact angles γ, the stable liquid volume decreases in proportion to the inverse square of the pressure. Equation (6.7) also applies to contact angles $\gamma \approx \pi$, where the family of possible unduloids reduces to the inscribed sphere. The capillary pressure of the inscribed sphere is $\tilde{p} = 4$, and its volume is $\tilde{V} = \pi/6$.

6.1.5 The Instability Due to the Minimum Volume (Due to $D\{2,0\}$)

Let us now turn to the stability limit due to by the minimum volume, within the family of symmetric unduloids. For an unduloid to fit between the parallel plates with contact angle γ, the axial distance z from the position of minimum radius to the position with slope γ must equal $L/2$. From

$$z = \int_{r_1}^{r(\gamma)} dr \tan \varphi = \frac{L}{2} \, , \tag{6.8}$$

we obtain, by the same substitutions as in Sect. 3.4,

$$\tilde{p} = 2 \int_0^{\xi_0} d\xi \frac{1 - \cos\psi \cos\xi}{\sqrt{1 - \cos\psi \cos\xi + \cos^2\psi}} \, . \tag{6.9}$$

The upper integration limit ξ_0 in (6.9) is given by

$$\xi_0 = \arcsin\left[\frac{\cos\gamma}{\cos\psi}\left(\sin\gamma \mp \sqrt{\cos^2\psi - \cos^2\gamma}\right)\right] \, . \tag{6.10}$$

Equations (6.9), (6.10) and those following apply formally to nodoids as well, although in that case one has to tolerate $\cos\psi > 1$, ψ imaginary. At the inflection point $\cos\psi = \cos\gamma$ of the unduloid, we obtain $\xi_0 = \gamma$.

Application of the same procedure to the volume of the unduloid considered leads to

$$\tilde{V} = \frac{2\pi}{\tilde{p}^3} \int_0^{\xi_0} d\xi \, (1 - \cos\psi \cos\xi) \sqrt{1 - 2\cos\psi \cos\xi + \cos^2\psi} \, . \tag{6.11}$$

128   6. Liquid Zones

To find the minimum volume within the family of symmetric surfaces given by (6.9)–(6.11) we require

$$\frac{dV}{dc} = 0, \tag{6.12}$$

where $c = \cos\psi$, which is equivalent to

$$3\frac{d}{dc}\log\left(\int_0^{\xi_0} d\xi \frac{1 - c\cos\xi}{\sqrt{1 - 2c\cos\xi + c^2}}\right)$$

$$= \frac{d}{dc}\log\left(\int_0^{\xi_0} d\xi\, (1 - c\cos\xi)\sqrt{1 - 2c\cos\xi + c^2}\right). \tag{6.13}$$

The differentiations required are straightforward, though somewhat tedious. One obtains

$$\left(\int_0^{\xi_0} d\xi \frac{\cos\xi}{\sqrt{1 - 2c\cos\xi + c^2}} \pm \frac{\sin\xi_0}{\sqrt{c^2 - \cos\gamma}}\right)$$

$$\times \int_0^{\xi_0} d\xi\, (1 - c\cos\xi)\sqrt{1 - 2c\cos\xi + c^2}$$

$$= \left(\int_0^{\xi_0} d\xi\, \cos\xi\sqrt{1 - 2c\cos\xi + c^2} \pm \frac{1}{3}\frac{c^2\sin^3\xi_0}{\cos^2\gamma\sqrt{c^2 - \cos^2\gamma}}\right)$$

$$\times \int_0^{\xi_0} d\xi \frac{\cos\xi}{\sqrt{1 - 2c\cos\xi + c^2}}. \tag{6.14}$$

Equation (6.14), together with (6.10), gives $\cos\psi$ as a function of $\cos\gamma$. Solving (6.14) numerically, we obtain, in particular,

$$\begin{array}{lll}
\cos\psi = 1.095085 & \text{for } \cos\gamma = 1, & \gamma = 0, \\
\cos\psi = 1 & \text{for } \cos\gamma = 0.966030, & \gamma = 14.976925°, \\
\cos\psi = \cos\gamma & \text{for } \cos\gamma = 0.855852, & \gamma = 31.146031°, \\
\cos\psi = 0 & \text{for } \cos\gamma = 0, & \gamma = 90°, \\
\cos\psi = -1 & \text{for } \cos\gamma = -1, & \gamma = 180°.
\end{array} \tag{6.15}$$

This means that at low contact angles γ the stability limit arises within the family of nodoids, that the catenoid represents the stability limit for $\gamma \approx 15°$ and that the stability limits due to the minimum volume and due to the bifurcation coincide for $\cos\gamma = 0.855852$. For contact angles larger than 31.146031°, the bifurcation represents the primary stability limit.

Substitution of $c = \cos\psi$ according to (6.14) into (6.9) and (6.11) gives the dimensionless pressure $\tilde{p}$ and the volume $\tilde{V}$ as a function of the contact angle γ. $\tilde{V}$ is plotted versus $\tilde{p}$ in Fig. 6.6. The asterisks and circles again indicate the contact angle γ in steps of $\pi/12$.

At the crossover point of the stability limits due to the minimum-volume and due to the bifurcation, i.e. for $c = \cos\psi = \cos\gamma$, (6.14) reduces to

$$\tilde{V} = \frac{\pi}{3}\frac{\sin^2\gamma}{\tilde{p}^2}. \tag{6.16}$$

Equation (6.16) is equivalent to the condition

$$(3 + \cos^2\gamma)\,E(\cos^2\gamma) - 2\sin^2\gamma K(\cos^2\gamma) = 4\cos\gamma. \tag{6.17}$$

Equation (6.17) is an implicit equation for $\cos\gamma$ in terms of complete elliptic integrals. Its solution is given by $\cos\gamma = 0.85585183$, $\gamma = 31.14603108°$.

The stability limit due to the minimum volume condition has been studied by Rositto et al. [1987]. By excluding from the start liquid surfaces with an inflection point between the plates, they find unstable surfaces for $\gamma < 20°$. This, apparently, is a rough estimate of the actual condition, $\gamma < 31.15°$. In a subsequent paper on the MASER-1 results, a correct stability diagram including the instability due to the inflection points is given [Passerone et al. 1988].

6.1.6 Differing Contact Angles

Up to now the contact angles of the liquid with the two parallel plates have been assumed to be equal. If the contact angle with plate 1 is γ_1 and that with plate 2 is $\gamma_2 > \gamma_1$, families of symmetric solutions no longer exist. The family of unduloids with no inflection point between the plates now joins with the family exhibiting an inflection point close to the plate with the smaller contact angle. At the same time, the family of unduloids with an inflection point close to the plate with the larger contact angle joins the family with two inflection points. The latter family, however, is not stable. The primary stability limit arises at the minimum volume within the former family of unduloids.

Figure 6.7 shows the volume $\tilde{V}$ versus the pressure $\tilde{p}$ for constant mean contact angles γ_0 and increasing contact angle differences, where $\gamma_1 = \gamma_0 - n\,\Delta\gamma$, $\gamma_2 = \gamma_0 + n\,\Delta\gamma$.

6.1.7 Gravity

A second important violation of symmetry results from gravity. Gravity tends to move the liquid to the lower side, i.e. to increase the radius of the liquid bridge in its lower region. We define the Bond number by

$$Bo = \frac{g\,\Delta\rho\,L^2}{\sigma}. \tag{6.18}$$

130 6. Liquid Zones

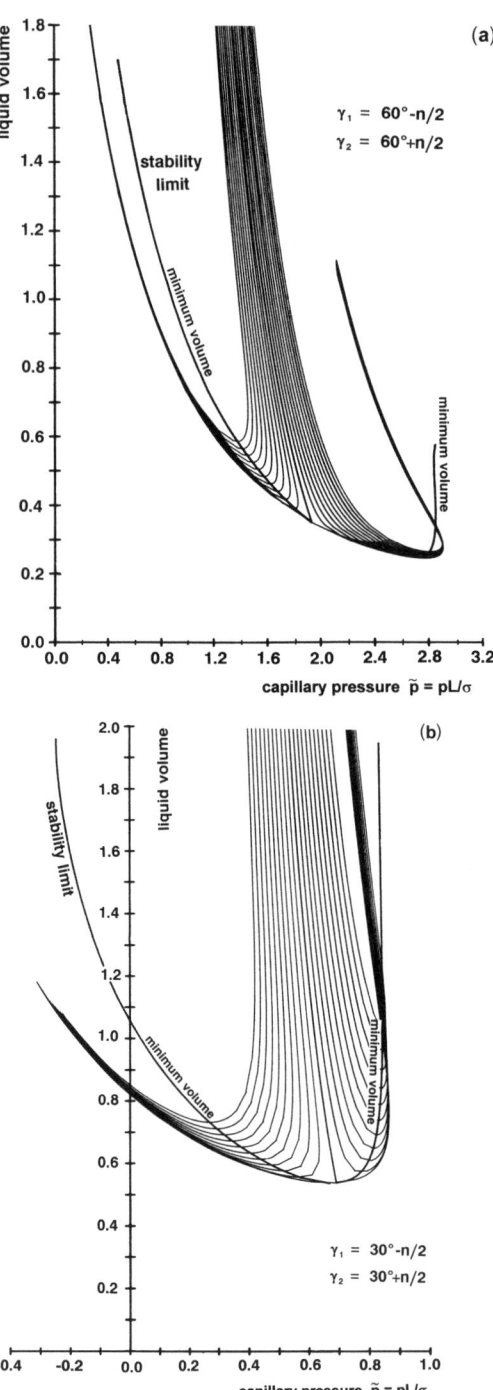

Fig. 6.7a,b. Volume of liquid bridges between parallel plates versus a capillary pressure for differing contact angles $\gamma_1 = \gamma_0 - n\,\Delta\gamma$, $\gamma_2 = \gamma_0 + n\,\Delta\gamma$, where $\Delta\gamma = 1°$

Figure 6.8 shows the volume $\tilde{V}$ of liquid bridges versus the pressure $\tilde{p}$ for increasing Bond number $Bo = 0, 1, 2, 3, \ldots$. As in the case of differing contact angles, there are two families of solutions. The first family exhibits a larger radius at the bottom plate than at the top plate. For the second family the situation is reversed. The radius at the top plate is larger than at the bottom plate. It is obvious that the latter family of solutions is always unstable. Exchanging top and bottom is equivalent to changing the sign of the Bond number. The second family of solutions, therefore, is marked by negative Bond numbers.

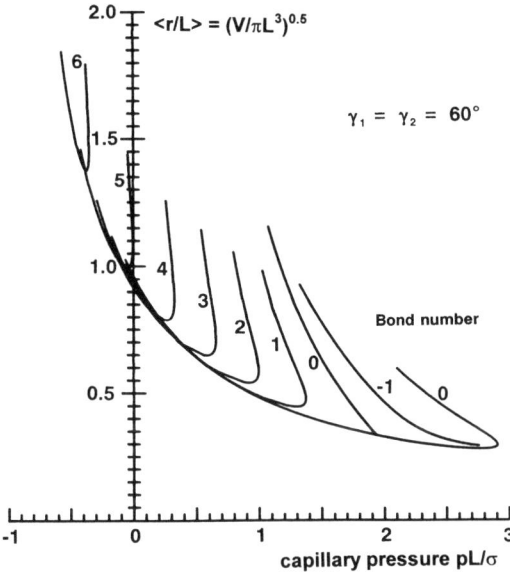

Fig. 6.8. Volume of liquid bridges between parallel plates versus capillary pressure for Bond numbers $Bo = -1$ to 6

Figure 6.9 shows the minimum liquid volume of stable liquid bridges between parallel plates versus the Bond number. The reciprocal square root of the liquid volume divided by π, that is, the reciprocal average radius $\tilde{r}$ of the bridge, is plotted on the abscissa. The various curves correspond to different pairs of equal contact angles $\gamma_1 = \gamma_2$. Figure 6.9 has been obtained numerically. Nevertheless, there are several exact limits. For $Bo = 0$ and $\gamma_1 = \gamma_2 = \pi/2$, the surface with the minimum stable volume is a cylinder with reciprocal radius $(\tilde{V}/\pi)^{-0.5} = \pi$. For $\gamma_1 = \gamma_2 = \pi$ the surface with the minimum stable volume is the inscribed sphere, yielding $(\tilde{V}/\pi)^{-0.5} = \sqrt{6}$.

Most importantly, there is an exact limit for large Bond numbers. Large Bond numbers generally require large liquid volumes. The minimum liquid volume required for obtaining a stable bridge increases with increasing Bond

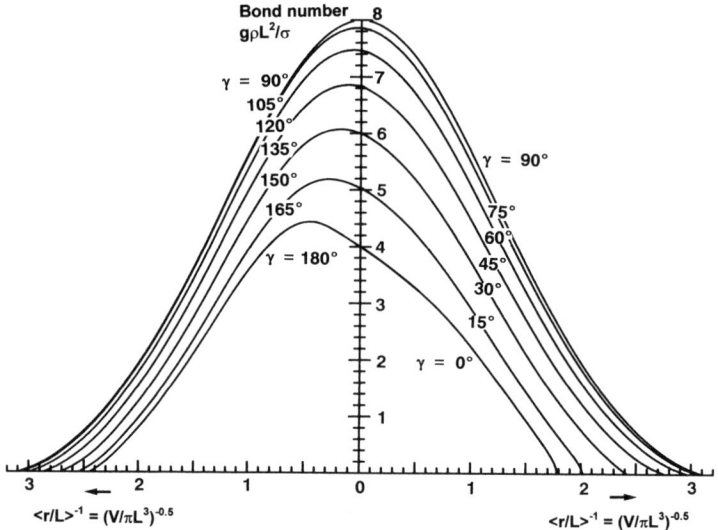

Fig. 6.9. Maximum allowed Bond number of stable liquid bridges between parallel plates versus reciprocal average radius (= square root of liquid volume divided by π): *right*, contact angles $\gamma \leq \pi/2$; *left*, contact angles $\gamma \geq \pi/2$

number. The corresponding liquid surfaces all exhibit a much smaller radius in the top region than in the bottom region. And the surface with the minimum liquid volume has a nearly horizontal tangent at the inflection point. By integrating the capillary equation in cylindrical coordinates once, we obtain

$$\frac{dr/dz}{\sqrt{1+(dr/dz)^2}} - \int_0^z \frac{dz}{r\sqrt{1+(dr/dz)^2}} = Bo\frac{z^2}{2} - \tilde{p}z + C, \qquad (6.19)$$

where C is the constant of integration. The boundary conditions on the contact angles at the plates lead to

$$\cos(\pi - \gamma_1) = C \qquad (6.20)$$

and

$$\cos\gamma_1 + \cos\gamma_2 = \frac{Bo}{2} - \tilde{p} + \int_0^1 \frac{dz}{r\sqrt{1+(dr/dz)^2}}. \qquad (6.21)$$

In the case of very large volumes, i.e. large radii of the liquid bridge, the last term in (6.21) can be neglected. The bridge has a horizontal tangent at the inflection point, yielding

$$Bo\, z = \tilde{p}, \qquad (6.22)$$
$$\tilde{p}^2 = 2Bo[1 + \cos(\pi - \gamma_1)]. \qquad (6.23)$$

From (6.21) and (6.23), we obtain

$$Bo = 4\left[\sin\frac{\gamma_1}{2} + \cos\frac{\gamma_2}{2}\right]^2 \tag{6.24}$$

$$\tilde{p} = 4\sin\frac{\gamma_1}{2}\left[\sin\frac{\gamma_1}{2} + \cos\frac{\gamma_2}{2}\right] = 2\sin\frac{\gamma_1}{2}\sqrt{Bo}. \tag{6.25}$$

Equation (6.24) gives the maximum allowed Bond number for the case of very large liquid volumes as a function of the contact angles γ_1 and γ_2. This is the value which the curves in Fig. 6.9 assume at the ordinate. The deviation from (6.24) and (6.25) for finite liquid volumes has an order of magnitude of $(\tilde{V}/\pi)^{-0.5} = 1/\tilde{r}$. This determines the slope of the curves at the ordinate in Fig. 6.9.

For equal contact angles $\gamma_1 = \gamma_2$, (6.24) yields the same Bond number for the complementary angles γ_1 and $\pi - \gamma_1$. In both cases nearly the same surface shape appears in the limit of very large volume (see Fig. 6.10). The inner liquid and outer liquid are just exchanged, as are the contact angles. By application of this principle also to the terms proportional to $1/\tilde{r}$ neglected in (6.19) and (6.21), we conclude that the slopes of the curves for complementary contact angles γ_1 and $\pi - \gamma_1$ differ in sign only. This is confirmed by the numerical calculations. This is the reason for plotting the stability curves for γ_1 and $\pi - \gamma_1$ in Fig. 6.9 with opposite signs of $1/\tilde{r}$ on the two sides of the ordinate.

By combining the stable surfaces with the minimum liquid volume for a sessile drop with contact angle γ_1 and a hanging drop with contact angle γ_2, we find, for the maximum allowed Bond number inclusive of terms of order $1/\tilde{r}$,

$$Bo = 4\left[\sin\frac{\gamma_1}{2} + \cos\frac{\gamma_2}{2}\right]^2 + \frac{4}{3\tilde{r}}\tan\frac{\gamma_1}{4}\left[1 + \cos\frac{\gamma_1}{2} + \cos^2\frac{\gamma_1}{2}\right]$$
$$- \frac{4}{3\tilde{r}}\cot\frac{\gamma_2}{4}\left[1 + \sin\frac{\gamma_2}{2} + \sin^2\frac{\gamma_2}{2}\right]. \tag{6.26}$$

The increase of the maximum allowed Bond number with decreasing liquid volume for contact angles such that $\gamma_1 + \gamma_2 > \pi$ is in agreement with the corresponding behavior of sessile drops: sessile drops with contact angle $\gamma_1 > \pi/2$ reach their maximum height at a finite liquid volume, when the radial curvature is still adding positively to the positive capillary pressure resulting from the axial curvature [Finn 1986]. The radial curvature further stabilizes the liquid column. The strength of the radial curvature decreases when the liquid volume is further increased. This principle also applies to hanging drops with contact angles $\gamma_2 < \pi/2$.

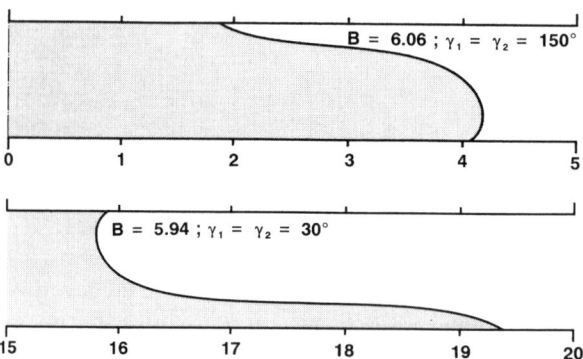

Fig. 6.10. Liquid bridges with complementary contact angles $\gamma_1 = \gamma_2 = 150°$ and $\gamma_1 = \gamma_2 = 30°$ in the limit of large liquid volume

6.1.8 Key Points

In the absence of gravity, liquid bridges between parallel plates are symmetric if the contact angles with the two plates are equal. The shapes are sections of nodoids, catenoids and unduloids. The unduloids become unstable if their inflection point coincides with the contact point, i.e. if the slope at the inflection point equals the contact angle. The unstable deformation is the antimetric mode $D\{1,0\}$. This, however, cannot be the principal instability. The liquid surface also becomes unstable if, within a family of solutions of the capillary equation, a minimum arises in the liquid volume. In that case the unstable deformation is the symmetric mode $D\{2,0\}$. There is actually a competition between the two instability criteria. For contact angles $\gamma_1 = \gamma_2 < 31.15°$ and $\gamma_1 = \gamma_2 > 31.15°$, the minimum-volume condition and the inflection point, respectively, lead to the principal instability. The minimum-volume condition also leads to the instability of the catenoids and the nodoids, which join the family of unduloids for zero and negative capillary pressure. The crossover point of the two stability criteria can be calculated explicitly in terms of complete elliptic integrals.

If the contact angles of the liquid bridge at the two plates differ, all bifurcations of the volume versus pressure curves, which correspond to different families of solutions, are resolved by a rearrangement of the families. In that case the minimum-volume condition is the only stability criterion applicable. Another common violation of symmetry is gravity. If the sum of the contact angles with the bottom plate and the top plate $\gamma_1 + \gamma_2$ exceeds π, it is not an infinite liquid volume which yields the highest allowed Bond number. The positive radial curvature of a bridge enclosing a finite liquid volume supports the bridge better than the zero radial curvature that occurs in the case of an infinite liquid volume.

The different stability criteria applied all share the same basic principle. A bifurcation or degeneracy of different families of liquid surfaces is generally associated with a symmetry operator. Any perturbation of the symmetry resolves the degeneracy, thus giving rise to extrema of the volume and to application of the extremum-volume stability criterion. This means that a bifurcation corresponds to a stability limit and that the first stability criterion given by Vogel [1987] can be subsumed into the second. In the case of parallel plates, equal contact angles and zero gravity, the symmetry operator is the reflection of the liquid surface in the middle plane. Differing contact angles and nonzero gravity break that symmetry. Rotation, on the other hand, maintains it.

Finally, it is worth mentioning that a liquid bridge between parallel plates is always in a metastable state. It may rest in any position. However, if the planes are tilted slightly, a wetting liquid drop moves in the direction of decreasing spacing until it is trapped in the wedge formed by the plates. It forms a drop if the sum of the contact angle γ and half the dihedral angle of the wedge, α, exceeds a right angle, i.e. $\gamma + \alpha > \pi/2$, but spreads along the wedge if $\gamma + \alpha \leq \pi/2$; see Sect. 9.1. A nonwetting liquid drop moves to that position where its surface takes a spherical shape. It forms a spherical bridge for $\gamma - \alpha > \pi/2$, but a blob in the wedge if $\gamma - \alpha \leq \pi/2$ [Concus & Finn 1998, Concus et al. 2001]. Being aware that planes are never perfectly parallel, experimenters usually give them a very small curvature in order to locate the liquid.

6.2 Double Float Zones

6.2.1 Introduction

A double float zone is composed of two coaxial liquid bridges supported on opposite ends by disks A and B and coupled in the middle by a freely floating disk C as shown in Fig. 6.11. Under microgravity conditions, the two liquid zones assume the shapes of sections of unduloids, catenoids or nodoids. The condition for equilibrium in a double float zone is that the forces which the two zones exert on the separating disk C must be equal. These forces consist of a repulsive pressure component and an attractive surface tension component. The net force is given by $2\pi\sigma r_1 r_2/(r_1+r_2)$, where R is the radius

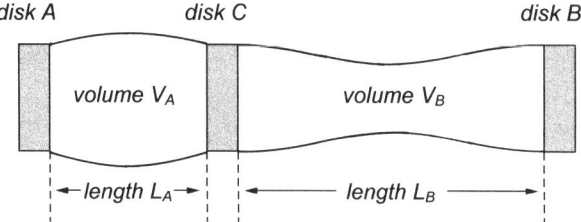

Fig. 6.11. Sketch of a double float zone

136   6. Liquid Zones

of the supporting disks, σ is the surface tension of the liquid, and r_1 and r_2 are the neck radius and the belly radius of the surfaces. Sections of unduloids (r_1 and $r_2 > 0$) always exert a positive, or attractive, force, spherical sections ($r_1 = 0$) exert no net force and nodoids ($r_2 < 0$) always exert a negative, or repulsive, force. The condition of equal forces on the separating disk thus excludes the coexistence of unduloids and nodoids.

A double-float-zone experiment was performed by Naumann [1994, 1995] on USML-1. The apparatus used consisted of a Lexan base with two 1 cm diameter Lexan support rods along a common axis as shown in Fig. 6.12. The clamps allowed the distance between the rods to be easily adjusted. Each rod was drilled along its axis to accept a stainless steel alignment wire. A Teflon liner was placed in the hole to prevent the working fluid (water) from wicking back into the hole as the alignment wire was withdrawn. A 10 mm diameter × 40 mm long Lexan rod served as typical floating piece. Other floating pieces consisted of rods of different lengths and geometries. The end faces of the rods were coated with EHEC CST103, a sucrose-like coating, to improve the wettability of the Lexan. Teflon shrink tube was placed over the cylindrical surfaces to act as an antispread barrier.

Double float zones of various lengths were established using the flat-ended rods. These configurations were remarkably stable even when they were intentionally perturbed by the Mission Specialist, Roger Chassay. With the

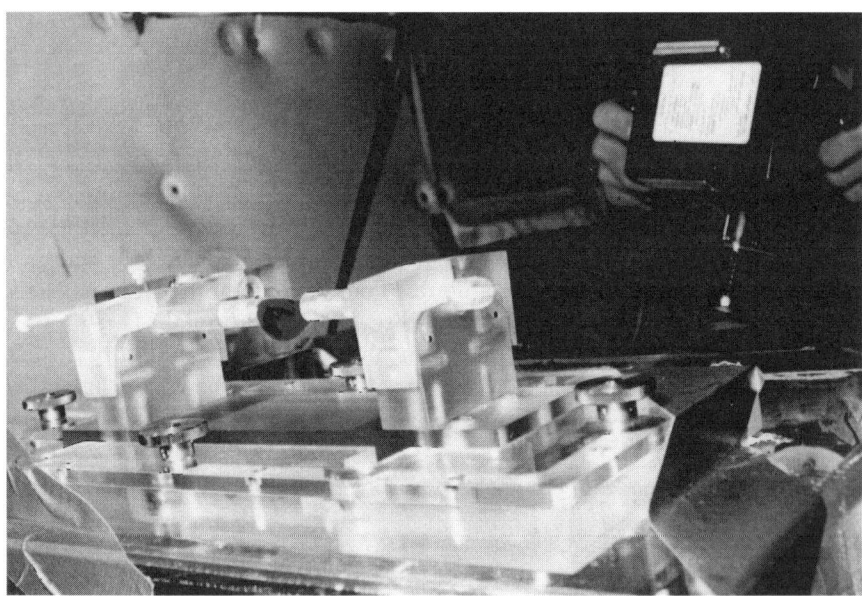

Fig. 6.12. Photograph of the double-float-zone apparatus during testing in a parabolic flight, where it could be configured for single float zones only, owing to the strong residual accelerations and the short times available

theory not yet having been developed, some phenomena came as a surprise. The central float seemed to be attracted more toward one support rod than the other. Adding more fluid to the short zone seemed to exacerbate the effect. In fact, at one point the short zone was bulging while the long zone was pinched in – totally defying the expected behavior! This was not a residual gravity effect since the residual acceleration was perpendicular to the direction of the offset and turning the apparatus around produced the same effect. The theoretical investigations presented below clearly support the observed behavior.

6.2.2 Unduloids and Nodoids

The properties and equations of unduloids and nodoids are discussed in Sects. 3.4 and 6.1. In Sect. 3.4 we were interested in the minimum-volume and in the period of the unduloids, in Sect. 6.1 we considered the minimum-volume and the inflection point. In the following we are primarily interested in the forces exerted on the floating disk C.

For the length z_1 of a section of an unduloid between two disks with radius R, we have

$$z_1 = 2\int_{r_1}^{R} dr \tan\varphi = 2\int_{r_1}^{R} dr \frac{r^2 + r_1 r_2}{\sqrt{[r(r_2+r_1)]^2 - [r^2 + r_2 r_1]^2}} \ . \tag{6.27}$$

The subscript 1, i.e. $i = 1$ in z_i, A_i, V_i, refers to a neck section (z_1 equals twice the length z_1 used in Sects. 3.4 and 6.1). In a belly section (subscript 2) the lower and upper integration limits r_1 and R have to be replaced by R and r_2, respectively. Substitution of

$$\left(\frac{2r}{r_2 + r_1}\right)^2 = 1 - 2\cos\psi\cos\xi + \cos^2\psi \tag{6.28}$$

leads to the following equations for the length, surface area and volume:

$$z_1 = (r_2 + r_1)\int_0^{\xi_0} d\xi \frac{1 - \cos\psi\cos\xi}{\sqrt{1 - 2\cos\psi\cos\xi + \cos^2\psi}} \ , \tag{6.29}$$

$$A_1 = 4\pi\left(\frac{r_1 + r_2}{2}\right)^2 \int_0^{\xi_0} d\xi \sqrt{1 - 2\cos\psi\cos\xi + \cos^2\psi} \ , \tag{6.30}$$

$$V_1 = 4\pi\left(\frac{r_1+r_2}{2}\right)^3 \int_0^{\xi_0} d\xi (1 - \cos\psi\cos\xi)\sqrt{1 - 2\cos\psi\cos\xi + \cos^2\psi} \ . \tag{6.31}$$

ξ_0 is obtained from (6.28) by substituting the disk radius R. For the belly sections z_2, A_2 and V_2 are obtained by replacing the lower and upper integration limits 0 and ξ_0 by ξ_0 and π, respectively. Note that the lengths, areas and volumes used here are equal to twice the lengths, areas and volumes considered in Sects. 3.4 and 6.1.

The condition for the coexistence of sections of different unduloids on sides A and B is that they exert the same force f on the separating disk C. This force is made up of the attractive force of the surface tension along the periphery $2\pi r \sin\varphi$ and the repulsive contribution $\pi R^2 p$ of the capillary pressure p, yielding the net attractive force

$$f_A = 2\pi\sigma_A \left[\frac{r_1 r_2}{r_1 + r_2}\right]_A , \quad f_B = 2\pi\sigma_B \left[\frac{r_1 r_2}{r_1 + r_2}\right]_B . \tag{6.32}$$

6.2.3 Branches of Solutions

The adaptation of the position of the separating disk C to the condition of equal forces allows for a large variety of double float zones. They are distinguished by the numbers of necks and bellies arising on the two sides A and B. Configurations containing at least one period of an unduloid on one side are unstable in all cases. A few of the symmetric double float zones are depicted in Fig. 6.13; some nonsymmetric double float zones are shown in Fig. 6.14. All configurations are at least twofold degenerate: zone A and zone B may simply be interchanged. Up to nine different configurations (and others not shown) may arise for the same liquid volume at different pulling forces. The different branches may be characterized as follows:

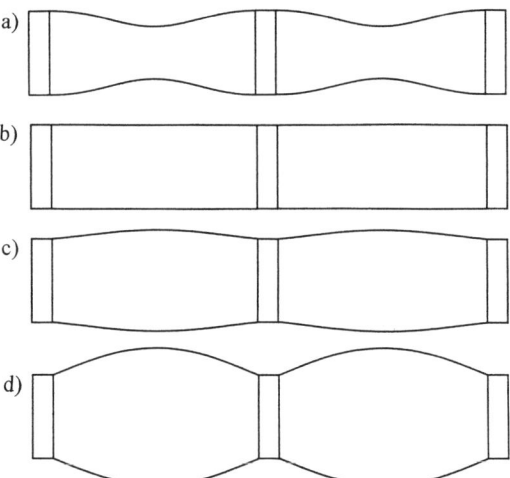

Fig. 6.13. Symmetric double float zones for $L_c = (L_A + L_B)/(4R) = 0.8\pi$ and increasing liquid volume $V_A = V_B$. (a) Necks on both sides at the stability limit of the individual zones: unstable. (b) Cylinders: unstable. (c) Bellies on both sides at the bifurcation with the nonsymmetric branch: unstable. (d) Bellies on both sides: stable

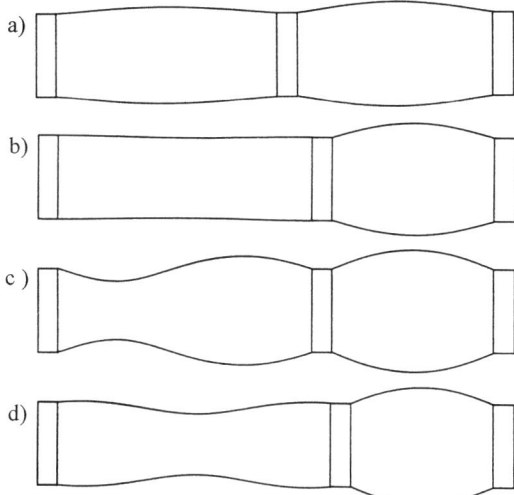

Fig. 6.14. Nonsymmetric double float zones for $L_c = 0.8\pi$ and $V_A = V_B$. (**a**) Bellies on both sides: stable. (**b**) Neck on side A, belly on side B: stability limit. (**c**) Full unduloid on side A, belly on side B: unstable. (**d**) Neck and two bellies on side A: unstable

Ia: in this branch both zones exhibit a belly. The diameter of the belly decreases with decreasing liquid volume V_c. Close to the bifurcation with branch IIa, zone A becomes cylindrical and then shows a neck rather than a belly.

Ib: for $V_c > 1$, both zones exhibit a belly; for $V_{rmc} < 1$, both zones exhibit a neck. This branch includes the case of cylindrical surfaces on both sides.

IIa: the zone with larger volume (zone B) shows a belly, and zone A a neck and a belly.

IIb: the zone with smaller volume (zone A) shows a belly, and zone B a neck and a belly.

IIIa: the zone with larger volume (zone B) shows a belly, and zone A a neck and two bellies.

IIIb: the zone with smaller volume (zone B) shows a belly, and zone B a neck and two bellies.

IV: both zones show a neck and a belly.

Va,b: one of the zones shows a neck and a belly, and the other one a neck and two bellies.

VI: both zones show a neck and two bellies.

A symmetric double float zone bifurcates with a nonsymmetric configuration if a surface deformation exists which maintains the length of the double float zone, satisfies the condition (6.32) of equal forces on the separating disk C and allows equal volumes V_A and V_B on both sides of disk C, yielding

140   6. Liquid Zones

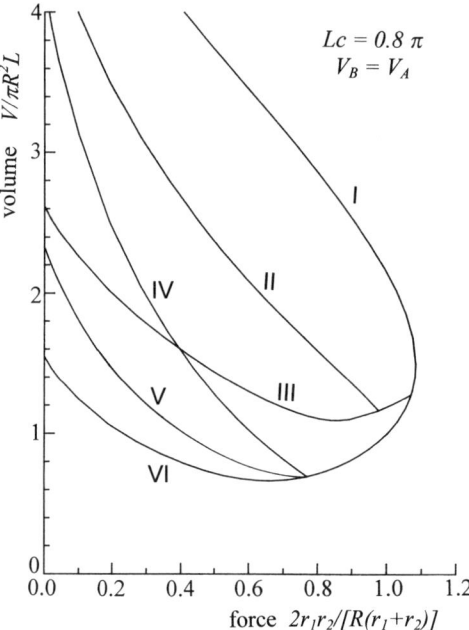

Fig. 6.15. The liquid volume $V_c = V/\pi R^2 L$ versus the net pulling force of the zones on the separating disk C. V equals the total volume $V = V_A + V_B$. L is half of the total length $L = (L_A + L_B)/2$. $V_B = 1.1 V_A$ and $L_c = 0.8\pi$ here

$$\frac{\partial f}{\partial r_1} dr_1 + \frac{\partial f}{\partial r_2} dr_2 = 0 \,, \tag{6.33}$$

$$\frac{\partial V}{\partial r_1} dr_1 + \frac{\partial V}{\partial r_2} dr_2 = 0 \,, \tag{6.34}$$

and

$$r_1^2 \frac{\partial V}{\partial r_1} = r_2^2 \frac{\partial V}{\partial r_2} \,. \tag{6.35}$$

In Figs. 6.15 and 6.16 the liquid volume $V_c = (V_A + V_B)/[\pi R^2 (L_A + L_B)]$ has been plotted versus the net pulling force of the zones on the separating disk C. In Fig. 6.15 $V_B = V_A$ and $L_c = 0.8\pi$, and in Fig. 6.16 $V_B = 1.1 V_A$, and $L_c = 0.8\pi$.

In the cases considered, namely $V_B = V_A$, $L_c = 0.8\pi$ and $V_B = 1.1 V_A$, $L_c = 0.8\pi$, the only stable branch is branch Ia. Its bifurcation with branch IIa leads to the stability limit. In Fig. 6.16 the minimum-volume of branches Ia and IIIa arises at a lower volume than does the bifurcation with branch IIa. For shorter zones this relation is inverted, i.e. the stability limit is given by the minimum-volume. This is fully analogous to the result found with single zones.

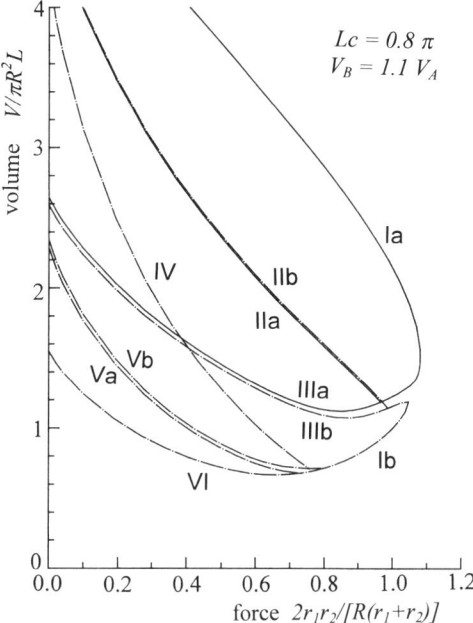

Fig. 6.16. The liquid volume $V_c = V/\pi R^2 L$ versus the net pulling force of the zones on the separating disk C. V equals the total volume $V = V_A + V_B$. L is half the total length $L = (L_A + L_B)/2$. $V_B = 1.1 V_A$ and $L_c = 0.8\pi$ here

6.2.4 Results of the Spacelab Experiments

Figure 6.17 shows the double float zones established during USML-1. The simple experimental equipment used did not allow one to establish double float zones with equal volumes $V_A = V_B$ on both sides. Experimentally, there was no symmetric configuration. The zones look much more symmetric for large liquid volumes than they look for small liquid volumes. It can be noted that it is always the left-hand side which becomes long and slim. In all cases the lengths of the zones and their central radii (neck radius r_1 or belly radius r_2, respectively) were measured. From these data the corresponding radius r_2 or r_1, the liquid volumes V_A and V_B, the forces on the separating disk C and the derivatives of these forces with respect to volume and length were computed. It appears that there was a systematic error in the measurement of the right-hand side due to parallax. The right-hand side was farther from the camera. The apparent rod diameters on the left-hand side and on the right-hand side were 42.5 pixels and 40.8 pixels, respectively. These values have been used to normalize the corresponding liquid bridge diameters. The contrast on the video on the bottom of the right-hand disk was very low and it was difficult to determine its position accurately. A re-examination of the apparatus showed that the Teflon on one side of the floating piece

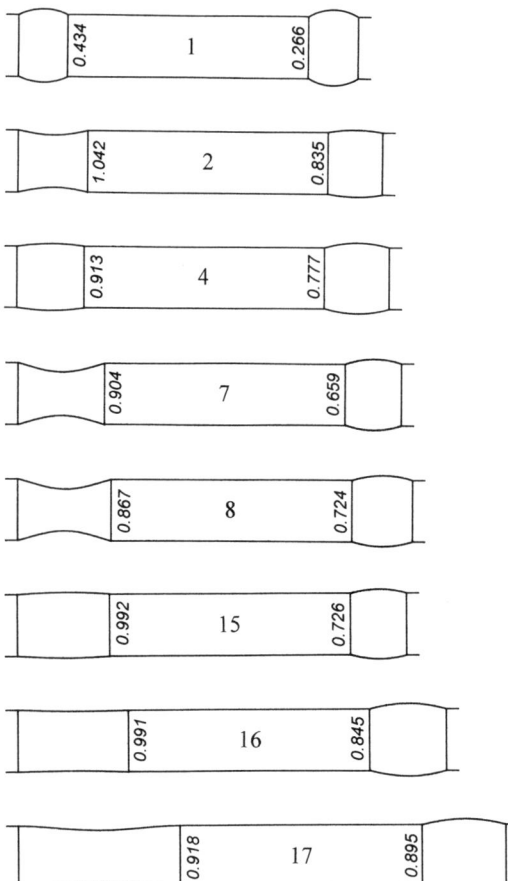

Fig. 6.17. Eight of the 17 double float zones established during USML-1 and the forces exerted on the separating disk C

was not exactly flush with the flat face, but extended approximately half a millimeter beyond it. This tends to prevent contact angles larger than $\pi/2$ on this face and might explain the reason for the asymmetry – or at least why the left-hand side always chose to extend and form the neck.

There is a variation of 5–6% in the computed volumes among those observations where no liquid was added or subtracted. This is consistent with the approximately 2% in uncertainty in the length and diameter measurements determined by the smallest possible measurement of the video analyzer (42.5 counts = 1 cm on the left-hand side, 40.8 counts = 1 cm on the right-hand side). In all cases the force on the left-hand side is higher than the force on the right-hand side. There is, however, a strong dependence of the force on the radius measured on the right-hand side. We find $\partial f/\partial r_\mathrm{B} < -2$ in all cases except the last two. This suggests that all radii measured on the right-hand side are too large.

The breakage of the left-hand zone, when further liquid was added to that zone in configuration 17, appears to be due to the antimetric instability $D\{1,0\}$. The aspect ratio of that zone was already $L_A/2R = 2.68$, and the neck radius was $r_1/R = 0.847$. At that aspect ratio a single column must break for a neck radius $r_1/R = 0.718$. Adding more liquid therefore would have been the correct countermeasure. The stability of configuration 17 will be reconsidered in the next section in view of the consideration of the stability of double float zones there.

6.2.5 The Stability Diagram

Double float zones with equal volumes $V_A = V_B$ are symmetric and exhibit a belly on both sides, as long as the liquid volumes are large compared with the cylindrical volume (fat zones). If the double float zone is lengthened or the liquid volume is reduced, the zone does not follow the symmetric branch of solutions, but rather switches to a nonsymmetric configuration, to a longer, slimmer zone and a shorter, fatter zone (from branch I to branch II).

If the liquid volumes on the two sides A and B differ, $V_A \neq V_B$, the symmetric branch of solutions discussed above is nonsymmetric from the start, i.e. the zone on the side where liquid is added becomes fatter than the other zone. Since the fat zones as a rule exert a lower force on the separating disk C than do the slim zones, the former tend to become even fatter and shorter, whereas the latter tend to become even slimmer and longer. The maximum force is obtained for zones slightly fatter than cylindrical. The degeneracy of the nonsymmetric branch and its bifurcation with the symmetric branch that exists for $V_A = V_B$ are resolved for $V_A \neq V_B$. The branch comprising the larger liquid volumes together with the corresponding modified section of the symmetric branch is stable, whereas the branch comprising the smaller liquid volumes is unstable.

The stable branch of solutions is found to exert the strongest pulling force on the floating disk C. Increasing the volume of one zone usually lowers the pulling force, such that the zone becomes shortened. This means a stable situation. In the case of equal volumes, $V_A = V_B$, the antimetric stability limit $D\{1,0\}$ (see Fig. 5.4) obtained for a single zone applies to a double zone as well. In a single zone, one half of the zone widens, and the other half shrinks accordingly. In a double float zone, one zone widens, and the other zone shrinks accordingly.

In Figs. 6.15 and 6.16 the liquid volume is plotted versus the force on the separating disk. The forces on disk C are equal on both sides, whereas the pressure is not. An advantage of these plots is that all spherical solutions, because they exert zero force on disk C, are represented by points on the volume axis. For any given length L (or aspect ratio), the force has a maximum as a function of volume. This maximum occurs for zones slightly fatter than

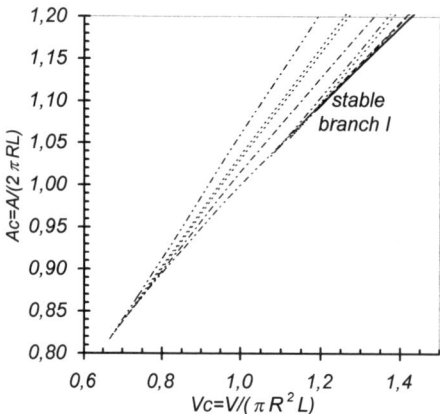

Fig. 6.18. Area A_c versus volume V_c for the branches shown in Fig. 6.16. The stable branch appears on the *upper right*, at $V_c = 1.18812$, $A_c = 1.089437$

cylindrical. In Fig. 6.18, the area A_c is shown versus the volume V_c for the branches shown in Fig. 6.16. The stable branch appears on the upper right, at $V_c = 1.18812$, $A_c = 1.089437$.

The stability diagram is obtained by plotting the minimum liquid volume and the volumes corresponding to the bifurcations versus the aspect ratio L_c. Figure 6.19 shows the stability diagram for equal volumes $V_A = V_B$. In that case one finds a stable symmetric branch, where both zones show a belly. However, before these nodoids shrink to cylinders with decreasing volume, the configuration bifurcates with a nonsymmetric branch, where one zone lengthens with decreasing neck radius, whereas the other zone shortens with increasing neck radius. Curve (a) represents this bifurcation between the symmetric branch and the nonsymmetric branch. Below curve (a), stable nonsymmetric zones exist. Curve (d) represents the stability limit due to the minimum-volume that arises in the nonsymmetric branch. Curve (e) shows the antimetric instability that arises when the longer zone extends over a full period.

The bifurcation with the nonsymmetric branch arises if both zones still exhibit a belly for $L_c > \pi/2$, but is postponed until both zones exhibit a neck for $L_c < \pi/2$. (For $V_B = 1.1 V_A$, this bifurcation arises at $L_c > 0.533\pi$ and $L_c < 0.533\pi$, respectively.) This property of double float zones is inherent in a pair of zones. The stability limit of single float zones, as shown in Fig. 5.5, cannot be achieved.

A single cylindrical zone becomes unstable if the length equals the circumference. For zones with approximately the cylindrical volume $V_c \approx 1$, one obtains (5.35) instead. The single zone switches into the antimetric mode and breaks. If the same zone is separated into two zones by an additional disk C, it likewise tends to switch into the antimetric mode. This, however, does not cause breakage, but leads to the two unequal zones described above. A small

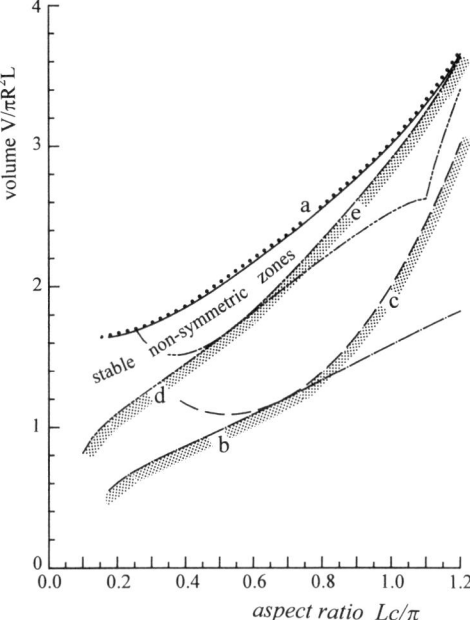

Fig. 6.19. Stability diagram of double float zone with equal volumes $V_A = V_B$. The liquid volumes at the bifurcations and the minimum liquid volumes have been plotted versus the aspect ratio $L/2\pi R$, $L = (L_A + L_B)/2$. The lower stability limit is given by curves (d) and (e). (a) Bifurcation between the symmetric and nonsymmetric solutions; (b) minimum-volume of the symmetric solutions; (c) bifurcation within the symmetric solutions; (d) minimum-volume of the nonsymmetric solutions; (e) bifurcation within the nonsymmetric solutions

hole in the separating disk C would lead to a condition of equal pressure rather than of equal force for the two zones. Liquid would flow from the slim zone to the fat zone and the double zone would break. This slightly increases the stability of the double float zone.

The stability diagram for unequal volumes $V_A \neq V_B$ looks similar to Fig. 6.19. However, all the doubly degenerate branches split up into two branches. One can add liquid either to the left-hand side or to the right-hand side. The last experimental configuration, 17, just falls outside the theoretical stability limit. However, it may be inside this limit if one takes account of experimental errors. It just had to break when further liquid was added to the left-hand zone.

6.2.6 Key Points

The stability of the nonsymmetric configurations was actually first concluded from experimental evidence. The zones established during USML-1 all satisfy the theoretical stability condition. The last experimental configuration, 17, falls slightly outside the theoretical stability limit, but may be inside this limit if one takes account of experimental errors. Subsequently, in the experiment, the double float zone broke.

A single cylindrical zone becomes unstable if the length equals the circumference (= Rayleigh instability, $D\{1,0\}$). For zones with approximately the cylindrical ratio between the volume V and length L, one finds (5.35) instead. The single zone switches into the well known amphora mode and breaks. Owing to the quadratic dependence of the liquid volume on the local radius, the increase in surface area at the side where the surface is bulging out is lower than the decrease in surface area at the side where the surface is curving in. If the same zone is separated into two zones by an additional disk C, it likewise tends to switch into the amphora mode. This, instead of just a liquid flow in the axial direction, requires a movement of the separating disk. This slightly increases the stability, i.e. a cylindrical double float zone with unequal volumes $V_A \neq V_B$ is stable for a total length $2L = L_A + L_B$ somewhat larger than the circumference $2\pi R$.

The key to the interpretation of these findings is the fact the net force on the separating disk is composed of an attractive contribution from the surface tension along the periphery, $2\pi R\sigma \sin\varphi$, and a repulsive contribution, $\pi R^2 p$, from the pressure p. Whereas the former contribution has an obvious maximum for $\varphi = \pi/2$, that is, for nearly cylindrical surfaces, we find the latter contribution to increase monotonically from the catenoid to the unduloids, the cylinder and the nodoids. The net attractive force hence shows a maximum for nearly cylindrical surfaces. The attractive force of fat zones decreases the fatter they become. The attractive force of slim zones decreases the slimmer they become.

Consider now a double float zone which has been established by putting together two equal single zones. If the separating disk is not placed exactly at the middle, which always will be the case, one of the zones becomes slightly shorter and fatter, whereas the other becomes slightly longer and slimmer. If the two single zones exhibit bellies, the attractive force of the shorter, fatter zone becomes smaller than that of the longer, slimmer zone. This is a stable situation, i.e. the longer zone decreases in length. On the other hand, if the two single zones exhibit necks, the attractive force of the shorter, fatter zone becomes larger than that of the longer, slimmer zone. The shorter zone decreases in length and a nonsymmetric double float zone is obtained. If this stable nonsymmetric configuration is further lengthened or the liquid volume is reduced, it will always be the slimmer zone which initiates the instability. This zone tends to become even slimmer and longer and eventually breaks.

References

1. Concus P, Finn R: Discontinuous behavior of liquids between parallel and tilted plates. Phys. Fluids **10** (1998) 39–43
2. Concus P, Finn R, McCuan J: Liquid bridges, edge blobs, and Sherk-type capillary surfaces. Indiana Univ. Math. J. **50** (2001) 411–441
3. Finn R: *Equilibrium capillary surfaces*. Grundlehren der mathematischen Wissenschaften, Vol. 284. Springer, Berlin, Heidelberg (1986)
4. Finn R, Neel RW: C-singular solutions of the capillary problem. J. reine angew. Math. **512** (1999) 1–25
5. Haynes JM: Capillary instabilities in $1\,g$ and $0\,g$. ESA SP-114 (1978) 467–471
6. Haynes JM: Kinetics of spreading of liquids in microgravity-experiment 1 ES 327. ESA SP-222 (1984) 43–46
7. Langbein D: Stability of liquid bridges between parallel plates. ESA SP-333 (1992) 85–93
8. Langbein D: Stability of liquid bridges between parallel plates. Microgravity Sci. Technol. **5** (1992) 2–11
9. Naumann RJ: USML-1 Glovebox experiments. Final report. NASA-38773, Marshall Space Flight Center, Huntsville AL (1995)
10. Naumann RJ: Stability of a double float zone. In: *Joint L+1 Science Review of USML-1 and USMP-1 with the Microgravity Measurement Group*, NASA Conference Publication 3272, Vol. 2, 691–700 (1994)
11. Passerone A, Rositto F, Sangiorgi R: Meniscus stability in immiscible metals, MASER-1 experiment. Appl. Microgravity Technol. **1** (1988) 62–66
12. Rositto F, Passerone A, Sangiorgi R, Minisini R: Liquid bridges formed by immiscible metals – a sounding rocket experiment. ESA SP-256 (1987) 215–220
13. Vogel TI: Stability of a liquid drop trapped between two parallel plates. SIAM J. Appl. Math. **47** (1987) 516–525
14. Vogel TI: Stability of a liquid drop trapped between two parallel plates. II: general contact angles. SIAM J. Appl. Math. **49** (1989) 1009–1028
15. Vogel TI: Numerical results on the stability of a drop trapped between parallel planes. Lawrence Berkeley Laboratory Technical Report LBL-30486, March 1991
16. Vogel TI: Types of instability for the trapped drop problem with equal contact angle. In: *Geometric Analysis and Computer Graphics*. P. Concus, R. Finn, D. Hoffman (eds.). Springer, Berlin, Heidelberg (1991b) 195–203

7. Canthotaxis/Wetting Barriers/ Pinning Lines

Under conditions of weightlessness, the shape of a capillary surface is determined by the container shape and the contact angle of the liquid with the adjacent solid faces. The location and handling of liquids require the application of wetting aids and wetting barriers such as sharp edges or coatings. At a wetting barrier, the liquid surface winds from the contact angle with the better-wetting face to the contact angle with the worse-wetting face. This effect has been termed "canthotaxis". Hitherto, canthotaxis has been an art rather than a science; exact results are scarce. Numerically, the length of the interval of canthotaxis (pinning interval) is proportional to the square of the angular interval in question. For surfaces with zero capillary pressure, the length of the interval of canthotaxis varies in proportion to the cube of the angular interval.

7.1 Introduction

Under the action of gravity, liquids generally rest on the bottom of their containers. The liquid surface is plane and horizontal. Small liquid volumes are likely to be found in wedges. In contrast, liquid volumes may be found almost anywhere under conditions of weightlessness. The shape of a capillary surface is determined by the shape of the container and the contact angle γ of the liquid with the adjacent solid faces. Wetting liquids ($\gamma < \pi/2$) try to maximize their area of contact with a solid, whereas nonwetting liquids ($\gamma > \pi/2$) try to minimize it. Special measures regarding the geometry and the wetting conditions have been taken successfully for manipulation and transport of liquids under low-gravity conditions.

The drops on a solid surface shown in Fig. 2.8 may rest in arbitrary positions if no measures are taken to locate them. For instance, if the solid surface has been treated so that the liquid wets it well in some positions and wets it badly in others, the liquid is most likely to be found in the former positions. Good and bad wetting may be achieved by different coatings or by joining different solids together. The most widely used method of locating liquids, however, is to introduce sharp edges. (A drop of water hanging on a window is quite atypical in that respect; pinning of the contact line overrules Young's boundary condition.)

7. Canthotaxis/Wetting Barriers/Pinning Lines

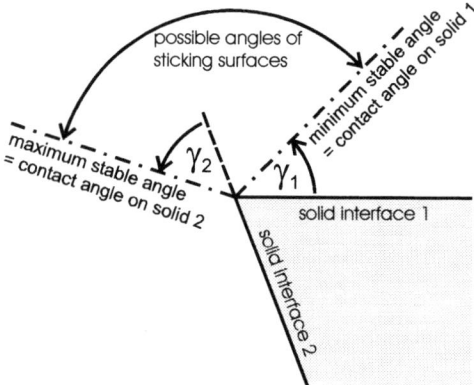

Fig. 7.1. A fluid interface sticks to a solid edge if a virtual displacement along one of the contacting solid faces 1 and 2 causes a restoring force of the interface tension. This implies that the angles of the fluid interface with faces 1 and 2 must be larger and smaller, respectively, than the respective contact angles γ_1, γ_2

A fluid interface sticks to a solid edge if a virtual displacement along one of the contacting solid faces 1 and 2 causes a restoring force of the interface tension (Fig. 7.1). This implies that the angles of the fluid interface with the solid faces 1 and 2 must be larger and smaller, respectively, than the contact angles γ_1, γ_2. The same condition arises along the contact line between two different solids or two different coatings. This phenomenon has been termed "canthotaxis". The Greek word *kanthos* means the angle made by the meeting of the eyelids. If the faces form a solid angle 2α, the range in which the contact angle is arbitrary given by

$$\gamma_1 < \gamma < \gamma_2 + \pi - 2\alpha \,. \tag{7.1}$$

Since the beginning of research under low-gravity conditions, canthotaxis has been used as a tool for locating liquids. It enables one to establish wetting barriers and to perform experiments in materials science and the life sciences. There are basically two cases of canthotaxis: the wetting barrier holds and, with increasing liquid volume, the liquid surface increasingly leans over to the other side; or else, the wetting barrier does not hold, and the liquid spreads across it sooner or later. In the latter case the liquid surface (more exactly, the normal to the liquid surface) must wind itself around the pinning line by at least the difference in contact angles $\Delta\gamma = \gamma_2 - \gamma_1$ across the wetting barrier, where $\gamma_2 > \gamma_1$ [$\Delta\gamma$ has to be supplemented by the angle between the planes forming the wetting barrier; see (7.1)). The length of the pinning interval basically depends on the angular difference $\Delta\gamma$. Numerically, it has been found to decrease to zero in proportion to $(\Delta\gamma)^2$ or even $(\Delta\gamma)^3$ [Mittelmann 1995, Mittelmann & Zhu 1996].

Canthotaxis is frequently used in microgravity experiments for keeping liquids at the intended position. Typical examples are the creation of plane liquid surfaces for investigations of thermocapillary convection or as the initial condition for studying liquid penetration into tubes or surface tension tanks [Dreyer et al. 1998, deLazzer et al. 1998], and the creation of axisymmetric liquid surfaces, in particular cylindrical liquid surfaces, for the purpose of studying their stability, their oscillations and, again, thermocapillary convection. Also, in all experiments on liquid columns between coaxial disks, the latter have been provided with extremely sharp edges. This applies, for example, to experiments in the Fluid Physics Module during the Spacelab-1, Spacelab D1 and Spacelab D2 missions. Rectangular configurations have been used in Spacelab D1 and Spacelab D2 (experiments conducted by Schwabe and Legros; see Sects. 13.11 and 13.12) and in the Bubble, Drop, and Particle Unit (BDPU) in IML-2 (experiments conducted by Koster and Legros; see Sect. 13.16).

Up to now, canthotaxis has been an art rather than a science, exact results are scarce. Numerical calculations on liquid surfaces involving canthotaxis performed using SURFACE EVOLVER have shown to lead to random normals along the wetting barrier. This and other codes based on finite elements minimize the energy of the liquid ($\sigma A_l - A_s \cos\gamma$), i.e. the boundary condition on the contact angle γ is accounted for by including the area of contact with a solid surface with the weight $\cos\gamma$. This, however, does not allow precise results for the direction of the normals.

7.2 Straight Wetting Barriers

In the following we shall discuss the basic properties of liquid surfaces governed by canthotaxis and the possible hysteresis effects of several well-wetting configurations on less well-wetting planes. We consider a wetting tile, a wetting stripe and a wetting cross. The contact angle of the liquid on the wetting configuration is denoted by γ_1 and that on the outer plane by γ_2, with γ_1 being smaller than γ_2. In all cases we successively increase the liquid volume through a virtual injection hole at the center of symmetry. Under conditions of weightlessness, the liquid drops will tend to form spherical caps, which are flattened in the presence of gravity. Gravity affects the sequence of liquid configurations presented below only quantitatively.

7.2.1 The Wetting Tile

Starting with a wetting tile, one will face the following situations:

- A small liquid drop injected in the center of the tile forms a spherical cap with contact angle γ_1, which increases in diameter with increasing liquid volume.
- The center of the drop, owing small wetting irregularities, may not strictly coincide with the injection hole (Fig. 7.2a).

- As soon as the periphery of the liquid drop touches one of the wetting barriers, i.e. one side of the tile, the contact line becomes pinned to that barrier and the drop is successively recentered.
- The pinning interval increases continuously and the liquid surface leans over to the other side (Fig. 7.2b).
- The liquid, however, will not wet the outer plane before the surface has reached the outer contact angle γ_2.
- Hence, if the angular difference $\Delta\gamma = \gamma_2 - \gamma_1$ is large, the pinning interval will reach the corners of the tile before the outer plane is wetted.
- In that case the shape of the liquid surface does not depend explicitly on the contact angles γ_1 and γ_2. The minimum angle $\gamma_{\min}$ of the liquid surface with the tile, which occurs at the corners, merely has to be larger than γ_1. And the maximum angle $\gamma_{\max}$, which occurs in the middle of the sides, has to be smaller than γ_2.
- This is the same family of surfaces that arises when a liquid is pressed out of a tube with a square cross section.
- On the other hand, if the angular difference $\Delta\gamma = \gamma_2 - \gamma_1$ is small, the outer plane is wetted before the pinning interval reaches the corners of the tile.
- This effect most likely will cause hysteresis, i.e. if the liquid volume is reduced, extrusion of liquid from the corners will not occur at the same volume as does penetration of liquid into the corners.

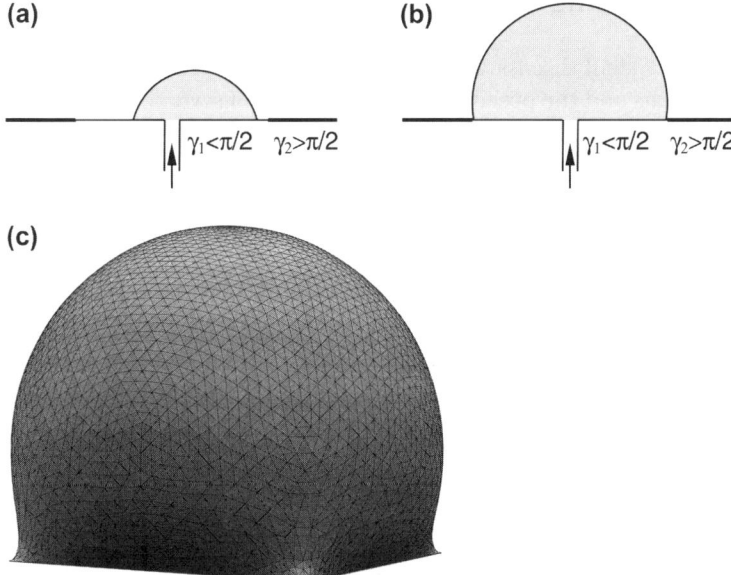

Fig. 7.2. (a) The center of a drop on a wetting tile, owing small wetting irregularities, may not always coincide with the injection hole. (b), (c) With increasing pinning interval, the liquid surface leans more and more over to the other side

- Once the liquid drop has crossed all four corners of the tile, it forms a spherical cap on the outer plane with contact angle γ_2, which, as before, is not necessarily centered on the injection hole.
- On the contrary, it appears most likely that the drop will stay pinned to one side of the tile even for large liquid volumes.
- It is of quantitative interest to know the length of the pinning interval at low volumes, the interval of hysteresis, and the energy gain as a result of the liquid staying pinned to the tile at large volumes as a function of the contact angles γ_1 and γ_2.

7.2.2 The Wetting Stripe

Next let us outline the corresponding configurations that occur at a wetting stripe (Fig. 7.3):

- For low liquid volumes, we find a spherical cap within the stripe, not touching the edges (wetting barriers). This is fully equivalent to the case of tile.
- For medium liquid volumes, the liquid surface touches the edges, but does not yet cross them. There is canthotaxis; the surface is pinned to the wetting barrier over a finite interval. The liquid volume looks like a snail. This case is of general interest for fluid handling, for liquid flow along a wetting stripe, etc.
- With increasing volume, the liquid surface increasingly leans over to the other side.
- Along the line of canthotaxis, owing to the vanishing axial curvature, the normal curvature is constant.
- For large liquid volumes, the liquid may cross the wetting barriers in the central region of the stripe.
- The question arises whether there is a bifurcation between the wetting barrier holding or not holding, i.e. whether both cases may arise for the same liquid volume depending on the initial conditions (including hysteresis).

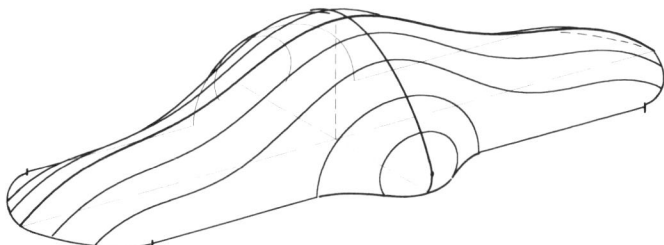

Fig. 7.3. Canthotaxis at a wetting stripe. For a medium liquid volume, the liquid surface touches the boundaries, but does not cross them. For a large liquid volume, the liquid crosses the boundaries in the central region

154   7. Canthotaxis/Wetting Barriers/Pinning Lines

- If so, at what energy (surface area) does the bifurcation arise? And how do the two branches of solutions behave as a function of the liquid volume?
- If the liquid does not cross the wetting barrier, the leaning over to the other side must asymptotically come to a stop, which means the "snail" gets longer and longer and reaches a constant profile in the middle. The surface reaches a maximum angle $\gamma_{\max}$.
- The shape of a surface with arbitrary length and maximum angle $\gamma_{\max}$ (γ_1) does not depend explicitly on the outer contact angle γ_2; it remains a solution if γ_2 is reduced to $\gamma_{\max}$. The effect of γ_1 is reduced to the shapes of the ends (head and tail of the snail).
- However, the Rayleigh instability has to be considered. The "snail" may break into droplets, as do surfaces of wetting liquids in wedges; see Sect. 8.1.
- Alternatively, the leaning over to the other side may not come to a stop, and the liquid may eventually spread to the outside with increasing volume. This happens, at the latest, if the maximum angle $\gamma_{\max}$ of the liquid surface reaches the outer contact angle γ_2.
- On the basis of energetic arguments, one may anticipate that this situation will eventually win the race. Nevertheless, once more hysteresis may be possible, i.e. there is an interval of volumes in which both cases may happen, depending on the initial conditions.

The stability of a liquid surface on a wetting stripe is similar in character to the stability of liquid zones between coaxial disks or parallel plates. Depending on the contact angles γ_1 and γ_2 and the liquid volume, the wetting barrier may not be crossed (= stability) or may be crossed (= instability). Large volumes favor instability. Corresponding stability diagrams, inclusive of the effect of the Bond number, may be constructed.

7.2.3 The Wetting Cross

The wetting cross has been treated numerically by Mittelmann (1995). The configurations that occur are the following (Fig. 7.4).

- For small liquid volumes the surface forms a spherical segment with contact angle γ_1, which does not touch the wetting barriers (corners).
- The liquid surface then reaches the corners, but does not cross them. Instead, it becomes recentered.
- Close to the corners, a limited interval of canthotaxis arises. The liquid leans over to the other side.
- The liquid surface crosses the barrier at the corners. The interval of canthotaxis increases.
- With increasing volume, the wetted solid area increases as (volume)$^{2/3}$. Since the region of good wetting increases in proportion to (volume)$^{1/3}$, whereas the region of bad wetting increases in proportion to (volume)$^{2/3}$, it may be anticipated that the liquid surface will prefer a stronger increase

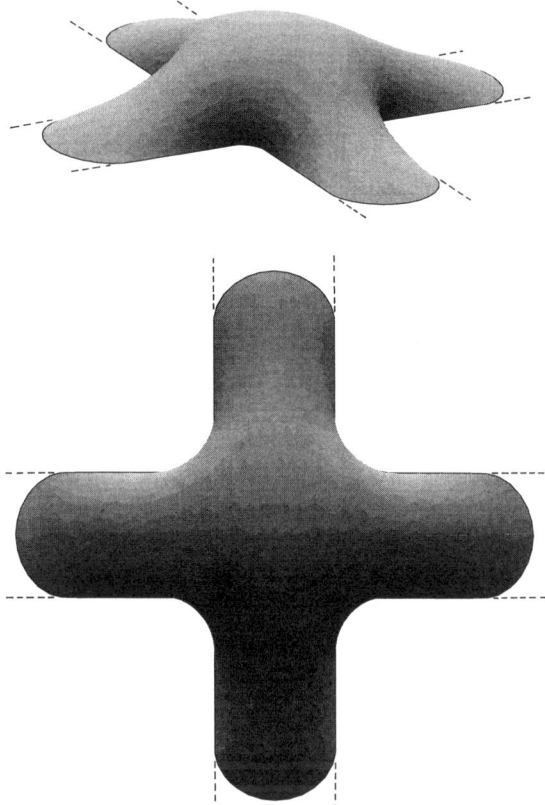

Fig. 7.4. Canthotaxis at a wetting cross [Mittelmann 1995]

of the region of good wetting, that is, a lengthening of the interval of canthotaxis.
- This gives rise to several physically important questions:
- Up to what volume do the corners prevent crossing of the wetting barriers?
- Does the length of the interval of canthotaxis increase with increasing volume? And if so, does it increase as (volume)$^{1/3}$ or as (volume)$^{2/3}$?
- The numerical tests by Mittelmann [1995] and Mittelmann and Zhu [1996] suggest that the interval of canthotaxis increases in proportion to $(\gamma_2 - \gamma_1)^2$ (Fig. 7.5)

7.2.4 Circular Tubes

Very similar configurations arise if a circular capillary is successively overfilled. The contact angle γ_1 inside the capillary, as before, is considered to be small. We assume the upper end of the capillary to be smooth, with contact

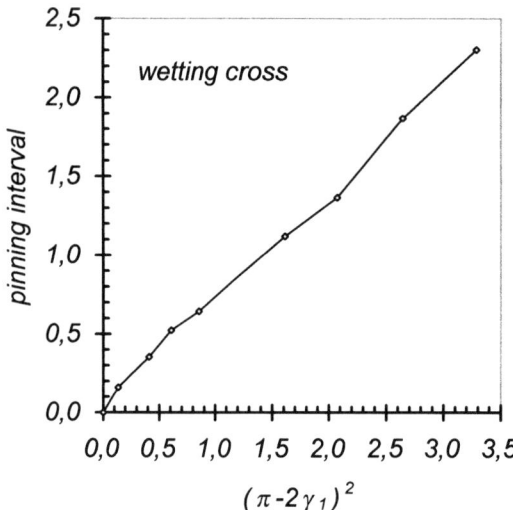

Fig. 7.5. Numerically, the pinning interval at the wetting cross increases in proportion to the square of the angular difference $(\gamma_2 - \gamma_1) = 2\pi - \gamma_1$ $(\gamma_2 = \pi - \gamma_1)$

angle γ_2. The contact angle with the outer wall is denoted by γ_3. If the capillary is only partly filled, the liquid forms a concave spherical surface. The condition for sphericity is that the pressure drop over the vertical extent of the surface $\Delta p = g\rho z$ is small compared with the capillary pressure, yielding

$$\Delta p = 2\sigma/r \geq g\,\Delta\rho\,r \sin\gamma\,, \tag{7.2}$$

where $r \cos\gamma = R$; R is the radius of the capillary, and r that of the spherical surface. The kinematics of filling due to capillary forces have been treated by Lucas [1917] and Washburn [1921]. The driving capillary underpressure works against gravity and against viscous friction of the intruding liquid. In the limit of low gravity, the filled section increases in proportion to the square root of time.

If, by an increase of the volume, the liquid surface is shifted upwards, it will reach the upper, inner periphery. Now canthotaxis enters the game. The contact line sticks to the periphery, the curvature of the surface decreases and the pressure jump across the surface decreases accordingly. A flat surface is reached. Subsequently, the liquid surface becomes convex, until it reaches the contact angle γ_2 with the upper end surface of the capillary. The surface now forms a spherical cap, which is not necessarily centered on the axis of the capillary, but may be shifted sideways. This depends on microscopic irregularities of the upper end of the capillary. However, as soon as the outer periphery is reached, the contact line becomes pinned there and the surface again assumes a symmetric position with increasing liquid volume.

The contact line is now pinned at the outer periphery and, with increasing volume, leans over more and more. In the absence of gravity, the liquid surface remains spherical. It is stable for arbitrary liquid volume. At best it may have a horizontal tangent at the periphery, but in that case it may drift away sideways as described in Sects. 5.2.3 and 5.2.4. (In the absence of gravity the term "horizontal" should be replaced by "radial".)

On the other hand, if gravity acts downwards, the liquid surface is bent downwards outside the capillary, too. The liquid becomes a sessile drop. If the surface reaches the contact angle γ_3 with the outer surface, the capillary surface may move down the outer surface and thus becomes unstable. Even more likely, a lateral deformation may give rise to a sideways movement; the liquid will spill down the outside of the capillary.

7.2.5 Large Liquid Volumes

A rough estimate of the stability of a liquid surface in the presence of a wetting stripe, a wetting cross or a figure exhibiting even more "tentacles" may be obtained on the basis of the fact that the surface are of a spherical liquid volume with constant contact angle γ_2 on the outer plane increases in proportion to (volume)$^{2/3}$, whereas the effective surface area of a cylindrical liquid volume on a wetting stripe or on a wetting tentacle increase in proportion to (volume)^1. One may anticipate that the spherical solution always has a lower surface energy for large liquid volume; the "snails" (= nearly cylindrical surface) may not become infinitely long.

If, with increasing volume, the "snail" becomes fatter and longer, however, with the angle with the outer plane being limited to γ_2, its volume V_{cyl}, the liquid surface area A_l, the wetted solid surface area A_s and the surface energy E_{cyl} are given by

$$V_{\mathrm{cyl}} = (\gamma_2 - \sin\gamma_2 \cos\gamma_2)\, R_{\mathrm{cyl}}^2 L\,, \tag{7.3}$$

$$A_\mathrm{l} = 2\gamma_2 R_{\mathrm{cyl}} L\,, \quad A_\mathrm{s} = 2\sin\gamma_2\, R_{\mathrm{cyl}} L\,, \tag{7.4}$$

$$E_{\mathrm{cyl}} = \sigma\,(A_\mathrm{l} - A_\mathrm{s}\cos\gamma_1) = 2\sigma\,(\gamma_2 - \sin\gamma_2\cos\gamma_1)\,R_{\mathrm{cyl}} L\,, \tag{7.5}$$

yielding

$$E_{\mathrm{cyl}} = 2\sigma V_{\mathrm{cyl}} \sin\gamma_2 \frac{\gamma_2 - \sin\gamma_2 \cos\gamma_1}{\gamma_2 - \sin\gamma_2 \cos\gamma_2}\,. \tag{7.6}$$

Here L is the width of the wetting stripe. On the other hand, if the liquid volume wets the outer plane and forms a spherical cap with contact angle γ_2, its volume V_{sph} and surface energy E_{sph} are given by

$$V_{\mathrm{sph}} = \frac{\pi}{3}(1-\cos\gamma_2)^2 (2+\cos\gamma_2)\, R_{\mathrm{sph}}^3\,, \tag{7.7}$$

$$A_\mathrm{l} = 2\pi(1-\cos\gamma_2)\, R_{\mathrm{sph}}^2\,, \quad A_\mathrm{s} = \pi\sin^2\gamma_2\, R_{\mathrm{sph}}^2\,, \tag{7.8}$$

$$E_{\mathrm{sph}} = \sigma\,(A_\mathrm{l} - A_\mathrm{s}\cos\gamma_2)\, R_{\mathrm{sph}}^2$$
$$= \sigma\pi(1-\cos\gamma_2)^2 (2+\cos\gamma_2)\, R_{\mathrm{sph}}^2 = 3\sigma V_{\mathrm{sph}}/R_{\mathrm{sph}}\,, \tag{7.9}$$

yielding

$$E_{\text{sph}} = 3\sigma V_{\text{sph}}^{2/3} \left[\pi \left(1 - \cos\gamma_2\right)^2 \left(\frac{2}{3} + \frac{1}{3}\cos\gamma_2\right) \right]^{1/3}. \qquad (7.10)$$

For large liquid volumes V, the spherical cap always has the lower energy. The energy E_{cyl} of the cylindrical "snail" is lower than the energy E_{sph} of the spherical cap for

$$L < \pi \frac{9\left(1 - \cos\gamma_2\right)^2 (2 + \cos\gamma_2)}{8\sin\gamma_2} \frac{(\gamma_2 - \sin\gamma_2 \cos\gamma_2)^2}{(\gamma_2 - \sin\gamma_2 \cos\gamma_1)^3} \approx \frac{9}{2\sin\gamma_2}. \qquad (7.11)$$

This result may be extended to the wetting cross or, quite generally, to figures with N "tentacles". There exist nearly spherical regions and nearly cylindrical regions of the liquid surface. In the case of a large liquid volume, the different "tentacles" may be treated as largely independent; they are cylindrical with an approximate radius $R_{\text{cyl}} = R_{\text{sph}}/2$. The liquid volume is distributed between the spherical region and the "tentacles" in such a manner that a balance of capillary pressure is reached.

7.3 Liquid Surfaces in Wedges

Having identified several situations where of canthotaxis occurs, let us now study the behavior of the liquid surfaces in more detail. We consider continuous, differentiable fluid surfaces in a wedge with dihedral angle 2α, as shown in Fig. 7.6. The wedge is formed by faces 1 and 2, with normals

$$\boldsymbol{n}_1 = ((1,0,0)\,, \quad \boldsymbol{n}_2 = (\cos(\pi - 2\alpha), \sin(\pi - 2\alpha), 0)\,. \qquad (7.12)$$

The angle between the normals is given by

$$\gamma_0 = \pi - 2\alpha\,. \qquad (7.13)$$

Let the contact angles of the fluid surface with the two faces 1 and 2 be γ_1 and γ_2, respectively. Then, the normal $\boldsymbol{n}_0$ of the fluid surface is located in a cone of angle γ_1 around $\boldsymbol{n}_1$ and in a cone of angle γ_2 around $\boldsymbol{n}_2$; see Fig. 7.6. Depending on the overlap of these cones, four different situations may arise:

A. The two cones do not overlap: $\gamma_1 + \gamma_2 < \gamma_0$. No parameterizable surface exists in the wedge; rather, the liquid penetrates into the wedge to infinity (provided the liquid volume is sufficient). This situation is treated explicitly in Chap. 8. It is the normal situation for liquids penetrating into wedges under terrestrial gravity.
B. The two cones partly overlap: $\gamma_1 + \gamma_2 \geq \gamma_0$. The liquid surface is intersected (or cut off) by the wedge. A liquid surface exists within the wedge. The direction of the normal to the surface is given by one of the intersections of the cones.

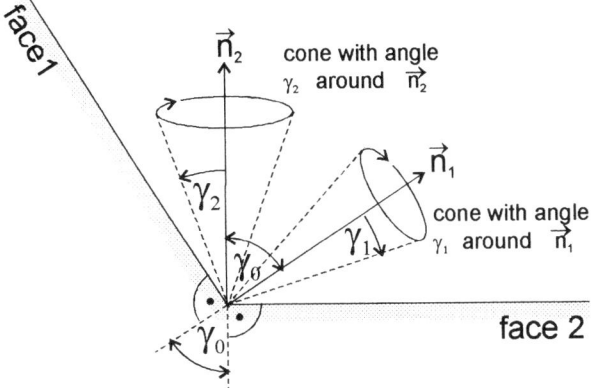

Fig. 7.6. A wedge with dihedral angle 2α between face 1 and face 2. The angle between the face normals equals $\gamma_0 \equiv \pi - 2\alpha$. The faces have contact angles γ_1, γ_2 with the liquid used

C. Cone 1 lies fully within cone 2 (or vice versa): $\gamma_2 > \gamma_0 + \gamma_1$ or $\gamma_1 > \gamma_0 + \gamma_2$. This is the case of canthotaxis. The liquid surface winds around the wedge from angle γ_1 to angle $\gamma_2 - \gamma_0$ in the interval of canthotaxis.

D. $\gamma_1 + \gamma_2 > 2\pi - \gamma_0$. This is equivalent to $(\pi - \gamma_1) + (\pi - \gamma_2) < \gamma_0$. The cones with the complementary angles $(\pi - \gamma_1)$ and $(\pi - \gamma_2)$ do not overlap. The complementary fluid penetrates into the wedge.

The resulting types of surfaces, as a function of the angles γ_1, γ_2 and γ_0, are presented in Fig. 7.7.

The transition from case B to case C, when the two cones just touch, deserves particular interest. In that case the contact lines approach tangentially the line of canthotaxis. The liquid surfaces on the two contacting solid faces join each other smoothly. However, as soon as $\gamma_2 > \gamma_0 + \gamma_1$, the liquid surface must wind around the axis by the defect angle $\Delta\gamma = \gamma_2 - \gamma_1 - \gamma_0$. The length of the interval of canthotaxis, according to numerical tests performed by Mittelmann using the computer code SURFACE EVOLVER, is proportional to $(\Delta\gamma)^2$.

The contact line with face 1 meets the axis of the wedge with a tangential slope, and the liquid surface there forms an angle γ_1 with face 1. The contact line with face 2 also meets the wedge axis with a tangential slope, and the liquid surface forms an angle γ_2 with face 2. Hence, along the interval of canthotaxis, the liquid surface must wind itself around the axis from γ_1 to $2\alpha - (\pi - \gamma_2)$. Distant from the wedge, for $r > 0$, the liquid surface is continuous and differentiable. The winding rate $[\partial z/\partial \varphi]_{r=0}$ along the interval of canthotaxis uniquely determines the behavior of $z(r, \varphi)$ for $r > 0$. The main task, therefore, is to find the winding rate $[\partial z/\partial \varphi]_{r=0}$ that allows constant contact angles γ_1, γ_2 along the faces.

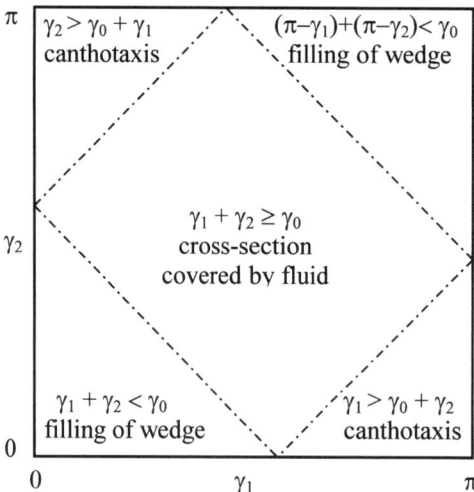

Fig. 7.7. The different types of surfaces that can occur in a wedge, as a function of the angles γ_1, γ_2 and $\gamma_0 \equiv \pi - 2\alpha$. *Lower left* and *upper right*: complete wetting of the wedge; *center*: liquid surface exists up to the wedge; *upper left* and *lower right*: canthotaxis. (From Concus and Finn 1994)

The behavior of the surfaces found for the wetting stripe and the wetting cross, namely the winding around the line of canthotaxis together with constant contact angles at the adjacent planes, cannot be described by an analytic function. Looking along the line of canthotaxis, the winding appears as a jump. An appropriate functional description must show a singularity at the ends of the interval of canthotaxis. There are, however, numerous alternative cases of canthotaxis which allow an analytical description. A few such situations, namely the helicoid, its additive extension to $z(r,\varphi) = c_0\varphi + z_1(r)$ and some others, will be studied in the following sections.

In the mathematical literature, liquid surfaces in wedges are commonly studied by looking along the axis of the wedge or wetting barrier. A surface exists in a wedge if it extends right into the wedge. In case A, when the wedge is filled with liquid, no surface exists in the wedge. In the case of canthotaxis the surface appears to jump between the ends of the interval of canthotaxis. In a paper by Lancaster and Siegel [1996], we read on this topic:

- Consider a bounded capillary surface defined on a two-dimensional region Ω that has a corner point O, with opening angle 2α.
- For $0 < \alpha < \pi/2$ the corner will be said to be convex, and for $\pi/2 \leq \alpha \leq \pi$ the corner will be said to be nonconvex.
- Rf is the radial limit of the capillary surface $Rf = \lim_{r \Rightarrow 0} f(r,\varphi)$.
- The limiting faces correspond to $\varphi = -\alpha$ and $\varphi = +\alpha$.

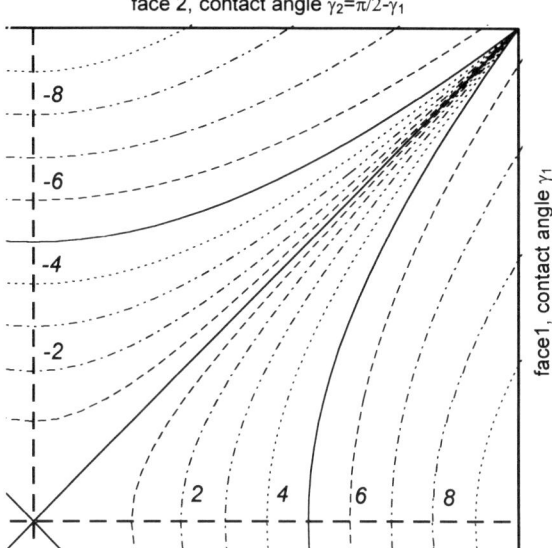

Fig. 7.8. Sketch of the height contours arising in a square cylinder with complementary contact angles γ_1 and $\gamma_2 = \pi - \gamma_1$ on neighboring faces. In the interval of canthotaxis, the liquid winds around the wedge axis from γ_1 to $\pi/2 - \gamma_1$. There exist 'fans' of directions adjacent to the wedge axes in which the radial limits are constant

- If the contact angle approaches limiting values as O is approached along each portion of the boundary, then there exist "fans" of directions adjacent to the two tangent directions at O in which the radial limits are constant.
- There exist fans of constant radial limits adjacent to each tangent direction at O: $\alpha_1 + \alpha \geq \gamma_0^-$ or $\pi - \gamma_0^-$. and $\alpha - \alpha_2 \geq \gamma_0^+$ or $\pi - \gamma_0^+$.
- There exist α_1 and α_2 such that $-\alpha \leq \alpha_1 < \alpha_2 \leq +\alpha$ and Rf is constant on $[-\alpha, \alpha_1]$ and $[\alpha_2, +\alpha]$ and strictly increasing or strictly decreasing on $[\alpha_1, \alpha_2]$.
- For a convex corner, the solution is continuous up to O when the limiting angles γ_0^+, γ_0^- satisfy $|\pi - \gamma_0^+ - \gamma_0^-| < 2\alpha$ and $2\alpha + |\gamma_0^+ - \gamma_0^a -| \leq \pi$.

The "fans of constant radial limits adjacent to each tangent direction" arise in regions close to faces 1 and 2, more exactly in the regions between face 1 and γ_1 and between $2\alpha - (\pi - \gamma_2)$ and face 2, where the contact line of the fluid surface with the solid faces approaches the wedge axis with tangential slope. If the axis is approached radially from angles smaller than γ_1 the upper end of the jump is reached, whereas if the axis is approached radially from angles larger than $2\alpha - (\pi - \gamma_2)$ the lower end of the jump is reached, or vice versa. The region where "Rf is strictly increasing or strictly decreasing" is the interval of canthotaxis, where the fluid surface winds around the wedge; see Fig. 7.8.

7.4 Taylor Expansions at Small Radii

There are numerous cases of canthotaxis which may be described by analytic functions. However, none of these cases allows constant contact angles γ_1, γ_2 on adjacent solid faces. A few of these cases are studied in the following. If a fluid surface exhibiting canthotaxis is analytical near the axis, it may be expanded into a power series in r, with the coefficients depending on the azimuth φ. Substitution of

$$z(r,\varphi) = \sum_{n=0}^{\infty} r^n c_n(\varphi) \tag{7.14}$$

into the capillary equation (3.68) and comparison of the coefficients of r^m leads to the following for the leading terms $n = 0, 1, 2, 3$:

$$2c_1(\varphi) = \widehat{p} c_0', \tag{7.15}$$

$$6c_2(\varphi) = -\frac{c_0''}{c_0'^2} + \widehat{p}^2 \frac{3}{4} c_0'', \tag{7.16}$$

$$12c_3(\varphi) = \widehat{p}\left(\frac{1}{c_0'} - \frac{c_0'''}{c_0'^2} + 2\frac{c_0''^2}{c_0'^3}\right) + \frac{\widehat{p}^3}{4}(c_0''' + c_0'), \tag{7.17}$$

$$20c_4(\varphi) = \left(\frac{2}{3}\frac{c_0''}{c_0'^4} + \frac{1}{6}\frac{c_0'''}{c_0'^4} - \frac{10}{9}\frac{c_0'''c_0''}{c_0'^5} + \frac{10}{9}\frac{c_0''^3}{c_0'^6}\right) + \widehat{p}^2[\ldots] + \widehat{p}^4[\ldots]. \tag{7.18}$$

By use of (7.15)–(7.18), all coefficients $c_n(\varphi)$ may be calculated from the winding rate along the axis $c_0' = [\partial z/\partial \varphi]_{r=0}$ and the pressure $\widehat{p}$.

Particular attention will be paid to symmetric situations (symmetric geometrically and with respect to the contact angle), in which zero capillary pressure of the liquid surface can be concluded by intuition or from the integral theorem (see Chap. 8). Examples are regular polygons with complementary contact angles $\gamma_2 = \pi - \gamma_1$ on adjacent faces; and wedges with dihedral angles 2α and $\pi - 2\alpha$, which may be symmetrically completed to form rhombi. Zero capillary pressure means that we are left with even powers of r in (7.14). In that case (7.15)–(7.18) reduce to

$$(c_{2m}'' + 4m^2 c_{2m}) + \sum_{n=0}^{m-1}(c_{2n}'' + 2nc_{2n}) \sum_{k=1}^{m-n} 2kc_{2k}(2m-2n-2k+2)c_{2m-2n-2k+2}$$
$$+ \sum_{n=1}^{m+1} 2n(4n-2m-1)c_{2n} \sum_{k=0}^{m-n+1} c_{2k}' c_{2m-2n-2k+2}' = 0. \tag{7.19}$$

7.4.1 Alternative Winding Functions

The best-known of the fluid surfaces that exhibit zero capillary pressure is the helicoid,

$$z(r,\varphi) = c_0\varphi = c_0 \arctan\left(\frac{y}{x}\right). \tag{7.20}$$

The angle γ between the helicoid and any plane through the axis is given by

$$\tan\gamma = \frac{r}{c_0}, \tag{7.21}$$

i.e. the helicoid requires a contact angle that increases linearly with the distance from the axis. One can easily experience this surface by descending a spiral staircase near its center. The helicoid allows a constant contact angle $\gamma = \pi/2$ on a coaxial circular cylinder.

The helicoid suggests separation of the capillary equation in cylindrical coordinates r and φ. Putting

$$z(r,\varphi) = c_0\varphi + z_1(r), \tag{7.22}$$

one obtains an ordinary differential equation for $z_1(r)$,

$$\frac{\partial^2 z_1}{\partial r^2} r(c_0^2 + r^2) + \frac{\partial z_1}{\partial r}\left[2c_0^2 + r^2\left(1 + \left(\frac{\partial z_1}{\partial r}\right)^2\right)\right]$$

$$= \widehat{p}\sqrt{c_0^2 + r^2\left[1 + \left(\frac{\partial z_1}{\partial r}\right)^2\right]^{-3}}, \tag{7.23}$$

Here $z_1(r)$ is an odd function in r. In the vicinity of the wedge axis, one has

$$z_1(r) = \sum_{n=0}^{\infty} c_{2n+1} r^{2n+1}, \tag{7.24}$$

$$c_1 = \frac{pc_0}{2}, \quad 6c_3 = c_0^{-2}\frac{pc_0}{2}\left[\left(\frac{pc_0}{2}\right)^2 + 1\right],$$

$$40c_5 = c_0^{-4}\frac{pc_0}{2}\left[\left(\frac{pc_0}{2}\right)^2 + 1\right]\left[3\left(\frac{pc_0}{2}\right)^2 - 1\right]. \tag{7.25}$$

The leading terms in c_{2n+1}, according to (7.25), are proportional to $(pc_0 2)^{2n+1}$. The coefficients $1, 1/6, 3/40$ lead to the arcsine function, i.e. in the limit of large pressures one obtains $z_1(r) \propto \arcsin(pc_0/2)$. In the limit of small radius the normal to the liquid surface approaches $\boldsymbol{n} \to (0,1,0)$. This normal has a component in the φ direction only, i.e. the contact line meets the wedge axis with contact angle zero. This is the same result as for the helicoid, (7.21).

It is also worth mentioning that an exact solution of the capillary equation may be achieved by separation in Cartesian coordinates. We obtain

$$z(x,y) = f(x) - f(y), \quad \text{where } f(x) = -\frac{1}{c}\log\left[\cos(cx)\right]. \tag{7.26}$$

This function, which is known as the Scherk function, does not, however, satisfy any plausible boundary condition on the contact angle with any nearby solid faces.

A further separation of the capillary equation in Cartesian coordinates is possible by putting

$$z(x,y) = yf(x). \tag{7.27}$$

One obtains

$$\frac{d^2 f}{dx^2}(1+f^2) - 2\left(\frac{df}{dx}\right)^2 f(x) = 0, \quad f(x) = \tan(c_1 x + c_0) \tag{7.28}$$

and

$$z(x,y) = y\cot x. \tag{7.29}$$

This solution does not contribute to the problem of canthotaxis either.

7.5 Liquid Surfaces in Square Cylinders, $\cos\gamma_1 + \cos\gamma_2 = 0$

Much attention has been paid to capillary surfaces in cylinders with polygonal cross section. Several microgravity experiments have been performed on such surfaces. Let us focus on square cylinders with complementary contact angles γ_1 and $\gamma_2 = \pi - \gamma_1$ on neighboring faces; see Figs. 7.8 and 7.9. From the integral theorem (see Sect. 8.2), we have

$$\cos\gamma_1 + \cos\gamma_2 = \widehat{p} = 0. \tag{7.30}$$

From the tetragonal symmetry, we conclude that $z(x,y)$ is an even function with respect to x and y and changes its sign on interchanging x and y:

$$z(-x,y) = z(x,y), \quad z(x,-y) = z(x,y), \quad z(y,x) = -z(x,y). \tag{7.31}$$

According to the general condition $\gamma_2 = \pi - 2\alpha - \gamma_1$, canthotaxis will arise in the edges between the faces of the square cylinder, i.e. the capillary surface will wind around the edge, if $\gamma_1 < \pi/4$. Figure 7.9 shows the capillary surfaces for $\gamma \equiv \gamma_1 = 60°$ and $\gamma \equiv \gamma_1 = 30°$ as computed by Mittelmann and Zhu [1996]. Numerically, the interval of canthotaxis increases in proportion to the third power of the angular interval of winding $\pi/2 - 2\gamma_1$; see Fig. 7.10.

7.5 Liquid Surfaces in Square Cylinders, $\cos\gamma_1 + \cos\gamma_2 = 0$

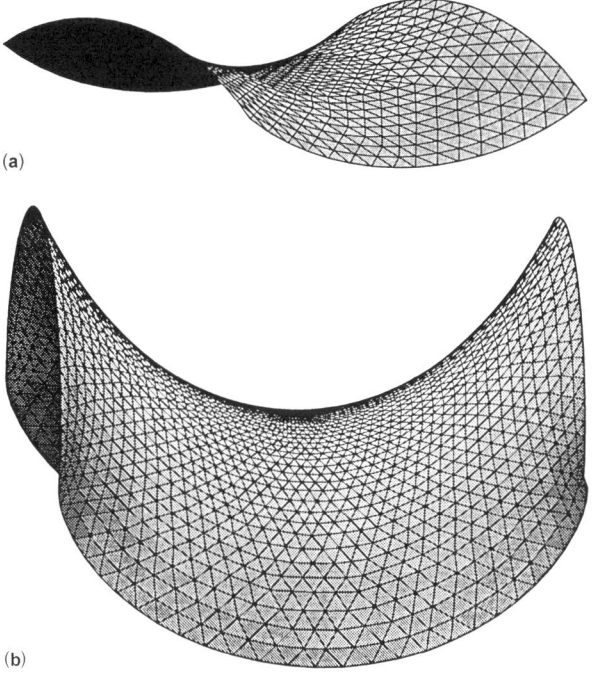

Fig. 7.9. The capillary surfaces that arise in a square cylinder with complementary contact angles γ_1 and $\gamma_2 = \pi - \gamma_1$ on neighboring faces, as computed by Mittelmann & Zhu [1996]; (**a**) $\gamma_1 = 60°$ (no canthotaxis) and (**b**) $\gamma_1 = 30°$ (canthotaxis)

For $\gamma > \pi/4$, the resulting capillary surfaces are analytical. They may be expanded into a Taylor series in x and y. The symmetry relations (7.31) reduce the relevant functions to $x^{2m}y^{2m}(x^{2n}-y^{2n})$, where $m = 0, 1, \ldots, \infty, n = 1, 2, \ldots, \infty$. Putting

$$z(x,y) = \sum_{m=0}^{\infty}\sum_{n=1}^{\infty} c_{m,n} x^{2m} y^{2m}(x^{2n} - y^{2n}) \,, \tag{7.32}$$

we anticipate convergence of (7.32) for $\gamma > \pi/4$. By substitution of (7.32) into the capillary equation for zero curvature and comparison of the coefficients of $x^{2M}y^{2M}(x^{2N} - y^{2N})$, one finds

$$c_{M,1} \text{ arbitrary for all } M \,, \tag{7.33}$$

$$(2M+2N+2)(2M+2N+1)c_{M,N+1} + (2M+2)(2M+1)c_{M+1,N-1}$$
$$= 8\sum_{i,j}\sum_{k,l}^{k \leq i}\sum_{m,n}^{m \leq k} A(i,j;k,l;m,n) c_{i,j} c_{k,l} c_{m,n} \,, \tag{7.34}$$

166   7. Canthotaxis/Wetting Barriers/Pinning Lines

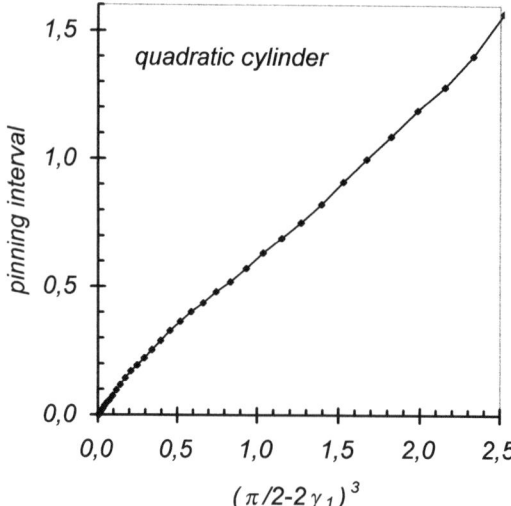

Fig. 7.10. Numerically, the interval of canthotaxis in a square cylinder increases in proportion to the third power of the winding interval $\pi/2 - 2\gamma_1$ ($\gamma_2 = \pi - \gamma_1$)

where $2(i+k+m)+j+l+n = 2M+N+2$. The coefficients $A(i,j;k,l;m,n)$ in (7.34) are integers. The leading terms up to $2M+N = 4$ are given in Table 7.1.

The coefficients $c_{M,1}$, which are arbitrary according to (7.33), can be used for satisfying the boundary condition on the contact angles at the faces $x = 1$ and $y = 1$. We obtain

$$\left.\frac{\partial z}{\partial x}\right|_{x=1} = \cot\gamma\sqrt{1 + \left(\frac{\partial z}{\partial y}\right)^2_{x=1}}. \tag{7.35}$$

Table 7.1. The leading terms $A(i,j;k,l;m,n)$ up $2M+N = 4$ in the trilinear expansion of (7.34). The repetition of products arising by permutation of the pairs of indices $(i,j), (k,l), (m,n)$ is taken into account

i	j	k	l	m	n	M	N	$A(i,j;k,l;m,n)$	M	N	$A(i,j;k,l;m,n)$
0	1	0	1	0	1	0	1	1			
0	1	0	1	0	2	0	2	4			
0	1	0	1	0	3	0	3	6	1	1	−15
0	1	0	1	1	1			−1			4
0	1	0	2	0	2			4			24
0	1	0	1	0	4	0	4	8	1	2	−28
0	1	0	1	1	2			−1			--4
0	1	0	2	0	3			12			36
0	1	0	2	1	1			−4			28
0	2	0	2	0	2			0			24

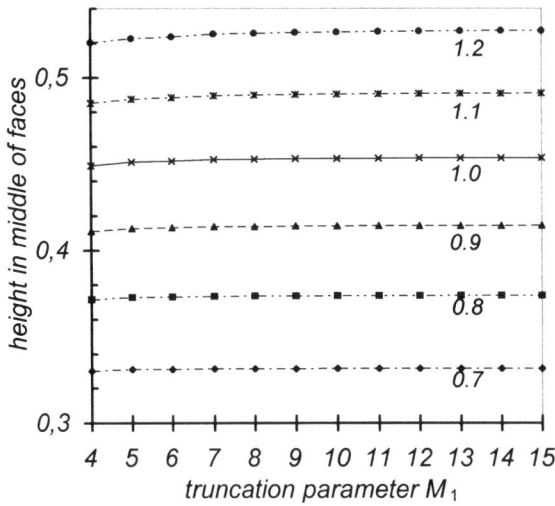

Fig. 7.11. Convergence of $z(1,0)$ in the center of a face with increasing truncation number M_1. The parameter of the various curves is $\cot\gamma$. Convergence is much slower for the forbidden values $\cot\gamma = 1.1$ and $\cot\gamma = 1.2$ than for the allowed values $\cot\gamma = 0.7$ to 1.2

Substitution of (7.32) and comparison of coefficients of y^{2M} leads to

$$\sum_{n=1}^{\infty} n c_{M,n} = \frac{1}{2}\cot\gamma \quad \text{for } M = 0, \tag{7.36}$$

$$\sum_{n=1}^{\infty} (M+n) c_{M,n} = \cot\gamma \left(\sum_{n=1}^{\infty} c_{1,n} - c_{0,1} \right)^2 \quad \text{for } M = 1, \tag{7.37}$$

$$\sum_{n=1}^{\infty} (M+n) c_{M,n} - \sum_{n=1}^{M-1} (M-n) c_{M-n,n}$$

$$= \cot\gamma \sum_{i+k=M+1} i \left(\sum_{j=1}^{\infty} c_{i,j} - \sum_{j=1}^{i} c_{i-j,j} \right) k \left(\sum_{l=1}^{\infty} c_{k,l} - \sum_{l=1}^{k} c_{k-l,l} \right)$$

$$+ \tan\gamma \sum_{i+k=M} \left(\sum_{j=1}^{\infty} (i+j) c_{i,j} - \sum_{j=1}^{i-1} (i-j) c_{i-j,j} \right)$$

$$\times \left(\sum_{l=1}^{\infty} (k+l) c_{k,l} - \sum_{l=1}^{k-1} (k-l) c_{k-l,l} \right) \tag{7.38}$$

for $M > 1$. The products of sums on the left-hand side of (7.38) result from $[\partial z/\partial y]_{x=1}$ and $[\partial z/\partial x]_{x=1}$.

168   7. Canthotaxis/Wetting Barriers/Pinning Lines

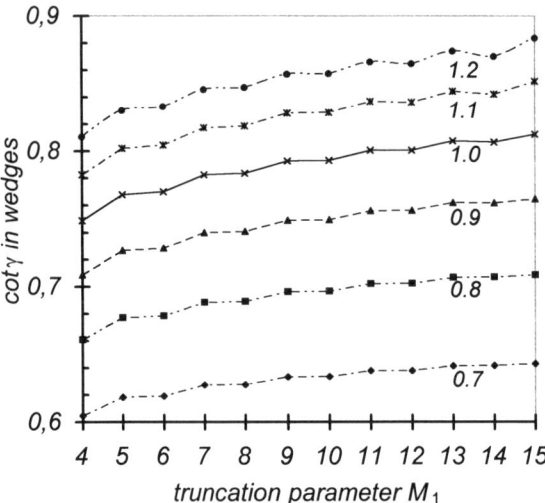

Fig. 7.12. The values of $\cot\gamma(1,1)$ reached in the edges. The approach of the value of $\cot\gamma$ obtained to the correct value clearly deteriorates with increasing $\cot\gamma$

Equation (7.34) includes a finite number of coefficients $c_{M,N+1}$ only, whereas (7.36)–(7.39) include an infinite number of coefficients. In order to achieve a convergent solution, it is convenient to truncate all powers $2m + n > M_1$ in (7.32) and to successively increase M_1. The absolute values of all coefficients $c_{m,n}$ increase monotonically with increasing truncation parameter M_1. According to (7.34), $c_{m,n}$ become very small for large n. The sums over n in (7.32) are hypergeometric series, which converge rapidly for $x \leq 1, y \leq 1$. The corresponding sums over m converge comparatively slowly.

For $\cot\gamma \ll 1$ we generally find

$$c_{0,1} \approx \frac{1}{2}\cot\gamma\,, \quad c_{0,2} \approx \frac{1}{12}\cot^3\gamma\,, \quad c_{m,n} \propto \begin{cases} \cot^3\gamma \text{ for } n \text{ odd} \\ \cot^5\gamma \text{ for } n \text{ even} \end{cases}. \quad (7.39)$$

The leading terms, proportional to $(\cot\gamma)^3$, in $c_{m,n}$ have been found numerically to converge in proportion to $1/M_1^3$. Going beyond (7.39), we find for $c_{m,1}$

$$c_{0,1} \approx 0.5\cot\gamma - 0.1373\cot^3\gamma \pm \ldots\,, \quad c_{1,1} \approx 0.1567\cot^3\gamma\,,$$
$$c_{2,1} \approx 0.0926\cot^3\gamma\,, \quad c_{3,1} \approx 0.1055\cot^3\gamma\,, \quad c_{4,1} \approx 0.1094\cot^3\gamma\,, \quad (7.40)$$

with the convergence clearly decreasing with increasing m. The terms for $n > 1$ are obtained from (7.34). The higher-order contributions, proportional to $(\cot\gamma)^5, (\cot\gamma)^7$, etc., can be determined in like manner.

Convergence of the iterative solution of the full system of (7.32) and (7.36)–(7.39) slows down strongly when $\cot\gamma = 1$ is approached. The iterative solution oscillates more strongly as higher values of the truncation parameter

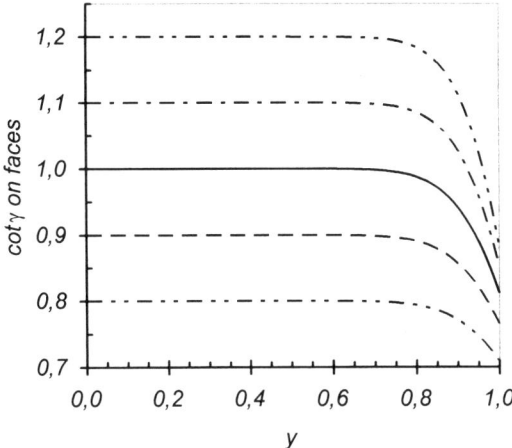

Fig. 7.13. $\cot \gamma(1, y)$ on the faces. If the solution was exact, $\cot \gamma(1, y)$ would have the same value as at the abscissa, independent of y. The representation of $z(x, y)$ by a power series simply cannot lead to a value of $\cot \gamma(1, 1)$ above 1

M_1 are chosen, i.e. more functions are taken into account. More functions mean satisfaction of the boundary condition on the contact angle γ up to higher powers y^{2M} and thus a better approximation to the test function (7.32). For $\gamma = \pi/4, \cot \gamma = 1$, the slopes $[\partial z/\partial x]_{x=1}$ and $[\partial z/\partial y]_{x=1}$ at the edge actually become infinite, which means a divergence of the power series used. For $\cot \gamma > 1$, a jump of the height z arises in the wedge.

Figure 7.11 shows the convergence of $z(1, 0)$ in the center of a face with increasing truncation number M_1. The parameter of the various curves is $\cot \gamma$. Convergence is much slower for the excessively high values $\cot \gamma = 1.1$ and $\cot \gamma = 1.2$. Figure 7.12 shows the actual values of $\cot \gamma$ in the edges $x = \pm 1, y = \pm 1$. The approach to the presumed correct values clearly deteriorates with increasing $\cot \gamma$.

Figure 7.13 shows $\cot \gamma(1, y)$ on the face $x = 1$. If the solution was exact, $\cot \gamma(1, y)$ would be constant. Figure 7.13 demonstrates the difficulties in realizing this requirement. The representation of $z(x, y)$ by a power series (7.32) together with the symmetry relations (7.31) simply cannot lead to a value of $\cot \gamma$ in the edge larger than 1.

7.6 Towards Modeling Canthotaxis

7.6.1 Helicoid and Catenoid

If γ_1 is smaller than $\pi/4$, i.e. if $\cot \gamma_1$ exceeds 1, the infinite slope of the contact line in a wedge axis is supplemented by a jump; an interval of canthotaxis arises. The liquid surface winds around the axis from γ_1 to $\pi/2 - \gamma_1$. Such a jump cannot be described by an analytic function; the surface must show a singularity at the wedge axis $r = 0$.

Looking for a starting point for modeling canthotaxis, one is reminded of the following facts:

- The helicoid $z(x, y) = \varphi = \arctan(y/x)$ winds around the axis and has curvature zero, i.e. is a minimal surface.
- The catenoid has zero curvature as well. It may be adjusted so that it has a contact angle γ_1 with face 1.
- The tangential planes of both surfaces coincide at the axis.
- The directions of the principal curvatures in these planes, however, are twisted by $\pi/4$.

The latter property makes it impossible to construct a surface from a helicoid along the axis and a catenoid at the solid surface. Along their line of intersection these surfaces form a concave angle, such that further liquid will be sucked into that region. This causes a negative capillary pressure in the azimuthal direction, which in turn must be balanced by a positive pressure in the radial direction. Thus, both surfaces, the helicoid and the catenoid, become strongly deformed.

7.6.2 Winding Rates $[\partial z(\varphi)/\partial \varphi]_{r=0} \propto [\cos(s\varphi)]^k$

As stated in Sect. 7.4, the winding rate along the line of canthotaxis fully defines the fluid surface in the neighborhood of the wedge axis. It is therefore worth studying various options for the winding rate. At the ends of the pinning interval, at the bounds of the angular interval in question, the winding rate $[\partial z/\partial \varphi]_{r=0}$ shows a singularity. A large variety of candidate functions can be obtained by considering winding rates of the form

$$c'_0 = \left[\frac{\partial z}{\partial \varphi}\right]_{r=0} = [\cos(s\varphi)]^k . \tag{7.41}$$

If k differs from an integer, the winding rate (7.41) has a singularity for $r = 0, \cos(s\varphi) = 0$, or $s\varphi = \pi/2$. Substitution of (7.41) into the set of equations (7.19) (or into the capillary equation) leads to a double series expansion in r and $\cos(s\varphi)$, (7.42), or else in r and $\sin(s\varphi)$, (7.43). The former series is clearly adapted to describing the behavior of the liquid surface at the ends of the pinning interval, at $\cos(s\varphi) \approx 0$, whereas the latter is applicable near the symmetry axis, where $\sin(s\varphi) \approx 0$. These series are

$$z(r,\varphi) = \int d\varphi [\cos(s\varphi)]^k + \sin(s\varphi) \sum_{n=0}^{\infty} [\cos(s\varphi)]^{2n+k+1} \sum_{m=n+1}^{\infty} a_{m,n} \hat{r}^{2m} ,$$

(7.42)

$$z(r,\varphi) = \int d\varphi [\cos(s\varphi)]^k + [\cos(s\varphi)]^{k+1} \sum_{n=0}^{\infty} [\sin(s\varphi)]^{2n+1} \sum_{m=n+1}^{\infty} b_{m,n} \hat{r}^{2m} ,$$

(7.43)

where

$$\hat{r} = \frac{r}{[\cos(s\varphi)]^{k+1}} .$$

(7.44)

In a variety of cases, the capillary surfaces described by (7.42) and (7.43) may be shown to be generalized hypergeometric functions. In these cases they can be summed exactly and thus may be transformed from small radii r to large radii r. The coefficients $a_{m,n}$ show a constant limiting ratio $a_0 = a_{m+1,n}/a_{m,n}$. This enables one to calculate the contact angles with the planes $s\varphi = \pm\pi/2$, i.e. at the azimuthal limits of the interval of canthotaxis. These planes differ in azimuth from the solid faces 1 and 2 by $\Delta\varphi = \pm\gamma_1$. However, except for the limiting cases $\gamma_1 = 0$ and $\gamma_1 = \pi$, results for the contact angles with the solid faces themselves are not obtained. This is in accord with the mathematical findings reported by Lancaster & Siegel [1996].

7.6.3 Winding Rate of Infinity

There is a class of winding functions that deserve particular interest: $0 \leq k \leq -1$ means that the winding rate approaches infinity, with the jump being integrable in the region $-\pi/2 \leq s\varphi \leq +\pi/2$. At the limits $\pm\pi/2$ the winding rate shows singularities, and the power series (7.42) with respect to r and $\cos(s\varphi)$ diverges. However, the generalized hypergeometric series arising can be transformed from small arguments to large arguments. If

$$k = -\frac{m-1}{m} ,$$

(7.45)

where m is an integer, one obtains for the radius of convergence

$$s\hat{r} \approx m \sin\left(\frac{\pi}{m}\right) (m+1)^{1/m} .$$

(7.46)

By explicitly summing all terms in (7.42), one finds

$$z(r,\varphi) \approx (m+1) \int d\varphi [\cos(s\varphi)]^{(m+1)/m}$$
$$- ms^{-1} \sin(s\varphi)[\cos(s\varphi)]^{1/m} \zeta + O\left[[\cos(s\varphi)]^{2+1/m}\right] ,$$

(7.47)

where ζ is given by the relations

Table 7.2. Coefficients in (7.49)

m	C_m	m	C_m	m	C_m	m	C_m
1	1.0000000	5	0.9974691	9	0.9531804	13	0.9304942
2	1.0471976	6	0.9831089	10	0.9461903	14	0.9265126
3	1.0339058	7	0.9712596	11	0.9402009	15	0.9229831
4	1.0146936	8	0.9614257	12	0.9350184	16	0.9198336

$$m = 2, \quad 1 - \zeta^{-2} = \frac{1}{12}(s\tilde{r})^2 \zeta^{-2}, \tag{7.48a}$$

$$m = 3, \quad 1 - \zeta^{-3} = \frac{1}{9}(s\tilde{r})^2 \zeta^{-2}, \tag{7.48b}$$

$$m = 4, \quad 1 - \sqrt{\frac{4}{5} + \frac{1}{5}\zeta^{-4}} = \frac{1}{80}(s\tilde{r})^2 \zeta^{-2}, \tag{7.48c}$$

$$m = 5, \quad 1 - \sqrt{\frac{16}{25} + \frac{9}{25}\zeta^{-5}} = \frac{3}{125}(s\tilde{r})^2 \zeta^{-2}. \tag{7.48d}$$

The fictive contact angle at the angular bounds of the interval of canthotaxis, i.e. the angle of the tangential planes of the fluid surface at the ends of interval, is given by

$$\tan\gamma(r,\varphi) = C_m \left(\frac{s}{\pi}\right)^{m-1} r^m \quad \text{for } m \geq 1. \tag{7.49}$$

The coefficients C_m are listed in Table 7.2.

Further analytical investigations are required, in particular with respect to the numerical finding that the length of pinning interval is proportional to the square of the defect angle $\Delta\gamma$.

7.6.4 Circular Tube with Complementary Contact Angles

In looking for a configuration which renders possible an analytical treatment of canthotaxis, one is led to consider a circular cylinder which shows complementary contact angles γ and $\pi - \gamma$ on opposite sides (Fig. 7.14). Explicitly, we assume a contact angle γ for $-\pi/2 < \varphi < +\pi/2$ and the complementary contact angle $\pi - \gamma$ for $\pi/2 < \varphi < 3\pi/2$. In this case, the liquid surface can be conveniently expanded into a Fourier series in the azimuth φ. The Fourier series contains only the terms $\cos(n\varphi)$, with n odd.

From the integral theorem introduced in Sect. 8.2, we conclude that the capillary pressure $\hat{p}$ of the resulting liquid surface vanishes. For $\gamma = \pi/2$, we obtain the plane $z(r,\varphi) = 0$. For $\gamma \leq \pi/2$, $\cos\gamma \approx 0$, it is possible to expand the surface into a Taylor series with respect to $\cos\gamma$ and to decompose the capillary equation accordingly. Putting

$$z(r,\varphi) = \sum_k c_1^{2k+1} z_{2k+1}(r,\varphi), \tag{7.50}$$

7.6 Towards Modeling Canthotaxis

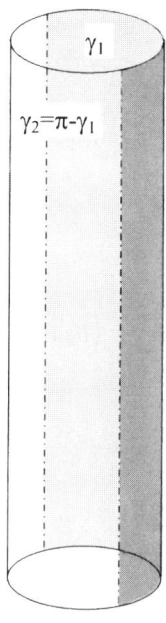

Fig. 7.14. Circular tube with complementary contact angles γ_1 and $\pi - \gamma$

where

$$c_1 = \frac{4\cos\gamma}{\pi}, \tag{7.51}$$

we obtain from (3.68)

$$\left(\frac{1}{r}\frac{\partial}{\partial r}r\frac{\partial}{\partial r} + \frac{\partial^2}{r^2\partial\varphi^2}\right)z_1(r,\varphi) = 0, \tag{7.52}$$

$$\left(\frac{1}{r}\frac{\partial}{\partial r}r\frac{\partial}{\partial r} + \frac{\partial^2}{r^2\partial\varphi^2}\right)z_3(r,\varphi) = -\frac{1}{r}\frac{\partial}{\partial r}r\frac{\partial z_1}{\partial r}\left(\frac{\partial z_1}{r\partial\varphi}\right)^2$$

$$-\frac{1}{r}\frac{\partial z_1}{\partial r}\left[\left(\frac{\partial z_1}{\partial r}\right)^2 - 2\frac{\partial z_1}{r\partial\varphi}\frac{\partial^2 z_1}{\partial\varphi\partial r} + \left(\frac{\partial z_1}{r\partial\varphi}\right)^2\right] - \frac{\partial^2 z_1}{r^2\partial\varphi^2}\left(\frac{\partial z_1}{\partial r}\right)^2. \tag{7.53}$$

Equation (7.52) has the homogeneous solutions $r^n \cos(n\varphi)$. Thus, we put

$$z_1(r,\varphi) = \sum_{n\,\text{odd}} a_{n0}^{(1)} r^n \cos(n\varphi). \tag{7.54}$$

Satisfaction of the boundary condition on the contact angle γ by means of the coefficients $a_{n0}^{(1)}$ of the homogeneous solutions is conveniently achieved by minimization of the effective surface area. Equating the radius R of the tube to 1, we obtain for the area A_l of the liquid surface

$$A_1 = \int_0^1 dr \int_0^{2\pi} r\, d\varphi$$
$$\times \sqrt{1 + \left(c_1 \sum_{n\,\text{odd}} a_{n0}^{(1)} n r^{n-1} \cos(n\varphi)\right)^2 + \left(c_1 \sum_{n\,\text{odd}} a_{n0}^{(1)} r^{n-1} n \sin(n\varphi)\right)^2}. \tag{7.55}$$

In the limit $c_1 \ll 1$, (7.55) can be reduced to

$$A_1 = \int_0^1 dr\, 2\pi r \left(1 + \frac{1}{2}c_1^2 \sum_{n\,\text{odd}} \left(a_{n0}^{(1)} n r^{n-1}\right)^2\right) = \pi + \frac{\pi}{2}c_1^2 \sum_{n\,\text{odd}} n \left(a_{n0}^{(1)}\right)^2. \tag{7.56}$$

The integral over the area of the solid surface is given by

$$A_s = c_1 \int_{-\pi/2}^{\pi/2} d\varphi \sum_{n\,\text{odd}} a_{n0}^{(1)} \cos(n\varphi) = c_1 \sum_{n\,\text{odd}} 2(-1)^{(n-1)/2} \frac{a_{n0}^{(1)}}{n}. \tag{7.57}$$

Hence, minimization of the effective area $A_1 - 2A_s \cos\gamma$ with respect to all $a_{n0}^{(1)}$ leads to

$$a_{n0}^{(1)} = \frac{(-1)^{(n-1)/2}}{n^2}. \tag{7.58}$$

The derivatives of $z_1(r, \varphi)$ and, hence, the normal to $z(r, \varphi)$ can be represented in closed analytic form:

$$\frac{\partial z_1(r,\varphi)}{\partial r} = \sum_{n\,\text{odd}} (-1)^{(n-1)/2} \frac{r^{n-1}}{n} \cos(n\varphi)$$
$$= \frac{1}{2r} \arctan\left(\frac{2r \cos\varphi}{1 - r^2}\right), \tag{7.59}$$

$$\frac{\partial z_1(r,\varphi)}{\partial \varphi} = \frac{i}{2} \sum_{n\,\text{odd}} (-1)^{(n-1)/2} \frac{1}{n} \left[(re^{i\varphi})^n - (re^{-i\varphi})^n\right]$$
$$= -\frac{1}{4} \log\left(\frac{1 + 2r\sin\varphi + r^2}{1 - 2r\sin\varphi + r^2}\right). \tag{7.60}$$

Whereas $\partial z/\partial r$ becomes a step function, $\partial z/\partial \varphi$ approaches infinity logarithmically at $\varphi = \pm\pi/2$, at the lines of canthotaxis. The normal to the surface becomes parallel to the azimuthal direction. At $r = 1$ we obtain, independent of the azimuth φ,

$$\left.\frac{\partial z(r,\varphi)}{\partial r}\right|_{r=1} = c_1 \left.\frac{\partial z_1(r,\varphi)}{\partial r}\right|_{r=1} = \pm \cos\gamma. \tag{7.61}$$

7.6 Towards Modeling Canthotaxis

Substitution of (7.59) and (7.60) into (7.53) yields products of order three in the terms $\cos(n\varphi)$, $\sin(n\varphi)$. These products can be reduced to linear terms by applying the addition theorem of the cosine. Three factors $\cos(n\varphi)$ or one factor $\cos(n\varphi)$ together with two factors $\sin(n\varphi)$, with n odd, reproduce $\cos(n\varphi)$ terms with n odd only. We obtain

$$\left(\frac{1}{r}\frac{\partial}{\partial r}r\frac{\partial}{\partial r} + \frac{\partial^2}{r^2\partial\varphi^2}\right) z_3(r,\varphi)$$
$$= 2 \sum_{n\,\text{odd}} (-1)^{(n-1)/2} \cos(n\varphi)$$
$$\times \sum_{m=1}^{\infty} \frac{(-1)^m r^{n+2m-2}}{(m+1)(m+n+1)} \begin{cases} (m+n)F_1(m) & m\text{ odd} \\ mF_1(m+n) & m\text{ even} \end{cases}.$$
(7.62)

The finite sums $F_1(m)$, $F_1(m+n)$ in (7.62) are given by

$$F_1(n) = \begin{cases} \log 2 + \frac{1}{2} + \frac{1}{4} + \ldots + \frac{1}{n} = \log 2 + \sum_{m=1}^{n/2} \frac{1}{2m} & \text{for } n \text{ even} \\ \frac{1}{1} + \frac{1}{3} + \ldots + \frac{1}{n} = \sum_{m=1}^{(n+1)/2} \frac{1}{2m-1} & \text{for } n \text{ odd} \end{cases}.$$
(7.63)

For large n, $F_1(n)$ may be approximated by

$$F_1(n) \approx \frac{1}{2}\{\log[2(n+1)] + c_2\},$$
(7.64)

where $c_2 = 0.57721566$ is Euler's constant.

We note that in (7.62), only terms $r^m \cos(n\varphi)$ with $m \geq n$ arise. This allows an analytic integration of the inhomogeneous capillary equation (7.62). We obtain

$$z_3(r,\varphi) = \sum_{n\,\text{odd}} \sum_{m=1}^{\infty} a_{nm}^{(3)} r^{n+2m} \cos(n\varphi),$$
(7.65)

where

$$a_{nm}^{(3)} = \frac{(-1)^{(n-1)/2+m}}{2(m+1)(m+n+1)} \begin{cases} F_1(m)/m & m\text{ odd} \\ F_1(m+n)/(m+n) & m\text{ even} \end{cases}.$$
(7.66)

The sums over m in (7.65) contain three factors m in the denominator. This guarantees rapid convergence. The sum over n for m odd, on the other hand, contains just one term $m+n+1$ in the denominator. This means slow convergence. For the largest term in $z_3(r,\varphi)$, we find at $r=1$

176   7. Canthotaxis/Wetting Barriers/Pinning Lines

$$z_3(1,\varphi) = -\frac{1}{2}\sum_{n\,\text{odd}}(-1)^{(n-1)/2}\frac{\cos(n\varphi)}{n}\sum_{m=1}^{\infty}\frac{F_1(m)}{m(m+1)}$$

$$= -\frac{\pi^2}{24}\sum_{n\,\text{odd}}(-1)^{(n-1)/2}\frac{\cos(n\varphi)}{n}, \tag{7.67}$$

where use has been made of

$$\sum_{m\,\text{odd}}^{\infty}\frac{F_1(m)}{m(m+1)} = \frac{\pi^2}{12}. \tag{7.68}$$

The formula (7.65) for $z_3(r,\varphi)$ is incomplete, insofar as the homogeneous solutions $a_{n0}^{(3)}r^n\cos(n\varphi)$ of order c_1^3 are still missing. As before, these terms must be determined by minimization of the effective surface area. We obtain

$$a_{n0}^{(3)} = -\sum_{m=1}^{\infty}a_{nm}^{(3)} + \frac{1}{4}\sum_{m\,\text{odd}}ma_{m0}^{(1)}$$

$$\times\left(\sum_{l=0}^{\infty}a_{n+2l0}^{(1)}a_{m+2l0}^{(1)}\frac{(n+2l)(m+2l)}{n+m+2l-1} + \sum_{l=1}^{(n-1)/2}a_{n-2l0}^{(1)}a_{m+2l0}^{(1)}\frac{(n-2l)(m+2l)}{n+m-1}\right). \tag{7.69}$$

The largest term in $a_{n0}^{(3)}$ results from the first sum on the right-hand side of (7.69). We obtain

$$a_{n0}^{(3)} \approx \frac{\pi^2}{24}\frac{(-1)^{(n-1)/2}}{n}. \tag{7.70}$$

The further contributions $a_{n0}^{(3)}$ in (7.65), and also the terms $a_{nm}^{(3)}$ with m even, generally decrease in proportion to $(\log n)^2/[n(n+1)]$. Equation (7.70) leads to

$$z_{3,0}(r,\varphi) \approx \frac{\pi^2}{24}\sum_{n\,\text{odd}}\frac{(-1)^{(n-1)/2}r^n\cos(n\varphi)}{n} = \frac{\pi^2}{48}\arctan\left(\frac{2r\cos\varphi}{1-r^2}\right). \tag{7.71}$$

Equation (7.70) means a jump (equal to the length of the interval of canthotaxis) given by

$$c_1^3\left[z_3(r,\varphi)|_{\pi/2-\varepsilon} - z_3(r,\varphi)|_{\pi/2+\varepsilon}\right] \approx \frac{c_1^3\pi^3}{48} = \frac{4}{3}\cos^3\gamma. \tag{7.72}$$

This jump is partly balanced by the terms $a_{nm}^{(3)}$ with m odd in (7.65). The difficulty that $\partial z/\partial\varphi$ approaches infinity logarithmically for $\varphi = \pm\pi/2$ does not become noticeable on substitution into the capillary equation (7.53), owing to the Fourier approach used here. However, one has to watch out for this difficulty during minimization of the effective area $A_l - 2A_s \cos\gamma$. Equations (7.70) and (7.71) must be considered as preliminary in that respect. In asymmetric cases, when the capillary pressure $\widehat{p}$ is nonzero, terms quadratic in c_1 rather than terms of order three arise on the right-hand side of (7.53). This causes a jump proportional to the square of the angular interval to be bridged.

References

1. Concus P, Finn R: Capillary surfaces in a wedge – differing contact angles. Microgravity Sci. Technol. **7** (1994) 152–155
2. de Lazzer A, Dreyer ME, Rath HJ: Capillary effects under low gravity. Part II: considerations on equilibrium capillary surfaces. Space Forum **3** (1998) 137–163
3. Dreyer ME, Gerstmann J, Stange M, Rosendahl U, Wölk G, Rath HJ: Capillary effects under low gravity. Part I. Surface settling, capillary rise and critical velocities. Space Forum **3** (1998) 87–136
4. Lancaster KE, Siegel D: Existence and behavior of the radial limits of a bounded capillary surface at a corner. Pacific J. Math. **176** (1996) 165–194; Pacific J. Math. **179** (1997) 397–402
5. Lucas R: Über das Zeitgesetz des kapillaren Aufstiegs von Flüssigkeiten. Kolloid-Zeitschrift **23** (1917) 15–22
6. Mittelmann HD: private communication (1995)
7. Mittelmann HD, Zhu A: Capillary surfaces with different contact angles in a corner. Microgravity Sci. Technol. **9** (1996) 22–27
8. Washburn EW: The dynamics of capillary flow. Phys. Rev. **17** (1921) 273–283

8. Cylindrical Containers

A liquid penetrates into wedge-shaped space between solid surfaces if the sum of the dihedral angle of the wedge and the contact angle with the faces is smaller than the sum π of angles in a triangle. If that sum exceeds π, a liquid volume pressed into the wedge by gravity or rotation breaks into droplets on reduction of the force. This is particularly important in long cylindrical vessels with a filling level sufficiently high for the end sections to be completely covered by liquid.

This situation allows an effective integral theorem: the cross-sectional integral over the curvature equals the integral over the cosine of the contact angle along the periphery. Cylindrical liquid volumes in edges or other narrowings are accounted for by the related reduction in cross section and periphery and an additional fluid boundary with contact angle zero. According to this theorem, the curvature of the fluid surface may be calculated without explicitly calculating the surface shape. The integral theorem is applicable whenever the forces act perpendicular to the symmetry axis. Its strength becomes fully obvious when cylindrical vessels spinning around their symmetry axis are considered. Various examples are given in this chapter.

8.1 Introduction

8.1.1 Fields of Application

In a zero-gravity environment, the configuration of a fluid interface is dictated by the shape of the container and the fluid/solid contact angle. Thus, by a judicious selection of these, one might be able to control, in a passive manner, the location of a bulk fluid. Cylindrical containers, whether circular, rhombic or adapted more specifically, offer great advantages in this respect.

The design of spacecraft tanks must ensure fuel outflow on demand. Since the fuel no longer rests on the bottom and pistons cannot safely prevent bubble intrusion after several months in orbit, so-called surface tension tanks are the obvious alternative. The surface energy of a wetting liquid generally decreases when it moves to concave regions of a container, in particular, to corners and edges. Various surface tension tanks based on longitudinal sections of circular cylinders and on rhombic cylinders have been successfully tested, exploiting the short-time microgravity environment provided by parabolic flights of a KC-135 aircraft [Bauer 1984, 1986; Langbein & Hornung 1988]. Other designs for longer-duration flights are being developed [Dreyer et al. 1998, de Lazzer et al. 1998a].

During materials processing under microgravity conditions, the wetting of the crucible by the liquid or melt used is of primary importance. Small changes in the contact angle often imply significant changes in the configuration of the liquid. There have been several attempts to produce finely dispersed mixtures of monotectic alloys under microgravity conditions. When a melt of such an alloy is cooled down into the miscibility gap, separation of the melt is very likely to start in corners and edges. First, the cooling generally has easy access to the corners and edges. Second, the free energy of nucleation of a melt that has a small contact angle is generally diminished in corners and edges. This applies in particular if there is complete wetting close to the critical point (or consolute point) [Cahn 1977, 1979], i.e. if one of the consolute melts wets the crucible so much better than the other melt that Young's boundary condition relating the interface tensions to the contact angle has no solution and the former melt spreads along the surface of the crucible anyway. However, with ongoing cooling and separation, the contact angle and thus the configuration of the liquid may change radically.

8.1.2 Liquids in Edges

By a "solid edge", we mean here either a wedge-shaped space (with dihedral angle $2\alpha < \pi$) (referrred to as a "wedge" below) or an edge in its more common sense (with dihedral angle $2\alpha > \pi$); see Fig. 8.1. An edge with dihedral angle $2\alpha = \pi$ is a fictitious straight line on a plane. Clearly, at such edges a family of cylindrical surfaces exists. It has been shown that these cylindrical surfaces represent the state with minimum energy if the sum of the dihedral angle 2α and the contact angles γ_1 and γ_2 with faces 1 and 2 is smaller than the sum π of the angles in a triangle [Concus & Finn 1974, Finn 1986], i.e.

$$2\alpha + \gamma_1 + \gamma_2 \leq \pi \,. \tag{8.1}$$

In Sect. 7.3, this situation is listed as case A. Equation (8.1) implies that the meniscus is concave, such that a capillary underpressure arises which favors penetration of the liquid into the wedge. On the ground and, even more so, under reduced gravity (at low Bond numbers), the liquid meniscus in such a wedge assumes a hyperbolic profile (Sect. 3.1).

On the other hand, it is hard to believe that the cylindrical surface shown in Fig. 8.2 might be stable. Experience tells us that it will break into drops. Whenever the sum of the dihedral angle 2α and the contact angles angles γ_1 and γ_2 with faces 1 and 2 exceeds a right angle,

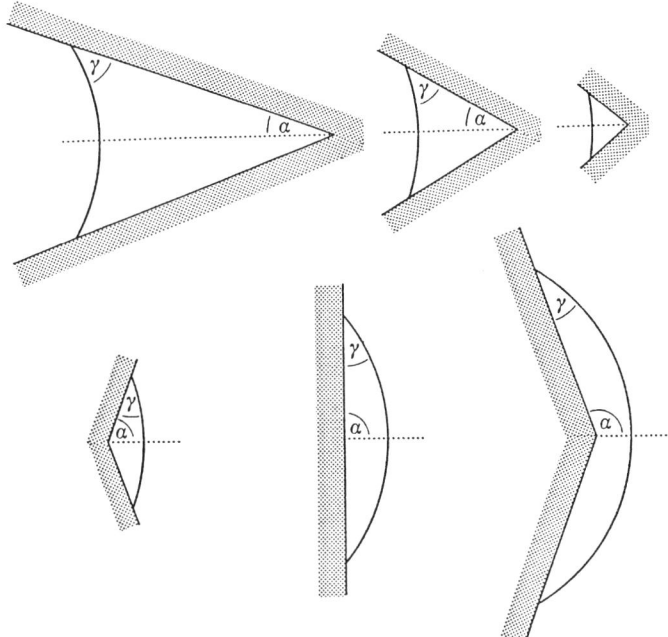

Fig. 8.1. Concave and convex liquid surfaces for $\alpha + \gamma \leq \pi/2$ and $\alpha + \gamma > \pi/2$, corresponding to negative and positive capillary pressure, at solid edges

$$2\alpha + \gamma_1 + \gamma_2 > \pi , \tag{8.2}$$

a convex meniscus causing a capillary overpressure arises. This corresponds to case B in Sect. 7.3. If the edge is sufficiently long, a liquid column on such an edge may lose energy by breaking into droplets. Stability criteria similar to those applying to free columns arise.

In the following sections we

- introduce an integral theorem about the pressure of fluid surfaces in cylindrical vessels
- study several of its consequences, in particular in rhombic and polygonal cylinders
- derive the stability criteria for convex surfaces in wedges
- report on a sounding-rocket experiment on the stability of fluid surfaces in rotating rhombic prisms.

182   8. Cylindrical Containers

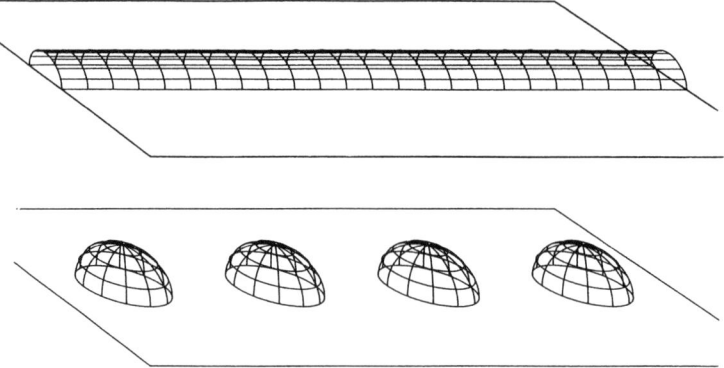

Fig. 8.2. Breakup of a cylindrical column with contact angle $\gamma = \pi/2$ on a plane (an edge with dihedral angle $2\alpha = \pi$)

8.2 The Integral Theorem for Cylindrical Vessels

8.2.1 Application of Divergence Theorem

Using cylindrical containers not only is convenient experimentally, but also offers great advantages regarding theory. We consider cylindrical vessels with a filling level sufficiently high for the end sections to be completely covered with liquid. This situation allows a general integral theorem; independent of whether the liquid penetrates into a wedge or does not, the curvature of the fluid surface may be calculated without explicitly calculating the surface shape. The proof of this principle has been given by de Lazzer et al. [1996, 1998b, de Lazzer 1998] on the basis of the following arguments:

- The capillary equation states that the curvature of the fluid surface equals the capillary pressure, or, more exactly, the ratio $p/\sigma \equiv \widehat{p}$ of the pressure to the surface tension.
- The curvature of the fluid surface is given by the divergence of its normal. (In the case of the capillary equation, the two-dimensional normal is considered. The same principle also applies to n-dimensional problems.)
- By applying the divergence theorem (Gauss theorem), the area integral of the curvature of the surface may be transformed into a boundary integral of the surface normal, or, more exactly, of the two-dimensional scalar product of the normals to the fluid surface and to the boundary.
- The scalar product of the normals equals the cosine of the contact angle.
- Hence, the cross-sectional integral of the pressure equals the boundary integral of the cosine of the contact angle.
- If the cross section and the boundary line are denoted by Ω and Σ, respectively (see Fig. 3.5), one has $\int_\Omega df\, \nabla . n = \int_\Sigma ds \cos\gamma$

- This theorem requires that the area of integration and the normal to the boundary are known; this is the case, for example, for surfaces completely covering the cross section of a cylindrical container.
- Liquid surfaces between two or more parallel cylindrical rods or between parallel solid stripes may be treated similarly.
- A quite important extension of this theorem can be applied to liquid surfaces that asymptotically approach a cylindrical shape in a wedge.
- In that case, the fluid surface itself becomes the boundary. The scalar product of the normals to the fluid surface and to the boundary equals one.
- This is true in particular for cylindrical surfaces arising in wetted wedges, which approach their final cylindrical shape exponentially; see Sect. 9.2.2. In the absence of external forces the boundary is a circle; in the presence of external forces normal to the wedge axis the boundary is the solution of an ordinary differential equation.
- On the other hand, it is not sufficient that the liquid surface has an axial slope just for reasons of symmetry, as for instance in the middle of a finite capillary. There must definitely be no curvature in the axial direction.
- It turns out that the exponential approach of cylindrical surfaces in wedges to their final shape renders possible an approximation along the above lines even for rather large liquid volumes.
- If the contact angle γ equals zero, the wetted and the nonwetted solid surfaces have the same energy. In that case, for statistical reasons, just one half of the total area should be wetted. This is confirmed by extrapolation of the formulae obtained for regular polygonal cylinders with N faces to the circular cylinder, $N \Rightarrow \infty$.

8.2.2 Minimization of Energy with Respect to Height

Let us present an alternative, more direct derivation of the integral theorem:

- The capillary equation is obtained by minimizing the energy of the liquid σA under the constraint of constant liquid volume V. The variation $\delta(\sigma A - pV)$ with respect to all parameters describing the capillary surface must vanish.
- The surface area A is given by the free liquid area A_l minus the wetted solid area $A_\text{s} \cos\gamma$, i.e. $A = A_\text{l} - A_\text{s}\cos\gamma$.
- In the particular case of a cylinder with arbitrary cross section, there exists one parameter the effect of which is known exactly, namely the absolute height h of the liquid surface. Since the fluid surface area A_l does not depend on h, we obtain

$$\frac{\mathrm{d}}{\mathrm{d}h}[(A_\text{l} - A_\text{s}\cos\gamma) + \widehat{p}V] = -\int_\Sigma \mathrm{d}s\cos\gamma + \widehat{p}\Omega = 0 \, . \tag{8.3}$$

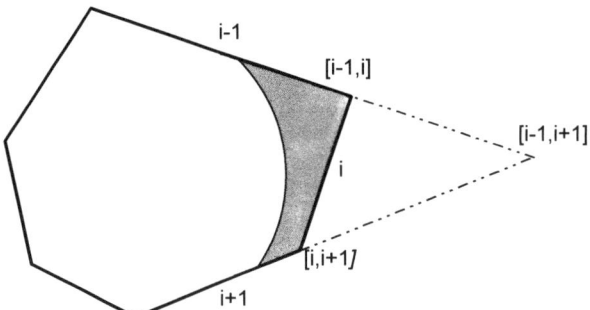

Fig. 8.3. Cylindrical vessel with polygonal cross section, faces 1 to N and wedges $[i, i+1]$. If one of the faces i is completely wetted, the liquid behaves as if faces $i-1$ and $i+1$ are neighboring faces, i.e. it completely ignores face i

Equation (8.3) allows the determination of the capillary pressure $\widehat{p}$ without explicitly solving the capillary equation. In cylindrical geometries, it is possible in a first step to drop the conservation of the liquid volume and to adjust the volume by means of the absolute height h afterwards.

Let us consider a cylindrical vessel with a polygonal cross section as depicted in Fig. 8.3. Face i has a contact angle γ_i with the liquid. If for any of the wedges $[ij]$ the dihedral angle and the contact angles satisfy the relation $2\alpha_{ij} + \gamma_i + \gamma_j < \pi$, the liquid will penetrate into that wedge up to infinity (or to the other end). Otherwise, only the ends of the vessel are covered by liquid. In a few symmetric cases the liquid surfaces are known to be spherical caps: this requires that an axis can be found which has a distance $R \cos \gamma_i$ from all faces i. Examples are vessels with triangular cross section, rhombic vessels with equal contact angles on all faces and, quite generally, regular polygons.

If there is no wetting of the wedges, application of the integral theorem to the geometry shown in Fig. 8.3 leads to

$$\widehat{p}\Omega = \sum_i L_i \cos \gamma_i . \tag{8.4}$$

On the other hand, if the wedge $[i, i+1]$ is wetted, i.e. if the wetting condition $2\alpha_{i,i+1} + \gamma_i + \gamma_{i+1} < \pi$ is satisfied, one has to subtract

- the wedge area Ω_1 covered by the liquid from the area integral on the left-hand side
- the wetted wedge section Σ_1 from the boundary line integral on the right-hand side.

In compensation, the length of the liquid arc Σ_2 with axial slope of the surface must be added, yielding $\cos \gamma = 1$ and

$$\widehat{p}\,(\Omega - \Omega_1) = \sum_i (L_i - \Sigma_1) \cos \gamma_i + \Sigma_2 . \tag{8.5}$$

If the filling ratio of the vessel is sufficiently high, another family of solutions of the capillary equation may arise, in which one of the faces i is fully wetted and thus does not behave as a solid face any more. The liquid instead "identifies" faces $i-1$ and $i+1$ as neighboring faces and completely wets this apparent wedge if $2\alpha_{i-1,i+1} + \gamma_{i-1} + \gamma_{i+1} < \pi$; see Fig. 8.3.

8.2.3 Evaluation of Wedge Contributions

Let us calculate the contributions $\Omega_1, \Sigma_1, \Sigma_2$ of the wetted wedges to the integral relation (8.5). Figure 8.4 shows a single wetted wedge, made up from faces 1 and 2 with dihedral angle 2α and contact angles γ_1, γ_2. The cylindrical liquid surface forms an arc of angle

$$2\delta = \pi - 2\alpha - \gamma_1 - \gamma_2 , \qquad (8.6)$$

yielding

$$\Sigma_2 = 2R\delta . \qquad (8.7)$$

For the angles α_1, α_2 which the line S from the central axis to the wedge axis forms with faces 1 and 2, one obtains

$$\alpha_1 + \alpha_2 = 2\alpha , \quad \sin\alpha_1 \cos\gamma_2 = \sin\alpha_2 \cos\gamma_1 , \qquad (8.8)$$

$$\tan\alpha_1 = \frac{\sin(2\alpha)\cos\gamma_1}{\cos(2\alpha)\cos\gamma_1 + \cos\gamma_2} , \quad \tan\alpha_2 = \frac{\sin(2\alpha)\cos\gamma_2}{\cos(2\alpha)\cos\gamma_2 + \cos\gamma_1} . \qquad (8.9)$$

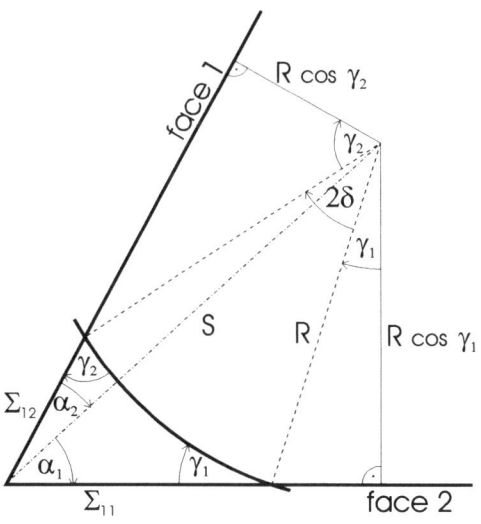

Fig. 8.4. Sketch of a circular cylindrical surface in a wedge made up from faces 1 and 2 with dihedral angle 2α and contact angles γ_1, γ_2

The distance S is given by

$$S = \frac{R}{\sin(2\alpha)} \sqrt{\cos^2 \gamma_1 + \cos^2 \gamma_2 + 2\cos(2\alpha) \cos \gamma_1 \cos \gamma_2} \ . \tag{8.10}$$

The wetted sections on faces $i = 1, 2$ are given by

$$\Sigma_{1i} = R(\cos \gamma_i \cot \alpha_i - \sin \gamma_i) = R \frac{\cos(\alpha_i + \gamma_i)}{\sin \alpha_i} \ , \tag{8.11}$$

$$\Sigma_{11} \cos \gamma_1 + \Sigma_{12} \cos \gamma_2$$
$$= 2R \frac{\sin \delta}{\sin(2\alpha)} \left[\sin(2\alpha + \delta) + \cos(\gamma_1 - \gamma_2) \sin \delta\right] \ . \tag{8.12}$$

Finally, for the wetted area, one finds

$$\Omega_1 = \frac{1}{2} RS \left(\Sigma_{11} \sin \alpha_1 + \Sigma_{12} \sin \alpha_2\right) - R^2 \delta \ , \tag{8.13}$$

$$\Omega_1 = R^2 \left(\frac{\sin \delta}{\sin(2\alpha)} \left[(\sin(2\alpha + \delta) + \cos(\gamma_1 - \gamma_2) \sin \delta\right] - \delta\right) \ . \tag{8.14}$$

Comparison of (8.14) with (8.7) and (8.12) leads to

$$\Omega_1 = \frac{1}{2} R(\Sigma_{11} \cos \gamma_1 + \Sigma_{12} \cos \gamma_2 - \Sigma_2) \ . \tag{8.15}$$

The area integral Ω_1 to be added to the left-hand side of (8.5) equals one half of the sum of the line integrals to be added on the right-hand side of (8.5). The line integrals are the derivatives of the area integral with respect to size. In effect, one may cancel the line integrals and add rather than subtract Ω_1 on the left-hand side. After multiplication by the "cylindrical" radius $R = 1/\widehat{p}$, we obtain

$$\widehat{p}^2 \Omega - \sum_i \widehat{p} L_i \cos \gamma_i + \sum_{[i,i+1]} \Omega_1 = 0 \ , \tag{8.16}$$

where the sum $[i, i + 1]$ runs over all wetted wedges. Equation (8.16) is a quadratic equation for the pressure $\widehat{p}$.

8.3 Examples

8.3.1 Ice Cream Cone

Let us first apply the above integral theorem to a cylindrical container composed of a circular arc with radius R and angle 2φ and a wedge with dihedral angle 2α, as shown in Fig. 8.5. Such a configuration has been termed an "ice cream cone" by Finn and Neel [1999]. Considering first the circular arc, we find for the cross-sectional area Ω_{arc} and the periphery Σ_{arc}

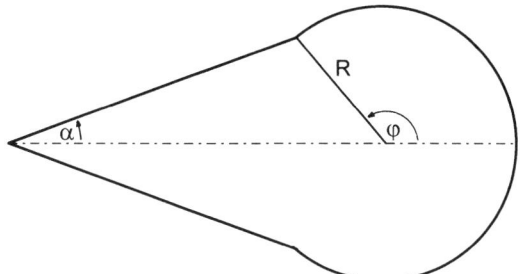

Fig. 8.5. The "ice cream cone": a cylindrical container composed of a circular arc with radius R and angle 2φ and a wedge with dihedral angle 2α

$$\Omega_{\text{arc}} = R^2(\varphi - \sin\varphi\cos\varphi), \quad \Sigma_{\text{arc}} = 2R\varphi. \tag{8.17}$$

If the arc is a full circle, i.e. for $\varphi = \pi$, one obtains from (8.3)

$$\widehat{p}R^2\pi = 2\pi R\cos\gamma, \quad \widehat{p} = \frac{2\cos\gamma}{R}. \tag{8.18}$$

This is the well-known capillary pressure in a circular tube. In the wedge region we obtain

$$\Omega_{\text{wdg}} = R^2\sin^2\varphi\cot\alpha, \quad \Sigma_{\text{wdg}} = 2R\frac{\sin\varphi}{\sin\alpha}. \tag{8.19}$$

If the wedge is not wetted, i.e. for $\delta = (1/2)\pi - \alpha - \gamma < 0$, substitution of (8.17) and (8.19) into (8.3) leads to

$$\widehat{p}R^2(\varphi - \sin\varphi\cos\varphi + \sin^2\varphi\cot\alpha) = 2R\cos\gamma\left(\varphi + \frac{\sin\varphi}{\sin\alpha}\right). \tag{8.20}$$

On the other hand, if the wedge is wetted, for $\delta = (1/2)\pi - \alpha - \gamma \geq 0$, the wetted area Ω_1 is given by

$$\widehat{p}^2\Omega_1 = \frac{\sin\delta}{\sin\alpha}\cos\gamma - \delta. \tag{8.21}$$

Substitution of (8.17), (8.19) and (8.21) into (8.16) leads to

$$(\widehat{p}R)^2(\varphi - \sin\varphi\cos\varphi + \sin^2\varphi\cot\alpha)$$
$$-2(\widehat{p}R)\cos\gamma\left(\varphi + \frac{\sin\varphi}{\sin\alpha}\right) + \left(\frac{\sin\delta}{\sin\alpha}\cos\gamma - \delta\right) = 0. \tag{8.22}$$

A typical result is shown in Fig. 8.6. The wetted region of the cone diminishes strongly with increasing dihedral angle 2α.

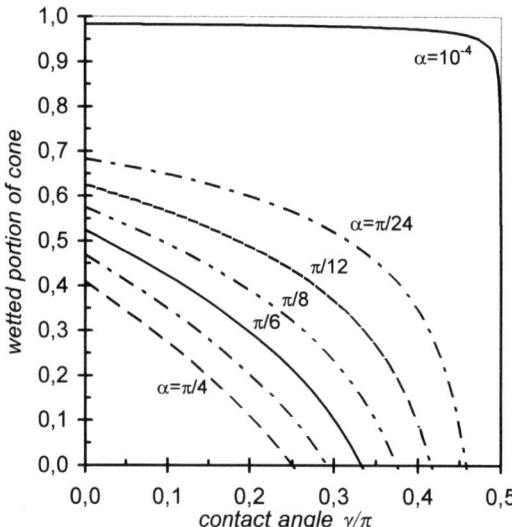

Fig. 8.6. The wetted portion of the wedge of the ice cream cone diminishes strongly with increasing dihedral angle 2α

8.3.2 Rhombic Cylinder

In the case of a *rhombic cylinder* (prism) with a dihedral angle 2α of the acute wedges and $\pi - 2\alpha$ of the obtuse wedges (as used in the Spacelab IML-2 experiment DYLCO and also in the corresponding MAXUS-2 experiment; see Sects. 8.5 and 11.3), we obtain from Fig. 8.7

$$\Omega = 2L^2 \sin\alpha \cos\alpha\,, \quad \Sigma = 4L\,, \tag{8.23}$$

$$\widehat{p}L \sin\alpha \cos\alpha = 2\cos\gamma\,. \tag{8.24}$$

If the acute wedge is wetted but the obtuse wedge is not wetted, i.e. if $\alpha + \gamma < \pi/2$ and $\alpha < \gamma$, (8.24) is replaced by

$$(\widehat{p}L)^2 \sin\alpha \cos\alpha - 2(\widehat{p}L)\cos\gamma + \left(\frac{\sin\delta}{\sin\alpha}\cos\gamma - \delta\right) = 0\,, \tag{8.25}$$

where $\delta = \pi/2 - \alpha - \gamma$, see Fig. 8.7b. If the obtuse wedge is also wetted, $\gamma > \alpha$, another term has to be added to the left-hand side of (8.25), yielding

$$\{\text{left-hand side of (8.25)}\} + \left(\frac{\sin(\alpha - \gamma)}{\cos\alpha}\cos\gamma - (\alpha - \gamma)\right) = 0\,. \tag{8.26}$$

Figure 8.8 shows the typical behavior of the pressure (curvature) versus the contact angle for $\alpha = \pi/4, \pi/6, \pi/12$. The side length L has been normalized to the constant cross-sectional area $\Omega = 1$, so that $L = (\sin 2\alpha)^{-0.5}$.

8.3 Examples   189

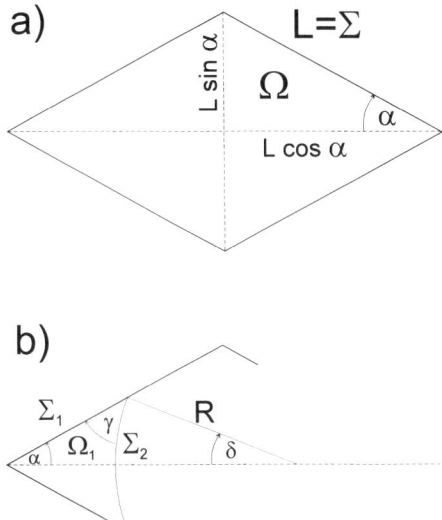

Fig. 8.7. (a) Cross section of a rhombic cylinder (prism) with dihedral angle 2α of the acute wedges and $\pi - 2\alpha$ of the obtuse wedges; (b) acute wedge filled

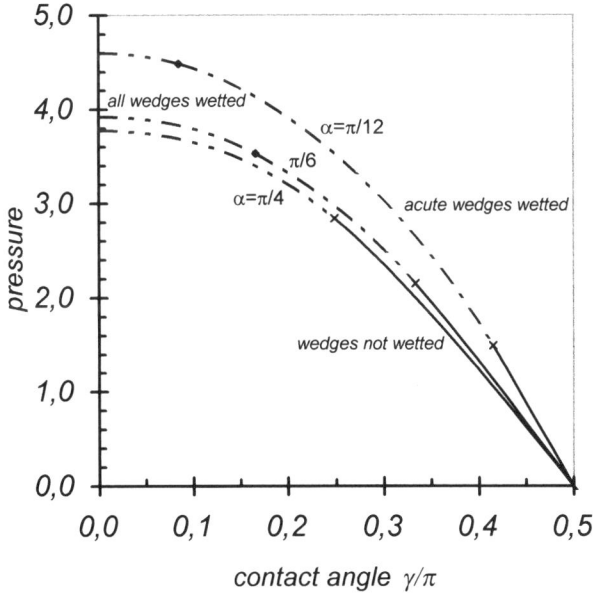

Fig. 8.8. The typical behavior of the pressure (curvature) versus the contact angle for $\alpha = \pi/4,\ \pi/6,\ \pi/12$. The side length L has been normalized to the constant cross-sectional area $\Omega = 1$, so that $L = (\sin 2\alpha)^{-0.5}$

8.3.3 Regular Polygon

Applying the integral theorem to a liquid surface in a cylinder whose cross section is a regular polygon with N sides and side length L, we obtain from Fig. 8.9

$$\alpha = \frac{\pi}{2} - \frac{\pi}{N}, \quad \delta = \frac{\pi}{2} - \alpha - \gamma = \frac{\pi}{N} - \gamma, \qquad (8.27)$$

$$\frac{\Omega}{N} = \left(\frac{L}{2}\right)^2 \tan \alpha, \quad \frac{\Sigma}{N} = L. \qquad (8.28)$$

If there is no wetting of the wedges, which is the case for $\delta < 0$, one has

$$\frac{1}{4}(\widehat{p}L) \tan \alpha = \cos \gamma. \qquad (8.29)$$

If the wedges are wetted, which is the case for $\delta > 0$ or $\pi/N > \gamma$, we find

$$\widehat{p}\frac{\Sigma_2}{N} = 2\delta, \quad \widehat{p}\frac{\Sigma_1}{N} = 2\frac{\sin \delta}{\sin \alpha}, \quad \widehat{p}^2\frac{\Omega_1}{N} = \left(\frac{\sin \delta \cos \widehat{p}\gamma}{\sin \alpha} - \delta\right), \qquad (8.30)$$

$$\frac{1}{4}(pL)^2 \tan \alpha - (\widehat{p}L) \cos \gamma + \left(\frac{\sin \delta \cos \gamma}{\sin \alpha} - \delta\right) = 0. \qquad (8.31)$$

Equation (8.31) leads to the following for the wetted portion Σ_1/Σ of the solid surface:

$$\frac{\Sigma_1}{\Sigma} = \left(\frac{2}{\widehat{p}L}\right)\frac{\sin \delta}{\sin \alpha}$$

$$= \frac{\sin \delta}{\cos \alpha \left(\cos \gamma + \sqrt{\cos^2 \gamma - \sin \delta \cos \gamma / \cos \alpha + \delta \tan \alpha}\right)}. \qquad (8.32)$$

Figure 8.10 shows the resulting surfaces, and Fig. 8.11 the wetted portion Σ_1/Σ of the solid surface. In the limit of the circular cylinder ($N \to \infty$) and

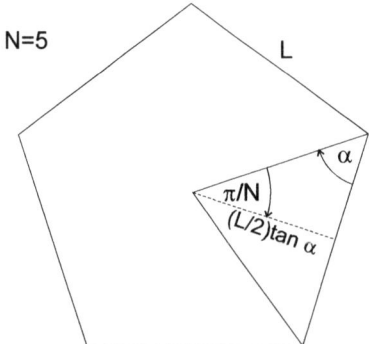

Fig. 8.9. A cylinder whose cross section is a regular polygon with N sides ($N = 5$ here) and side length L

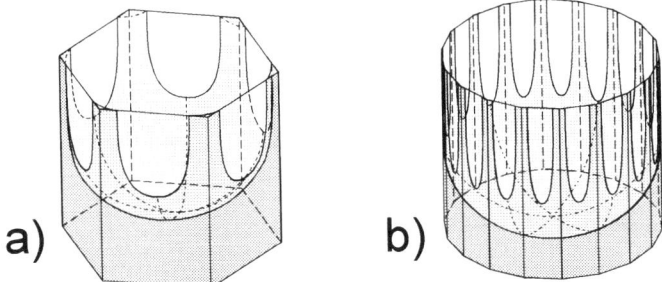

Fig. 8.10a,b. Fluid surfaces in a regular polygon with N sides

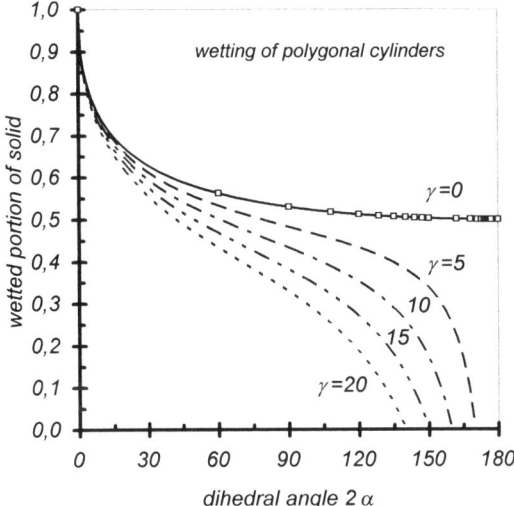

Fig. 8.11. The wetted portion Σ_1/Σ of the solid surface

for a contact angle $\gamma = 0$, we see that just one half of the upper, usually not wetted region of the circular cylinder is wetted. This unusual result is plausible energetically, since for a contact angle $\gamma = 0$ wetting of the upper region does not require energy. The difference betweeen the interface energies σ_{13} and σ_{23} of fluids 1 and 2 with the container material 3 vanishes, i.e. $\sigma_{13} = \sigma_{23}$.

In the limit of zero contact angle $\gamma = 0$, we are left with $\cos\alpha = \sin\delta$ and

$$\frac{\Sigma_1}{\Sigma} = \frac{\sin\delta}{\sin\delta + \sqrt{\delta}\sin\delta\cos\delta} = \frac{1}{1 + \sqrt{\delta}\cot\delta}, \tag{8.33}$$

and in the limit $N \to \infty$,

$$\frac{\Sigma_1}{\Sigma} = \frac{1}{1 + \sqrt{1 - \delta^2/3}} = \frac{1}{2} + \frac{\delta^2}{24}. \tag{8.34}$$

Complete wetting means, in contrast to the situation for a contact angle $\gamma = 0$, that the contact angle does not show hysteresis. A completely wetting phase creeps between the solid and any other phase. The circle is a polygon, too. No crucible is ideally round and free of roughness. In the case of complete wetting, a circular cylinder is wetted on just one half of its area.

8.3.4 Liquid in a Rotating Wedge

The strength of the integral theorem becomes fully obvious if a cylindrical vessel spinning around its symmetry axis is considered [de Lazzer et al. 1996, 1998, de Lazzer 1998]. Once more, it is not necessary to explicitly calculate the surface shape. The integral theorem is applicable whenever the forces act perpendicular to the symmetry axis. The energy of the liquid includes the potential of these forces, i.e. in (8.3) one has to replace pressure times cross-sectional area term by the integral of the pressure over the cross-sectional area.

Let us first consider concave liquid surfaces that arise for $\alpha + \gamma < \pi/2$. These are stable in the absence of rotation. If the center of the resting circular surface lies on the same side of the rotation axis as the surface itself,

- the distance from the axis of rotation decreases with increasing azimuth φ
- the negative curvature of the liquid surface increases with increasing azimuth φ
- constancy of γ requires that the negative curvature on the symmetry axis $\varphi = 0$ lessens
- the liquid surface bends outwards with increasing angular frequency, as depicted in Fig. 8.12a.

In the opposite case, if the center of the resting circular surface lies on the side opposite to the surface,

- the distance from the axis of rotation increases with increasing azimuth φ
- the negative curvature of the liquid surface is reduced with increasing azimuth φ
- constancy of γ requires that the negative curvature on the symmetry axis $\varphi = 0$ increases
- the liquid surface bends inwards with increasing angular frequency, as depicted in Fig. 8.12b.

The general trend is to approach a circular cylinder around the rotation axis and to leave satisfaction of the boundary condition on the contact angle to the regions near the boundary.

The main interest in rotation is related to the case $\alpha + \gamma > \pi/2$, when a liquid volume in a wedge is unstable at rest, but owing to rotation is pressed into and becomes stable in the wedge. The surface in such a case has positive curvature; it is convex at zero angular frequency,

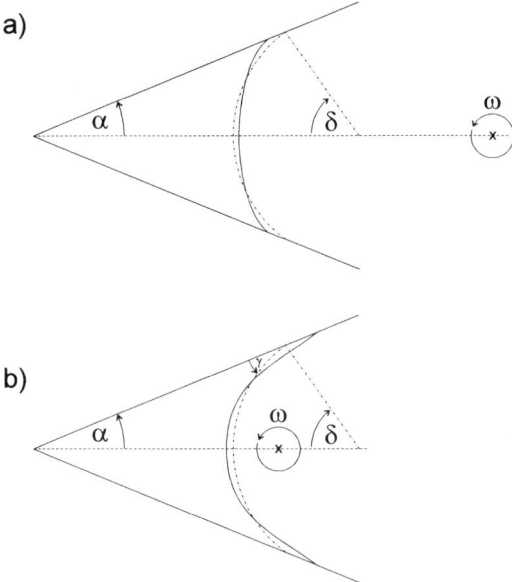

Fig. 8.12. (a) If the center of a resting stable circular surface is located on the same side of the rotation axis as the surface, the surface bends outwards with increasing rotation number. (b) If the center of the resting stable circular surface is located on the opposite side, the surface bends inwards with increasing rotation number

- its distance from the axis of rotation increases with increasing azimuth φ
- the positive curvature of the liquid surface becomes stronger with increasing azimuth φ
- constancy of γ requires that the positive curvature on the symmetry axis $\varphi = 0$ is lessened
- the liquid surface bends inwards with increasing angular frequency, as depicted in Fig. 8.13.

At high rotation numbers the curvature on the symmetry axis $\varphi = 0$ may even become negative; the surface approaches a circle around the axis of rotation and only bends outwards close to the faces in order to satisfy the contact angle boundary condition. In all these cases the precise shape of the surface may be obtained by a Runge–Kutta integration.

8.3.5 No Wetting of Wedge

It is inherent in the integral theorem that the geometrical parameters of the container and of the liquid volume in a wedge may be treated independently in a first step. For both, the container and the liqud volume, the integral of the pressure over the cross section and the projected lengths of the contact lines

8. Cylindrical Containers

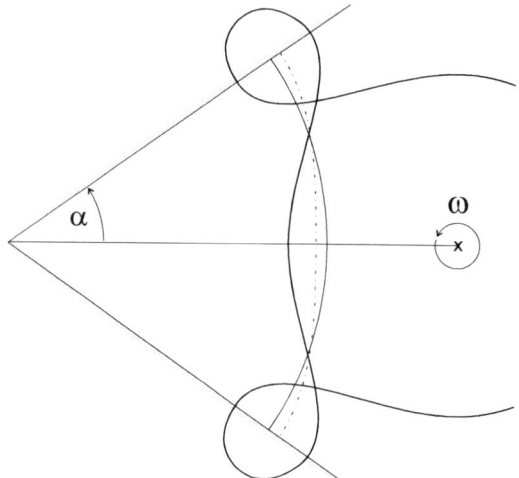

Fig. 8.13. Spinning up a liquid volume with $\delta = \alpha + \gamma - \pi/2 > 0$ in a wedge. The liquid surface bends inwards with increasing angular frequency

with the boundaries are required. Starting with the container (see Fig. 8.7), we have for the pressure (curvature)

$$\widehat{p} = \widehat{p}_0 + \widehat{p}_2 r^2, \quad \widehat{p}_0 = \frac{p}{\sigma}, \quad \widehat{p}_2 = \frac{\Delta \rho \omega^2}{2\sigma}. \tag{8.35}$$

The integral of the pressure (8.35) over the cross-sectional area leads to two contributions, the cross-sectional area itself

$$\Omega_0 = s^2 \tan \alpha, \tag{8.36}$$

where $s = L \cos \alpha$, and the moment of inertia

$$\Omega_2 = 2 \int_0^s \mathrm{d}x \int_0^{(s-x)\tan\alpha} \mathrm{d}y \, (x^2 + y^2) = \frac{s^4 \sin \alpha}{6 \cos^3 \alpha}. \tag{8.37}$$

The length of the projection of the contact line is given by

$$\Sigma = 2 \frac{s}{\cos \alpha}. \tag{8.38}$$

According to (8.3), this leads to

$$\widehat{p}_0 \Omega_0 + \widehat{p}_2 \Omega_2 = \Sigma \cos \gamma, \quad \widehat{p}_0 s + \widehat{p}_2 \frac{s^3}{6 \cos^2 \alpha} = 2 \frac{\cos \gamma}{\sin \alpha}. \tag{8.39}$$

In the case of no wetting of the wedges, a linear relation between $\widehat{p}_0$ and $\widehat{p}_2$ arises.

8.3.6 Liquid Volume Pressed into a Wedge

Wetting of the wedge reduces the contact line Σ by Σ_1 and creates a free fluid surface Σ_2. On the left-hand side of (8.39), one has to subtract the wetted volume and the corresponding angular momentum, and on the right-hand side one has to subtract the wetted surface $\Sigma_1 \cos\gamma$ and to add the free surface Σ_2. The following parameters are determined by the experimental conditions:

- the distance s of the wedge axis from the rotation axis
- the dihedral angle 2α of the wedge
- the filling ratio of the vessel, i.e. the wetted cross-sectional area
- the contact angle γ between the liquid and the wall
- the rotation number Rn, which is conveniently defined by means of the distance s of the wedge axis from the rotation axis: $Rn \equiv p_2 s^3 = \Delta\rho\omega^2 s^3/2\sigma$.

Adjustable parameters are

- the distance r_1 of the surface from the axis of rotation in the symmetry plane
- the capillary pressure $\widehat{p}_0$ at the rotation axis.

The capillary surfaces satisfying these conditions are conveniently obtained by means of a Runge–Kutta integration. Figure 8.14 shows the liquid volume (the wetted cross-sectional area) versus the pressure $\widehat{p}_0$ at the rotation axis for a half dihedral angle $\alpha = 40°$ and a contact angle $\gamma = 55°$. The parameter of the various curves is the rotation number Rn. The curve for zero rotation $Rn = 0$ comprises the well-known circular surfaces; their volume is proportional to $1/\widehat{p}_0^2$, where $\widehat{p}_0$ is positive. The curves corresponding to small positive rotation numbers $Rn > 0$ exhibit an inflection point. With increasing $\widehat{p}_2$ the inflection point reaches a slope of infinity, followed by the appearance of a maximum and a minimum of the pressure $\widehat{p}_0$ at the rotation axis. The stable fluid surfaces are those between these extremes. Outside that stable region, the surfaces break into droplets longitudinally. At the extremes of the pressure, $dp_0/dv = 0$, the wavenumber q of breakage equals zero; the wavelength becomes infinite. The maxima of the liquid volume as a function of the pressure $\widehat{p}_0$, $dv/dp_0 = 0$, give rise to a lateral instability: the liquid surface may spread further along the faces. This lateral instability, however, appears in regions where the longitudinal instability clearly prevails.

Small wedge volumes mean high capillary pressures, which are difficult to counteract by rotation. The minimum stable volume at constant rotation number Rn is, consequently, inversely proportional to Rn. The minimum rotation number Rn required for a stable liquid surface in a wedge with parameters α, γ arises when the inflection point has a vertical slope. At this singular point Rn varies in proportion to δ (see Figs. 8.7 and 8.13).

Numerically, one finds that all quantities entering the integral theorem, namely the pressure integral over the wetted area, the free fluid surface Σ_2

196   8. Cylindrical Containers

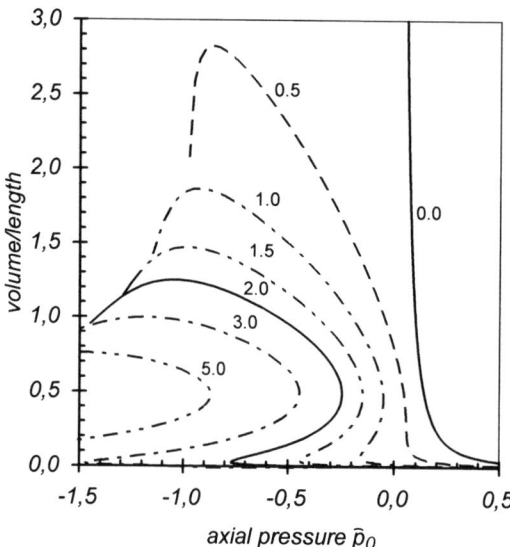

Fig. 8.14. The liquid volume (wetted cross-sectional area) versus the axial pressure $\widehat{p}_0$ for a half dihedral angle $\alpha = 40°$ and a contact angle $\gamma = 55°$

and the projection Σ_1 of the contact line, may be expanded convergently into power series in $(\delta/Rn)^{0.5}$:

$$\Sigma_1 = \sum_{n=1}^{\infty} c_{1,n} \left(\frac{\delta}{Rn}\right)^{n/2}, \quad \Sigma_2 = \sum_{n=1}^{\infty} c_{2,n} \left(\frac{\delta}{Rn}\right)^{n/2}, \tag{8.40}$$

$$\int_{\Omega_1} d\boldsymbol{r}\, (\widehat{p}_0 + \widehat{p}_2 r^2) = \delta \sum_{n=1}^{\infty} c_{3,n} \left(\frac{\delta}{Rn}\right)^{n/2}. \tag{8.41}$$

Small deviations from the stability limit $\delta = \alpha + \gamma - \pi/2 \geq 0$ require correspondingly low rotation numbers. In that case analytical approximations are possible, which lead to

$$\Sigma_1 = \frac{1 - (\delta/6)\tan\alpha}{\sqrt{2}\sin\alpha\cos\alpha} \left(\frac{\delta}{Rn}\right)^{0.5} + \frac{1 + 2\cos^2\alpha}{12\sin\alpha\cos^2\alpha} \left(\frac{\delta}{Rn}\right) + \ldots, \tag{8.42}$$

$$\Sigma_2 = \sqrt{\frac{\sin\alpha}{2\cos\alpha}} \left(1 - \frac{\delta}{6}\tan\alpha\right) \left(\frac{\delta}{Rn}\right)^{0.5} + \frac{1 + 2\cos^2\alpha}{12\cos^2\alpha} \left(\frac{\delta}{Rn}\right) + \ldots$$
$$\approx \Sigma_1 \sin\alpha,$$

$$\tag{8.43}$$

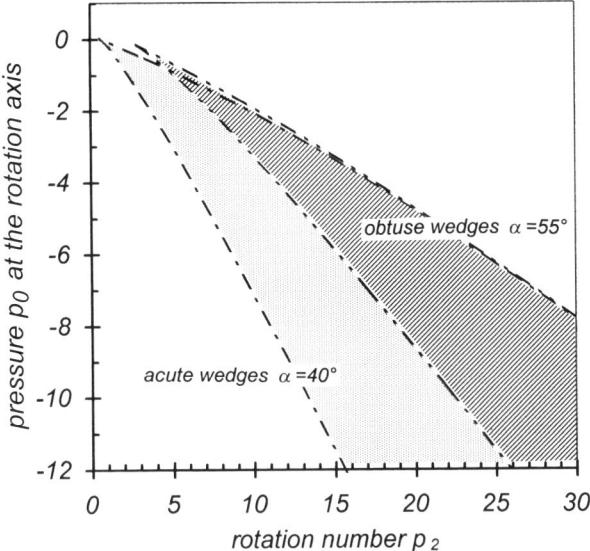

Fig. 8.15. The stable regions of the axial pressure of the liquid surface in rotating complementary wedges with $\alpha = 40°$ and $\alpha = 50°$ for a contact angle $\gamma = 55°$

$$\int_{\Omega_1} d\mathbf{r}\, (\widehat{p}_0 + \widehat{p}_2 r^2)$$
$$= \delta \left[\frac{4}{3} \sqrt{\frac{\cos\alpha}{2\sin\alpha}} \left(\frac{\delta}{Rn}\right)^{0.5} + \frac{1 + 2\cos^2\alpha}{8\sin\alpha\cos\alpha} \left(\frac{\delta}{Rn}\right) + \dots \right]. \tag{8.44}$$

Figure 8.15 shows the stable regions of the pressure $\widehat{p}_0$ of the liquid surface in rotating complementary wedges with $\alpha = 40°$ and $\alpha = 50°$ and contact angle $\gamma = 55°$. All surfaces outside these regions are unstable; they break in the longitudinal direction with wavenumber q (see Fig. 8.18). The surfaces in the obtuse wedge with $\alpha = 50°$ clearly are stable at higher rotation numbers only.

In Fig. 8.16, the straight line given by (8.39), i.e. corresponding to no wetting of the wedges, has been added. As soon as this curve intersects the region of stable surfaces in the acute wedges with $\alpha = 40°$, the corresponding terms given by (8.40) and (8.41) must be added to the integral theorem, and the $\widehat{p}_0, Rn$ curve bifurcates to the curve indicated by triangles. An equivalent situation happens a second time when the region of stable surfaces in the obtuse wedges with $\alpha = 50°$ is reached. The corresponding relation between $\widehat{p}_0$ and Rn is indicated by squares (slightly below the triangles). In both cases, wetting of the wedges slightly lowers the axial pressure $\widehat{p}_0$ at any given rotation number Rn.

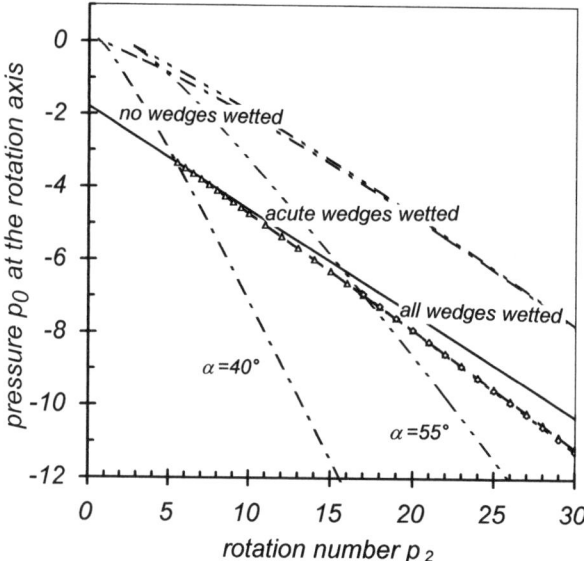

Fig. 8.16. Stability diagram of liquid surfaces in a rotating rhombic cylinder with dihedral angles $2\alpha = 80°$ and $100°$ and contact angle $\gamma = 55°$. At low rotation numbers only the cylinder ends are covered with liquid (up to about $\widehat{p}_2 = 5$). The $\widehat{p}_0, \widehat{p}_2$ curve then enters the region of stable surfaces in the acute wedges with $\alpha = 40°$, i.e. the corresponding terms have to be added to the integral theorem, and the $\widehat{p}_0, \widehat{p}_2$ curve bifurcates to the curve indicated by *triangles*. An equivalent situation happens a second time when the region of stable surfaces in the obtuse wedges with $\alpha = 50°$ is reached

Numerical analysis shows that the bifurcation between the cases of no wetting and wetting of the acute wedges (and of the obtuse wedges) arises slightly within the corresponding region of stable surfaces. There is indeed no argument which would suggest that the sum of the wedge contributions should vanish just at that crossing point. The contributions of the container and of the wetted regions to the quantities that appear integral theorem are basically independent. The axial pressure $\widehat{p}_0$ is only adjusted to the overall surface afterwards. If instead of a rhombus we had chosen an ice cream cone, the contribution of the container would have been different anyway.

The sum of the wedge contributions becomes small if $\delta = \alpha + \gamma - \pi/2$ only slightly exceeds the stability limit at rest, $\delta = 0$. In that case (8.42)–(8.44) apply. The largest term, the cross section integral, becomes proportional to $\delta^{1.5}$.

Real containers never have a length of infinity. In finite containers the region of stable surfaces is usually larger than in infinite containers, since the stability limit is based on perturbations with wavenumber $q = 0$, i.e.

wavelength infinity. In finite containers, only perturbations with a wavelength shorter than the dimensions of the container may arise.

8.4 Stability of Convex Cylindrical Surfaces

8.4.1 Longitudinal Normal Deformations

The stability limits shown in Fig. 8.15 correspond to the minimum and maximum wedge volumes for constant rotation number Rn when $\widehat{p}_0$ is variable (Fig. 8.14). When these volumes are exceeded or fallen short of, the liquid column in the wedge breaks in the longitudinal direction. The bifurcations depicted in Fig. 8.16 are due to the additional wetting of the obtuse wedges. In the following we shall study the wavelength of the normal longitudinal deformations $D\{1,0\}$.

The stability limit $L = 2\pi R$ of free cylindrical surfaces is conveniently obtained by embedding them into the family of unduloids. Liquid surfaces in wedges may be treated in similar manner [Langbein 1990]. In a first step, we identify a family of azimuthally modified unduloids. The family of regular unduloids does not contain solutions which would fit a solid edge with constant contact angle γ along the contact line. This condition requires nonaxisymmetric deformations of the liquid surface instead. Therefore, in Sect. 8.4.2 we construct a family of such nonaxisymmetric capillary surfaces. We consider small deviations d from the cylindrical shape and expand the liquid surface into a Fourier series with wavenumber q in the axial direction. The coefficients of this expansion become Fourier series with wavenumber s in the azimuthal direction.

The position of the z axis of the deformed surfaces under consideration does not need to be fixed before their adaptation to a solid edge. This is done only in Sect. 8.4.3, by requiring that the z axis coincides with the axis of the cylindrical solutions that have the same liquid volume per length. Any other choice of the axis would considerably impede the calculations. Subsequently, the equation for the contact line with the solid edge is derived, and constancy of the contact angle is required. This condition relates the wavenumber of the surfaces considered to the contact angle γ and the dihedral angle 2α of the edge.

The liquid volume per unit length is calculated in Sect. 8.4.4. It turns out not to depend on the azimuthal deformation to first order in the waviness parameter d if the liquid column considered has infinite length or extends over an integer number of wavelengths $L = 2\pi/q$. This means that the minimum-volume condition is satisfied, i.e. a stability limit has been reached. The really important limit is that corresponding to a single wavelength. A longer column is unstable anyway.

8.4.2 Axially Periodic Meniscus Shapes

Using cylindrical coordinates r, φ, z, with $r(\varphi, z)$ being the dependent variable, one obtains for the surface area A and the volume V enclosed by the liquid surface (see Sect. 3.7.3)

$$A = \int dz \int r\, d\varphi \sqrt{1 + \left(\frac{\partial r}{r\, \partial \varphi}\right)^2 + \left(\frac{\partial r}{\partial z}\right)^2}\,, \quad V = \frac{1}{2} \int dz \int r\, d\varphi\, r\,. \tag{8.45}$$

The capillary equation is given by (3.64). Being interested in small deviations from the cylindrical shape, we restrict ourselves to terms bilinear in $\partial r / r\partial \varphi$, $\partial r / \partial z$, etc. The capillary equation thus reduces to

$$\widehat{p} = \frac{1}{r}\left[1 + \frac{1}{2}\left(\frac{\partial r}{r\, \partial \varphi}\right)^2 - \frac{1}{2}\left(\frac{\partial r}{\partial z}\right)^2\right] - \frac{\partial^2 r}{(r\, \partial\varphi)^2} - \frac{\partial^2 r}{\partial z^2}\,. \tag{8.46}$$

Breakage of convex surfaces is due to axially periodic deformations, which may be expanded into a Fourier series with axial wavenumber q:

$$r(\varphi, z) = \sum_{m=-\infty}^{+\infty} r_m(\varphi) \exp(imqz)\,. \tag{8.47}$$

Owing to the structure of (8.46), $r_m(\varphi)$ is found to be of order m in the deviation d from the cylindrical shape. In zero order, $m = 0$, we obtain

$$r_0(\varphi) \equiv R = 1/\widehat{p}\,. \tag{8.48}$$

In the following all lengths are nondimensionalized by means of $R = 1/\widehat{p}$. The first-order terms with $m = \pm 1$ in (8.47) yield

$$-\frac{d^2 r_{\pm 1}(\varphi)}{d\varphi^2} = (1 - q^2) r_{\pm 1}(\varphi)\,. \tag{8.49}$$

Equation (8.49) can be solved as

$$r_{\pm 1}(\varphi) = d_{\pm 1} \cos(s\varphi)\,, \tag{8.50}$$

where

$$q^2 + s^2 = 1\,. \tag{8.51}$$

In (8.50), the sine term has been omitted, since we are interested in even functions of φ only. Odd functions do not fit the edges considered. In order for the deviation from the cylindrical shape to be real and small, we require

$$d_{+1} = d_{-1} \equiv d \ll 1\,. \tag{8.52}$$

8.4 Stability of Convex Cylindrical Surfaces

By substituting (8.48) and (8.50) into the capillary equation (8.46) and by Fourier decomposition with respect to z, we obtain, for the terms of order $m = \pm 2$ in the waviness d,

$$-\frac{\partial^2}{\partial \varphi^2}[r_0(\varphi) - 1]$$
$$= [r_0(\varphi) - 1] + \frac{1}{2}d^2\left[(3 - 4q^2)\cos(2s\varphi) + (1 - 2q^2)\right], \tag{8.53}$$

$$-\frac{\partial^2}{\partial \varphi^2} r_{\pm 2}(\varphi)$$
$$= (1 - 4q^2)r_{\pm 2}(\varphi) + \frac{1}{4}d^2\left[3(1 - 2q^2)\cos(2s\varphi) + (1 - 4q^2)\right]. \tag{8.54}$$

The terms $\cos(2s\varphi)$ in the azimuthal direction arise from products of first-order terms. The inhomogeneous solutions of (8.53) and (8.54) are given by

$$r_0(\varphi) - 1 = \frac{1}{2}d^2\left[\cos(2s\varphi) + (1 - 2s^2)\right], \tag{8.55}$$

$$r_{\pm 2}(\varphi) = \frac{1}{4}d^2\left[(1 - 2s^2)\cos(2s\varphi) + 1\right]. \tag{8.56}$$

The homogeneous solutions of (8.53) and (8.54) are not considered further here, since they represent

- a displacement of the column normal to its axis, which is fixed only in the next section
- an axial deformation of the column with wavenumber $2q$, which if this is the principal deformation, is included in the present investigations by replacing $2q$ by q.

By repeated substitution of the lower-order terms into the capillary equation and evaluation of all products and square roots, one obtains for $r(\varphi, z)$ the double Fourier series

$$r(\varphi, z) = \sum_{k=0}^{\infty} \sum_{l=0}^{k/2} \sum_{m=0}^{k/2} C(k, l, m) d^k \cos\left[(k - 2l)s\varphi\right] \cos\left[(k - 2m)qz\right]. \tag{8.57}$$

8.4.3 Adjustment to Fit Solid Edges

In order for the surfaces under consideration to fit a solid edge, the contact angle γ must be constant along the contact line. Let the distance of the z axis from the planes forming the edge be h; see Fig. 8.17. Then we obtain the condition for the contact line

$$r\sin(\varphi - \alpha) = h = \sin\left(\gamma - \frac{\pi}{2}\right). \tag{8.58}$$

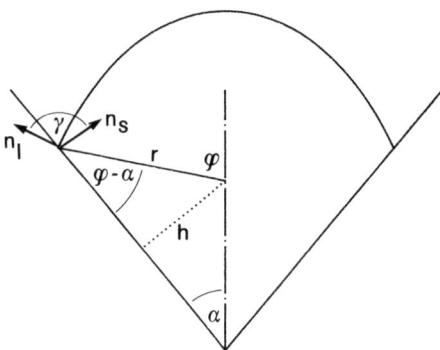

Fig. 8.17. Sketch of the calculation of the contact line and the contact angle

Substituting (8.50) into (8.58) and using

$$\varphi(z) = \delta = \alpha + \gamma - \frac{\pi}{2} \tag{8.59}$$

to zeroth order, we find, to first order in the waviness d,

$$\varphi(z) = \delta - 2d \tan\left(\gamma - \frac{\pi}{2}\right) \cos(s\delta) \cos(qz) \,. \tag{8.60}$$

The vector normal to one of the planes forming the solid edge equals

$$\mathbf{n}_s = (-\sin(\varphi - \alpha), -\cos(\varphi - \alpha), 0) \,. \tag{8.61}$$

The vector normal to the liquid meniscus is given by

$$\mathbf{n}_l = \frac{(1, -\partial r/r\,\partial\varphi, -\partial r/\partial z)}{\sqrt{1 + (\partial r/r\,\partial\varphi)^2 + (\partial r/\partial z)^2}} \,. \tag{8.62}$$

Hence, we obtain for the cosine of the contact angle

$$\mathbf{n}_s \cdot \mathbf{n}_l = \cos\gamma = \frac{-\sin(\varphi - \alpha) + \cos(\varphi - \alpha)(\partial r/r\,\partial\varphi)}{\sqrt{1 + (\partial r/r\partial\varphi)^2 + (\partial r/\partial z)^2}} \,. \tag{8.63}$$

Substituting (8.60) to obtain the contact line, we are left with

$$\tan\left(\gamma - \frac{\pi}{2}\right) = s\tan(s\delta) + O(d^2) \,. \tag{8.64}$$

A periodic deformation with constant contact angle γ along the contact line exists if (8.64) is satisfied. The second-order contributions to the meniscus shape given by (8.56) do not affect this condition, whereas the third-order terms cause correction terms of order d^2. The azimuthal wavenumber s is limited by the condition $s^2 \leq 1$. It becomes imaginary for large axial wavenumbers q, so that the tangent on the right-hand side of (8.64) turns into a hyperbolic tangent.

8.4.4 Volume and Energy

In Sect. 4.4.1 we showed that axisymmetric liquid surfaces become unstable if their length equals their period. In order to show that the same principle applies to the nonaxisymmetric surfaces considered here, we have to prove that they preserve volume and energy. For the liquid volume, we obtain from Fig. 8.17

$$V = \int dz \left(\int_0^{\varphi(z)} d\varphi\, r^2 + hr\cos[\varphi(z) - \alpha] + h^2 \cot\alpha \right). \tag{8.65}$$

Here $\varphi(z)$ is the azimuth along the contact line as given by (8.60). The integration over φ in (8.65) yields, inclusive of terms of second order with $m = \pm 2$ in the waviness parameter d,

$$V = \int dz \Big\{ \Phi + 4d\frac{\sin(s\delta)}{s}\cos(qz) + 2d^2(1-s^2)\left[\delta + \frac{\sin(2s\delta)}{2s}\right]$$

$$+ 4d^2\left[s\tan(s\delta) + \cot\gamma\right]\cos^2(s\delta)\cos^2(qz) \Big\}, \tag{8.66}$$

where

$$\Phi = \delta - \frac{\cos\gamma}{\sin\alpha}\sin\delta \tag{8.67}$$

is the cross-sectional area of the cylindrical liquid volume that fits into the edge under consideration. The terms of first order in the waviness parameter d do not change the liquid volume if the column has infinite length or extends over an integer number of wavelengths $L = 2\pi/q$. In that case we are left with

$$V = L\left\{ \Phi + 2d^2(1-s^2)\left[\delta + \frac{\sin(2s\delta)}{2s}\right] + 2d^2[s\tan(s\delta) + \cot\gamma]\cos^2(s\delta) \right\}. \tag{8.68}$$

For the liquid volume not to change to second order in the waviness parameter d, the basic radius r_0 has to be adjusted accordingly, i.e. (8.48) has to be replaced by

$$\frac{r_0}{R} = 1 - \frac{d^2}{\Phi}\left\{ 1 - s^2\left[\delta + \frac{\sin(2s\delta)}{2s}\right] + [s\tan(s\delta) + \cot\gamma]\cos^2(s\delta) \right\}. \tag{8.69}$$

Taking into account (8.64) with a constant contact angle, we find that the basic radius r_0 generally decreases and, equivalently, that the pressure $\widehat{p}$ increases with increasing waviness d.

The conservation of the liquid volume by the nonaxisymmetric periodic deformations considered means that this branch of solutions bifurcates with

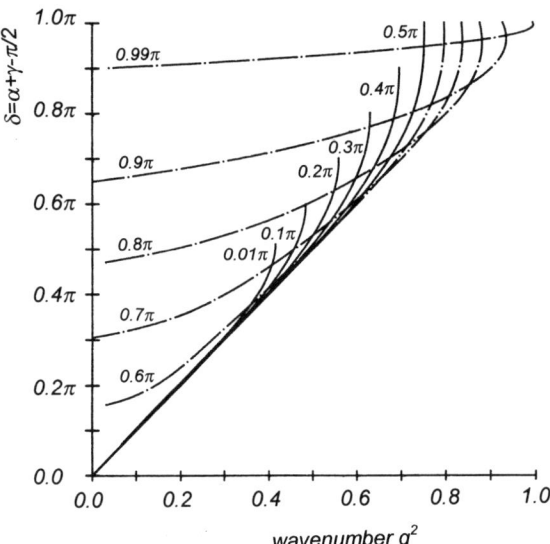

Fig. 8.18. Squared wavenumber q^2 of breakage of liquid columns with cross-sectional area $\Phi = \pi$ in a solid wedge versus $\delta = \alpha + \gamma - \pi/2$. The parameter of the various curves is half the dihedral angle of the wedge, α

the branch of cylindrical surfaces in the wedge. The bifurcation point is determined by (8.64) and (8.51).

When the cross-sectional area Φ is normalized to the volume π of a free cylindrical column with radius 1, i.e. $\Phi = \pi$, Fig. 8.18 is obtained for the squared wavenumber q^2 of breakage versus $\delta = \alpha + \gamma - \pi/2$. The parameter of the various curves is half the dihedral angle, α. The largest wavenumber $q = 1$, i.e. the shortest stable length $L/2\pi R = 1$, arises in the case of a free cylindrical column into which a virtual half-plane has been introduced up to the axis. This case is covered by the present investigations if we take $\alpha = \pi, \gamma = \pi/2$. At edges with $\alpha > \pi/2$ and small contact angles $\gamma < \alpha - \pi/2$, the cylindrical surfaces are actually intersected by the faces forming the edge. This however does not affect the analytical calculations presented if the vertex of the edge is truncated accordingly.

From Fig. 8.18, it is obvious that there is a lower limit for the stable length that depends on $\delta = \alpha + \gamma - \pi/2$ only:

$$L/2\pi R = \sqrt{\pi/\delta} \ . \tag{8.70}$$

The effective surface area of the liquid volume is made up by the free liquid surface A_l and the contact area A_s with the solid edge times $\cos\gamma$. One obtains

$$A = A_1 - A_s \cos\gamma ,\tag{8.71}$$

$$A = 2\int dz \left(\int_0^{\varphi(z)} r\, d\varphi \sqrt{1 + \left(\frac{\partial r}{r\,\partial\varphi}\right)^2 + \left(\frac{\partial r}{\partial z}\right)^2} \right.$$
$$\left. - \sin\gamma\{r\cos[\varphi(z) - \alpha] + h\cot\alpha\} \right)\tag{8.72}$$

and, after performing the required integrations over φ and z in a similar manner to that for the volume V,

$$A = 2L\left\{\Phi - d^2\left[s\tan(s\delta) + \cot\gamma\right]\cos^2(s\delta)\right\} .\tag{8.73}$$

The expression in the brackets in (8.73) corresponds to (8.64). The effective surface area is not changed to second order in the waviness parameter d if the condition (8.64) for constancy of the contact angle along the contact line is satisfied. Equation (8.64) is the stability condition. More generally, the effective surface energy of the cylindrical surface is diminished by any deformation the axial wavenumber of which exceeds that given by (8.64), even if the deformation does not satisfy the condition of constant contact angle. Deformations with wavelengths longer and shorter than that corresponding to (8.64) are stable and unstable, respectively.

8.4.5 Rotating Wedges

When the linear stability analysis presented above is applied to rotating wedges, the rotational terms have to be taken into account additionally in (8.50)–(8.64). This lengthens the calculations without changing the reasoning.

Figure 8.19 shows the result of a linear stability analysis of liquid surfaces in an infinite wedge with dihedral angle $2\alpha = 80°$. The squared longitudinal wavenumber q^2 of breakage is plotted versus the rotation number Rn for contact angles $\gamma = 20°$ to $160°$ in steps of $10°$. It is striking that, over a wide range of frequencies, q^2 depends linearly on Rn. This is analogous to the linear dependence of the stability of a free liquid column on the rotation number according to (5.2) and (5.3). A negative squared wavenumber means an imaginary wavenumber and a stable surface.

8.5 The MAXUS Experiment DYLCO

The stability of liquid surfaces in rotating vessels has been studied experimentally in the MAXUS-2 experiment DYLCO (launched on 24 November 96 in Kiruna, Sweden). The MAXUS rocket provides twelve minutes of microgravity. The acronym DYLCO stands for *Dy*namics of *L*iquids in Edges and *Co*rners.

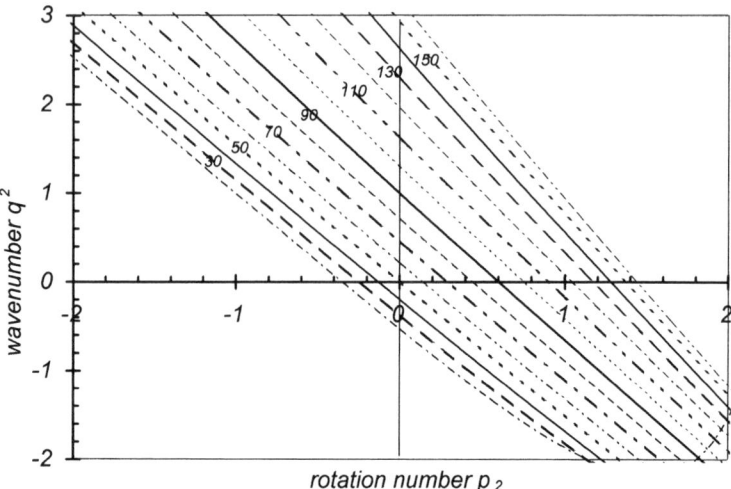

Fig. 8.19. Squared longitudinal wavenumber q^2 of breakage of liquid surfaces in a wedge with dihedral angle $2\alpha = 80°$ as a function of the rotation number Rn for contact angles $\gamma = 20°$ to $160°$ in steps of $10°$. Negative values of q^2 mean stable surfaces. The data were obtained from linear stability analysis

The aim of the MAXUS experiment was the controlled observation of the stability and instability of fluid surfaces in solid wedges [de Lazzer et al. 1998b, de Lazzer 1998]. In the wedges of rhombic prisms, stable surfaces were generated by rotation. Then their breakage in the wedges or else their retreat to the cylinder ends was studied by reducing the rotation rate stepwise. The experimental sequence for the MAXUS flight shown in Fig. 8.20 was defined on the basis of tests during preceding parabolic flights. The rotation rates were chosen so as to achieve wetting of the wedges in a reasonable time and to resolve the onset of unstable surface deformation as precisely as possible by stepwise spinning down. The liquids used were

- Cargille 50350, with contact angle $\gamma \approx 67°$ versus FC-724
- Dow Corning silicone fluid DC 200, 10 cSt, $\gamma \approx 60°$ versus FC-724
- distilled water, $\gamma \approx 60°$ versus PMMA.

The dihedral angles 2α of the acute wedges of the prisms were $75°$, $80°$, $85°$ and $90°$; see Fig. 8.21. Together with three different filling ratios of 10, 20 and 30 volume percent, this meant a total of twelve vessels. In order to obtain contact angles $\gamma > \pi/2 - \alpha$, the PMMA vessels were coated with the surface modifier FC-724 for the experiments with the Cargille and Dow Corning fluids. The vessels were grouped together in two rotation units, each containing six vessels. These units were rotated in opposite directions in order to minimize the overall angular momentum. One of the groups comprised the $75°|105°$ (i.e. with dihedral angles of $75°$ and $105°$) and $90°|90°$ vessels, the other the $80°|100°$ and $85°|95°$ vessels. The six vessels were imaged side by

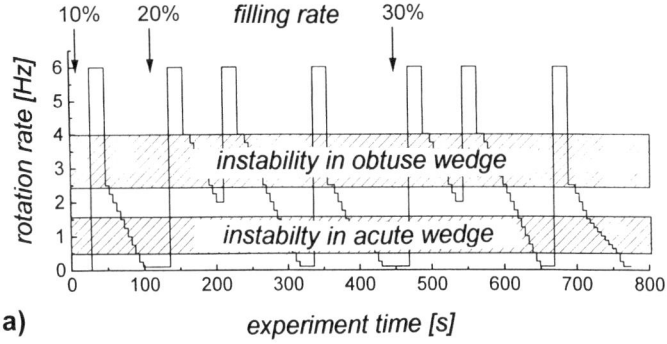

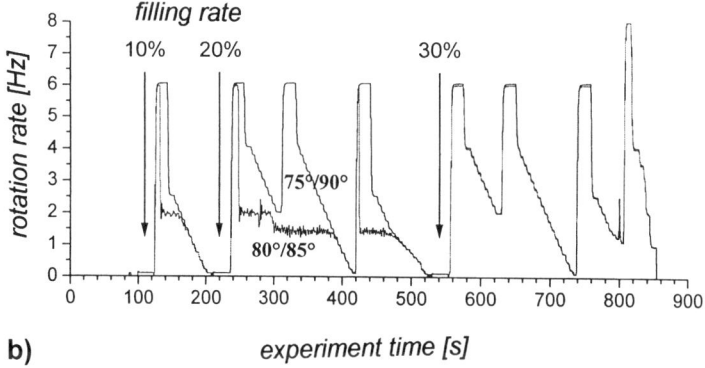

Fig. 8.20. The experimental sequence for the MAXUS flight, defined on the basis of results of preceding parabolic flights. The rotation rates were chosen so as to achieve wetting of the wedges in a reasonable time and to resolve the onset of unstable surface deformation as precisely as possible by stepwise spinning down. (**a**) planned; (**b**) flown

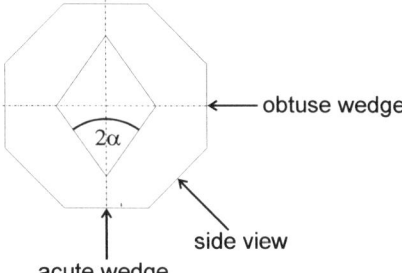

Fig. 8.21. The dihedral angles 2α of the acute wedges of the prisms were $75°$, $80°$, $85°$ and $90°$

side by a CCD camera, such that a single frame showed the vessels in the order Dow Corning|water|Cargille||Cargille|water|Dow Corning.

Changing the filling ratio of the vessels does not directly affect the cross-sectional area in the wetted wedges. The latter is given by the overall pressure obtained from the integral theorem. However, the filling ratio affects the length of the wetted region and thus may enable or prevent breakage, depending on the theoretical wavelength. One has to bear in mind also that the surfaces show an exponential transition from the region covering the vessel ends to the cylindrical regions in the wedges, see Sect. 9.2. The transition region favors retreat of the liquid volume to the vessel ends during or after spinning down because of the symmetric normal deformation $D\{2,0\}$. With low filling ratios there is the risk that a dry spot may appear at the vessel ends. In that case the theory presented in the preceding sections does not apply.

In general, the theoretical investigations suggest the following configurations during spinning up:

- at low rotation rate, the liquid wets the bottom and/or the top of the prisms
- increase of the rotation rate forces the liquid into the acute wedges
- further increase of the rotation rate forces the liquid into the obtuse wedges
- eventually, all wedges and the vessel ends will be wetted.

At high rotation rates, dry spots at the vessel ends may arise.

The behavior expected and found during spinning down clearly shows the existence of at least four concurrent instabilities:

- There is the Rayleigh instability $D\{1,0\}$ of a liquid surface in a wedge of infinite length. A convex liquid surface breaks if its length exceeds its period. The wavelength is proportional to the square root of the wetted cross-sectional area.
- There is the retreat of the liquid to the cylinder ends due to the symmetric normal deformation $D\{2,0\}$. At rest, the stable surface covers the cross section only.
- Retreat of liquid to the vessel ends can dynamically induce breakage of the surface distant from the ends. The flow of liquid along the wedge may just not be fast enough to follow the deformation. Although the normal deformation $D\{3,0\}$ does not initiate this behavior, that is the outcome.
- Likewise, if the liquid flow is not able to follow the symmetric deformation $D\{2,0\}$, a single drop is left in the middle of the wedge. This may be ascribed to $D\{4,0\}$.

The periodic deformation $D\{1,0\}$ requires a low filling ratio, whereas retreat to the ends requires a high filling ratio. As a rule, one wavelength is sufficient for the periodic deformation $D\{1,0\}$ to show up.

The most unambiguous results were obtained on Cargille 50350 fluid. They confirmed quantitatively the calculated stability limits and the transition from the Rayleigh breakage $D\{1,0\}$ to the retreat $D\{2,0\}$ of the liquid to the ends. The experiments on water showed little accuracy and reproducibility, owing to the strong contact angle hysteresis of that liquid. Examples of results obtained with Cargille fluid are depicted in Figs. 8.22 and 8.23, which

Table 8.1. Comparison of experimental and theoretical rotational frequencies for breakage of Cargille 50350 fluid in the 80°|100° and the 85°|95° prisms. $D\{2,0\}$ indicates breakage according to the symmetric normal deformation, $D\{1,0\}$ indicates breakage according to the Rayleigh instability

	Cargille 50350, 80° prisms							
	Acute wedges, 80°				Obtuse wedges, 100°			
	Theory		Experiment		Theory		Experiment	
Filling ratio (%)	Type	f (Hz)	Type	f (Hz)	Type	f (Hz)	Type	f (Hz)
10	$D\{2,0\}$	1.9	$D\{2,0\}$	1.9	–	–	–	–
20	$D\{2,0\}$	1.9	$D\{2,0\}$	1.9	$D\{1,0\}$	1.9	$D\{1,0\}$	1.9
20	$D\{2,0\}$	1.4	$D\{2,0\}$	1.4	$D\{1,0\}$	1.4	$D\{1,0\}$	1.4
30	$D\{2,0\}$	2.5	–	–	$D\{2,0\}$	≤ 4	$D\{2,0\}$	3.5
30	$D\{2,0\}$	2.5	$D\{2,0\}$	2.5	$D\{2,0\}$	≤ 4	$D\{2,0\}$	3.75
30	$D\{2,0\}$	2.5	$D\{2,0\}$	2.5	$D\{2,0\}$	2.5	$D\{2,0\}$	2.5

	Cargille 50350, 85° prisms							
	Acute wedges, 80°				Obtuse wedges, 100°			
	Theory		Experiment		Theory		Experiment	
Filling ratio (%)	Type	f (Hz)	Type	f (Hz)	Type	f (Hz)	Type	f (Hz)
10	$D\{1,0\}$	1.9	$D\{1,0\}$	1.9	$D\{1,0\}$	1.9	$D\{1,0\}$	1,9
20	$D\{2,0\}$	1.9	$D\{2,0\}$	1.9	$D\{2,0\}$	1.9	$D\{2,0\}$	1.9
					$D\{1,0\}$	1.9	$D\{1,0\}$	1.9
20	$D\{2,0\}$	1.4	$D\{2,0\}$	1.4	$D\{*,0\}$	1.4	$D\{2,0\}$	1.4
30	$D\{2,0\}$	3.0	$D\{2,0\}$	2.5	$D\{2,0\}$	≤ 3.75	$D\{2,0\}$	3.5
30	$D\{2,0\}$	3.0	$D\{2,0\}$	2.5	$D\{2,0\}$	≤ 3.75	$D\{2,0\}$	3.75
30	$D\{2,0\}$	2.5	$D\{2,0\}$	2.5	$D\{2,0\}$	2.5	$D\{2,0\}$	2.5

show breakage of surfaces in the 80°|100° and 85°|95° prisms at 10% and 20% filling ratios.

Observation was performed through a face of the vessels, as indicated by "side view" in Fig. 8.21. The acute wedges are located on the lelft-hand side at the front (and on the right-hand side at the rear); the obtuse wedges are located on the right-hand side at the front (and on the left-hand side at the rear). Immediately after reduction of the rotation frequency, the deformation $D\{2,0\}$ is observed near the lower end of the obtuse wedges. A little later, the same deformation appears near the lower end of the acute wedges. Whereas the cylindrical surfaces in the acute wedges are basically stable, the surfaces

210   8. Cylindrical Containers

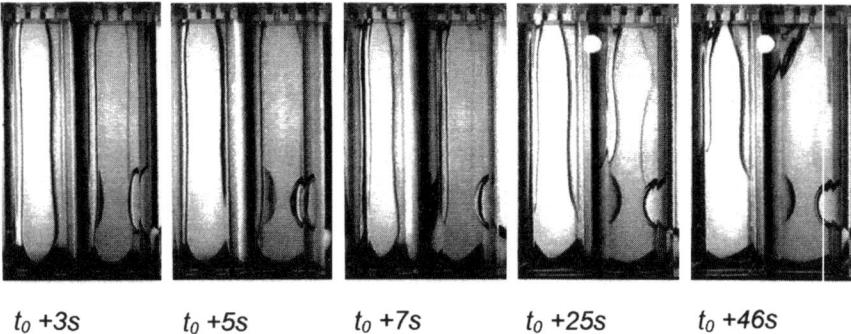

t_0 +3s t_0 +5s t_0 +7s t_0 +25s t_0 +46s

Fig. 8.22. Breakage of surfaces of Cargille fluid in the 80°|100° and 85°|95° prisms at 10% filling ratio after spinning down. Observation was performed through a face of the vessels, as indicated by "side view" in Fig. 8.21. The acute wedges are located on the left-hand side at the front (and on the right-hand side at the rear); the obtuse wedges are located on the right-hand side at the front (and on the left-hand side at the rear)

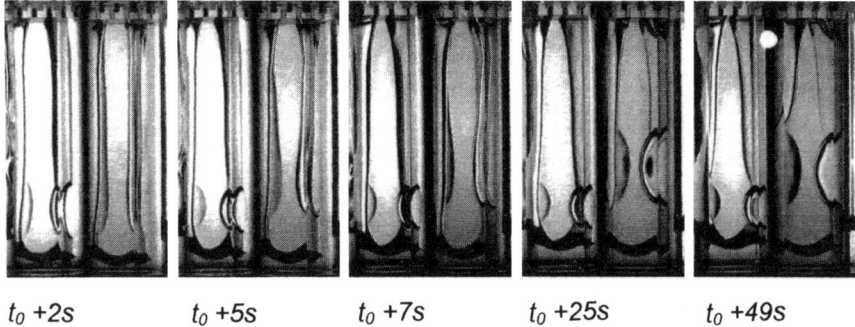

t_0 +2s t_0 +5s t_0 +7s t_0 +25s t_0 +49s

Fig. 8.23. Breakage of surfaces of Cargille fluid in the 80°|100° and 85°|95° prisms at 20% filling ratio after spinning down

in the obtuse wedges of the 85°|95° prisms break according to $D\{1,0\}$. The obtuse wedges of the 80°|100° prism have been wetted only halfway. This leads to immediate formation of a drop during spinning down. The upper ends of the prisms are not wetted in the sequences presented (the increased dynamic contact angle has prevented wetting). The $D\{2,0\}$ instability, therefore, cannot arise; the cylindrical surface in the wedge is stabilized.

Figure 8.22 shows that during the initial rotation, the acute wedges were wetted completely, whereas the wetting of the obtuse wedges remained incomplete owing to the low filling ratio and the short time of rotation. It is obvious that the surfaces in the acute wedges, in spite of having reached the top, do not wet the upper vessel ends. It is also worth noting that the surface in the obtuse wedge of the 85° prism, in spite of having moved only halfway up the wedge, does not simply retreat but breaks, forming a drop.

The observed breakage of the surfaces agrees well with the theoretical predictions. In Table 8.1, the experimental and theoretical stability limits in the 80°|100° and 85°|95° prisms are compared. The theoretical values given refer to the surface configuration at the end of the interval of fast rotation. The accuracy of the angular frequencies is limited by the preprogrammed steps of spinning down.

The influence of the vessel ends becomes negligible only for low filling ratio and fast spinning down into the instability region. In all other cases, the symmetric instability $D\{2,0\}$ dominates breakage. In several instances $D\{1,0\}$ is not observed, because the length of the wedge falls short of the wavelength.

References

1. Bauer HF: Natural damped frequencies of an infinitely long column of immiscible viscous liquids. Zeitschrift Angewandte Mathematik und Mechanik **64** (1984) 475–490
2. Bauer HF: Free surface and interface oscillations of an infinitely long liquid column. Acta Astronautica **13** (1986) 9–22
3. Cahn JW: Critical point wetting. J. Chem. Phys. **66** (1977) 3667–3672
4. Cahn JW: Monotectic composite growth. Metall. Trans. **10A** (1979) 119–121
5. Concus P, Finn R: On capillary free surfaces in the absence of gravity. Acta Math. **132** (1974) 77–198
6. de Lazzer A, Langbein D, Dreyer M, Rath HJ: Mean curvature of liquid surfaces in cylindrical containers of arbitrary cross-section. Microgravity Sci. Technol. **9** (1996) 208–219
7. de Lazzer A, Dreyer ME, Rath HJ: Capillary effects under low gravity. Part II: considerations on equilibrium capillary surfaces. Space Forum **3** (1998a) 137–163
8. de Lazzer A: Zum Verhalten kapillarer Flüssigkeitsgrenzflächen in Ecken und Kanten. Dissertation, University of Bremen (1998) 1–120. dto: Progress Reports VDI (*V*erein *D*eutscher *I*ngenieure), Series 7, flow technology, No. 345
9. de Lazzer A, Langbein D, Dreyer ME, Rath HJ: Stabilität von Grenzflächen in rotierenden rhombischen Prismen. Bilanzsymposium Forschung unter Weltraumbedingungen. M.H. Keller, P.R. Sahm (eds.). Norderney, 21–23 September 1998b, 42–49
10. Dreyer ME, Gerstmann J, Stange M, Rosendahl U, Wölk G, Rath HJ: Capillary effects under low gravity. Part I. surface settling, capillary rise and critical velocities. Space Forum **3** (1998) 87–136
11. Finn R: *Equilibrium Capillary Surfaces*, Grundlehren der Mathematischen Wissenschaften, Vol. 284. Springer, Berlin, Heidelberg (1986) 1–244
12. Finn R, Neel RW: Singular solutions of the capillary problem. J. Reine Angew. Math. **512** (1999) 1–25
13. Langbein D: The shape and stability of liquid menisci at solid edges. J. Fluid Mech. **213** (1990) 251–265
14. Langbein D, Hornung U: Liquid menisci in polyhedral containers. *Proceedings of the Workshop on Differential Geometry, Calculus of Variations and Computer Graphics*, MSRI Book Series. Springer, Berlin, Heidelberg (1988)

9. Liquid Surfaces in Polyhedral Containers

Corners formed by three or more solid surfaces offer better wall contact to liquids than do wedges ("wedge" is used here in the same sense as in Chap. 8, i.e. a wedge-shaped space between solid surfaces). Corners lower the free energy of wetting liquids more than do wedges. A liquid volume in a tripod forms a spherical surface, i.e. no liquid penetrates into the wedges, if for each of the three wedges forming the tripod the sum of the dihedral angle and the contact angles with the adjacent faces exceeds π. The volume and surface area of such a drop may be calculated by subtracting the three segments outside the faces from the spherical cap given by the intersection of the sphere with the wedges.

If the sum of contact angles falls short of π, the liquid drop is sucked into the wedges owing to the resulting capillary underpressure. Nevertheless, a surplus volume is left in the corner. Far from the corner, the liquid surface assumes a cylindrical shape. The corner volume extends into the wedges exponentially. The surface of a liquid in a polyhedron may thus be pasted together from cylindrical surfaces in the wedges and surplus volumes piled up in the corners. The corresponding similarity relation is based on decomposing the liquid volume into portions associated with the corners, the wedges and their intersection at the corners. Although the specific corner volume requires numerical simulation, this approach has the advantage that this volume has to be determined only once for a given geometry.

9.1 Spherical Surfaces at Edges and Corners

9.1.1 Nonwetting Drops

A liquid volume completely wets a wedge if the sum of the dihedral angle 2α of the wedge and the contact angles γ_1 and γ_2 of the liquid with the solid faces 1 and 2 is smaller than π, the sum of the angles in a triangle; see Sect. 7.3 and (8.1). The concave shape gives rise to a capillary underpressure, which sucks the liquid into the wedge. If several wedges are interconnected, the same capillary underpressure, i.e. the same curvature $\widehat{p}$, occurs in all wedges.

214   9. Liquid Surfaces in Polyhedral Containers

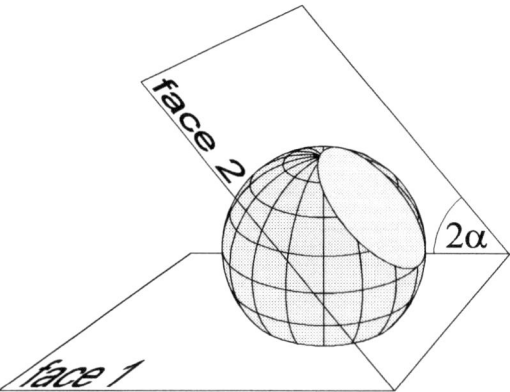

Fig. 9.1. Nonwetting spherical drop distant from a wedge

In the opposite case, if the sum of half the dihedral angle α and the contact angle γ exceeds the sum of the angles in a triangle, (8.2), there exist cylindrical solutions of the capillary equation as well. These surfaces, however, are convex, which means a capillary overpressure. The corresponding instabilities are treated in Chap. 8. The wavenumber q of breakage is given by (8.51) and (8.64). Breakage results in the formation of spherical drops in the wedge.

Eventually, with nonwetting liquids it may happen that the spherical drop does not wet the wedge any more, and instead forms a spherical bridge distant from the wedge. This situation arises for

$$2\alpha + (\pi - \gamma_1) + (\pi - \gamma_2) < \pi \,, \quad 2\alpha < \gamma_1 + \gamma_2 - \pi \,; \tag{9.1}$$

see Fig. 9.1 and also Sect. 6.1.8. The distance of the drop's center from the wedge vertex is given by (8.10). The case of a spherical drop sitting in a wedge and of a spherical bridge distant from the wedge both allow calculation of the volume, surface area and curvature of the liquid by elementary integrations.

A drop in a tripod forms a spherical surface, i.e. no liquid penetrates into the wedges, if for each of the three wedges $i = 1, 2, 3$ forming the tripod the sum of the dihedral angle $2\alpha_i$ and the contact angles γ_j, γ_k with the adjacent faces j, k exceeds π. The volume and surface area of such drops may be calculated by subtracting the three segments outside the faces from the spherical cap given by the intersection of the sphere with the wedges.

9.1.2 Drops in Planar Wedges

A liquid sphere of radius R forms a contact angle γ with a solid plane if the distance of the sphere's center from the plane equals $R \cos \gamma$. If several planes $1, 2, \ldots, N$ with contact angles $\gamma_1, \gamma_2, \ldots, \gamma_N$ are present, we obtain a solution of the capillary equation inclusive of the boundary

9.1 Spherical Surfaces at Edges and Corners   215

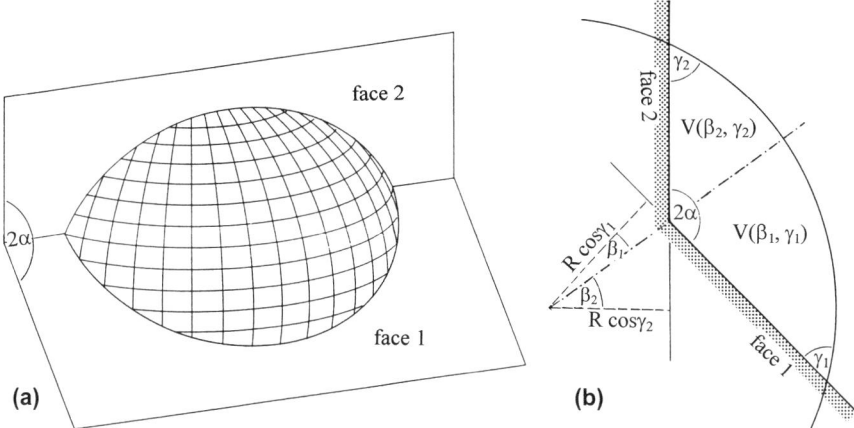

Fig. 9.2a,b. A spherical liquid drop in a wedge, which has a contact angle γ_1 with face 1 and a contact angle γ_2 with face 2. The volume and surface area of the drop are conveniently obtained by dissecting it along the plane through its center and the edge

conditions on the contact angles if the sphere of radius R has distances $R\cos\gamma_1, R\cos\gamma_2, \ldots, R\cos\gamma_N$ from the respective planes; see Fig. 3.6.

By dissecting the system into several parts, we can use this principle to calculate the volume and surface area of [Langbein 1994, 1995a, 1995b]

- a liquid drop in a straight wedge
- a liquid drop in a wedge formed by touching spheres
- a liquid drop in a polygonal cylinder
- a liquid drop in a tripod
- a liquid drop in a regular N-pod.

Let us consider a spherical drop in a wedge, which has contact angles γ_1 and γ_2 with faces 1 and 2, respectively (see Fig. 9.2a). In order to find the volume of the spherical drop, we can dissect it into volumes $V(\beta_1, \gamma_1)$ and $V(\beta_2, \gamma_2)$ by means of the plane through the sphere's center and the edge; see Fig. 9.2b. By elementary integrations, we find the following results for the volume $V(\beta_i, \gamma_i)$ and the surface area $A(\beta_i, \gamma_i)$ of the two sections $i = 1, 2$:

$$V(\beta, \gamma) = R^3 \left[\frac{2}{3} \arccos\left(\frac{\sin\beta}{\sin\gamma}\right) - \left(\cos\gamma - \frac{1}{3}\cos^3\gamma\right) \arccos\left(\frac{\tan\beta}{\tan\gamma}\right) \right.$$
$$\left. + \frac{1}{3}\cos^3\gamma \tan\beta \sqrt{\tan^2\gamma - \tan^2\beta} \right], \tag{9.2}$$

$$A(\beta, \gamma) = 2R^2 \left[\arccos\left(\frac{\sin\beta}{\sin\gamma}\right) - \cos\gamma \arccos\left(\frac{\tan\beta}{\tan\gamma}\right) \right]. \tag{9.3}$$

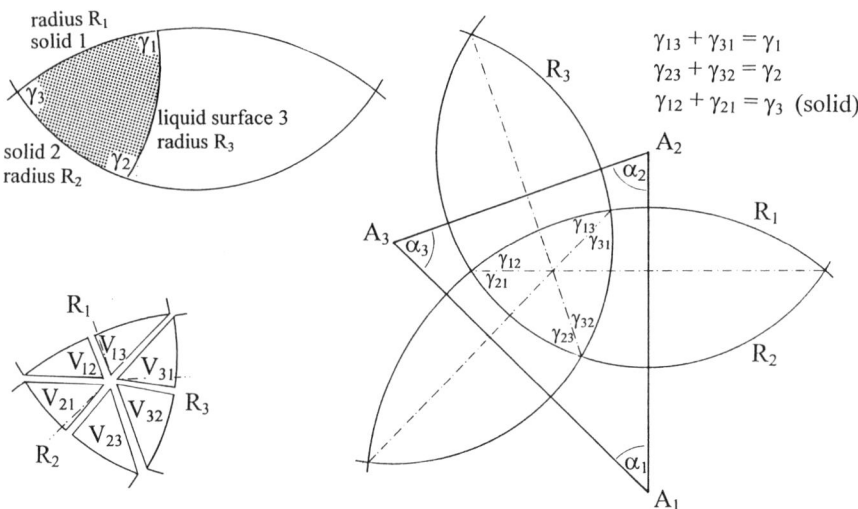

Fig. 9.3. A spherical liquid drop in a spherical wedge. The volume and surface area of the drop are obtained by using six auxiliary cuts. First, cutting along the intersections of spheres 1, 2 and 3, one obtains three drops in three plane wedges. Second, each drop is further cut by a plane through its center and the corresponding wedge

The angles β_1, β_2 between the faces and the auxiliary cut through the sphere's center and the edge are given by

$$\cos\beta_1 \cos\gamma_2 = \cos\beta_2 \cos\gamma_1 , \tag{9.4}$$
$$2\alpha + \beta_1 + \beta_2 = \pi . \tag{9.5}$$

Any inconvenience resulting from the fact that the arcos functions in (9.4), (9.5) are multivalued can be avoided by using the addition theorem

$$\arcos\left(\frac{\sin\beta_1}{\sin\gamma_1}\right) + \arcos\left(\frac{\sin\beta_2}{\sin\gamma_2}\right) = \arcos\left(\frac{\cos\gamma_1 \cos\gamma_2 + \cos 2\alpha}{\sin\gamma_1 \sin\gamma_2}\right). \tag{9.6}$$

9.1.3 Drops in Spherical Wedges

Turning to liquid drops in a spherical wedge (a space between two spherical surfaces; see Fig. 9.3), we can conveniently apply six auxiliary cuts. Let the radii of the solid surfaces 1 and 2 be R_1 and R_2 and the distance between the centers of the spheres be a_{12}. We denote the radius of the drop's surface by R_3 and the distances of its center from the centers of spheres 1 and 2 by a_{13} and a_{23}, respectively.

First we cut the drop into three segments with the planes given by the intersections of spheres 1 and 2, 2 and 3, and 3 and 1. This leaves us with three drops in three auxiliary planar wedges, in which the total liquid volume and the sums of the contact angles

$$\gamma_{13} + \gamma_{31} = \gamma_1, \quad \gamma_{23} + \gamma_{32} = \gamma_2, \quad (\gamma_{12} + \gamma_{21} = \gamma_3) \tag{9.7}$$

rather than the contact angles themselves are given. Second, we cut each of the three drops, in a manner similar to that used for the drop in the planar wedge, with the plane through the center of the spherical surface and the auxiliary edge. R_1, R_2 and a_{12} reflect the properties of the solid wedge. R_3, a_{13} and a_{23} refer to the drop's surface, and have to be determined iteratively from the contact angles γ_1 and γ_2.

9.1.4 Liquid Drops in a Tripod

Corners formed by solid surfaces offer an even better wall contact to liquid volumes than do wedges. In the case of wetting liquids, this further lowers their free energy. There exists a region of contact angles and dihedral angles where the cylindrical liquid surfaces in wedges are convex while the corresponding spherical surfaces in corners are still concave.

Starting with a liquid drop in a corner formed by three faces, we apply arguments similar to those used in Sect. 9.1.2 and the same equations. The drop again has a spherical shape. Let the contact angles of the liquid drop with the three faces 1, 2 and 3 be γ_1, γ_2 and γ_3, respectively. The center of this sphere is at a distance $R \cos \gamma_1$ from face 1, $R \cos \gamma_2$ from face 2 and $R \cos \gamma_3$ from face 3.

Let the normals to the three faces forming the corner be $\boldsymbol{n}_1, \boldsymbol{n}_2, \boldsymbol{n}_3$. We find the following for the position $\boldsymbol{a}$ of the center of the liquid sphere (see Fig. 9.4):

$$\boldsymbol{a} = R \frac{(\boldsymbol{n}_2 \times \boldsymbol{n}_3) \cos \gamma_1 + (\boldsymbol{n}_3 \times \boldsymbol{n}_1) \cos \gamma_2 + (\boldsymbol{n}_1 \times \boldsymbol{n}_2) \cos \gamma_3}{\boldsymbol{n}_1 \cdot (\boldsymbol{n}_2 \times \boldsymbol{n}_3)}. \tag{9.8}$$

The condition for the spherical drop to satisfy the boundary conditions is that the sphere fully occupies the corner, i.e. that it is intersected by all three edges or that the distance of the edges from the sphere's center $\boldsymbol{a}$ is smaller than R. This is equivalent to

$$2\alpha_i + \gamma_j + \gamma_k > \pi. \tag{9.9}$$

Here $2\alpha_i$ is the dihedral angle of edge i, which is given by the intersection of faces j, k, i.e. we reobtain the general nonwetting condition for wedges, (8.2) (i, j, k denote cyclic permutations of $1, 2, 3$). If condition (9.9) is not satisfied, the liquid cannot stay in the corner, but penetrates into the wedges.

9.1.5 Regular N-Pods

Let us consider in detail a regular corner made up from N edges. Figure 9.5 depicts a regular tripod with $N = 3$.

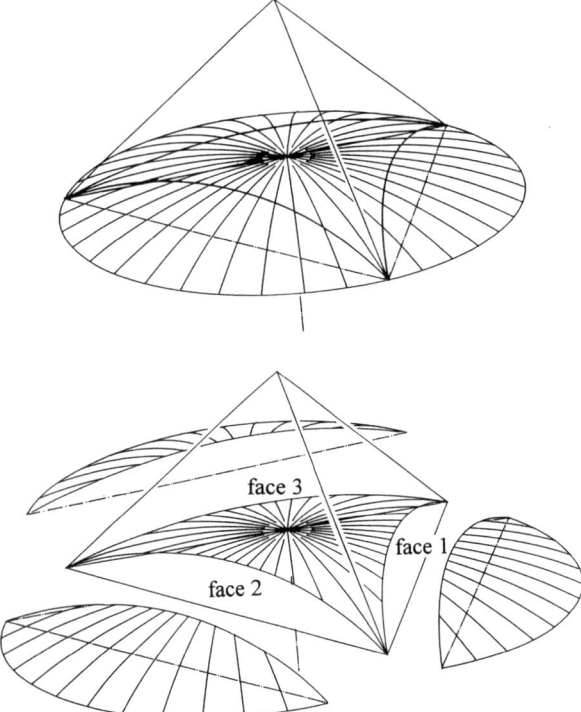

Fig. 9.4. A spherical liquid drop in a corner. The contact angles of the drop with the three faces forming the corner are $\gamma_1, \gamma_2, \gamma_3$. The volume and surface area of the drop can be calculated by subtracting the sections outside the faces from the spherical cap given by the drop's intersection with the edges

The angles of importance are the following:
- ϑ_1 is the polar angle between the space diagonal and a face diagonal
- ϑ_2 is the polar angle between the space diagonal and an edge
- ζ is the angle between a face diagonal and an edge
- α is half the dihedral angle between the faces.

These angles are interrelated by

$$\sin \zeta = \sin \vartheta_2 \sin(\pi/N) \,, \tag{9.10}$$
$$\cos \vartheta_2 = \cos \zeta \cos \vartheta_1 \,, \tag{9.11}$$
$$\sin \vartheta_1 = \sin \vartheta_2 \sin \alpha \,, \tag{9.12}$$
$$\cos(\pi/N) = \cos \zeta \sin \alpha \,. \tag{9.13}$$

Figure 9.6 shows a liquid sphere of radius $R = 1$, which has a distance $\cos \gamma_1$ from all three faces, i.e. it represents a liquid with contact angle γ_1. The distance of the sphere's center from the corner therefore equals

$$l = R \frac{\cos \gamma_1}{\sin \vartheta_1} \,. \tag{9.14}$$

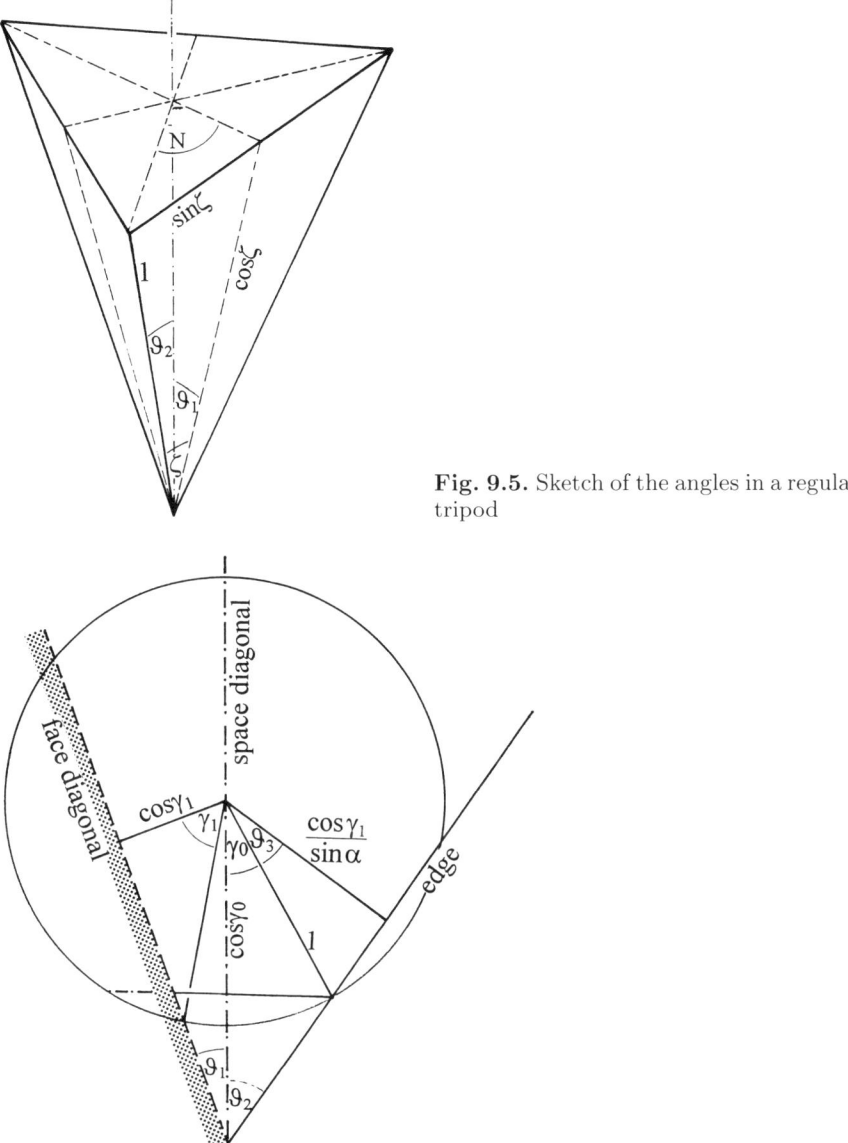

Fig. 9.5. Sketch of the angles in a regular tripod

Fig. 9.6. A liquid drop of radius $R = 1$, which has a contact angle γ_1 with the faces of a regular N-pod. The intersections with the face diagonal (*left-hand side*) and with the edge (*right-hand side*) lie in the same plane for N odd and in different planes for N even

We are interested in the liquid volume within the corner. This volume is obtained in several steps. From the volume of the N-sided pyramid given by the sphere's intersection with the edges, we subtract the volume of the spherical cap at the corner side of this intersection diminished by its N sections outside the faces; see Fig. 9.4. The plane through the intersections of the edges with the sphere has a distance $\cos\gamma_0$ from the sphere's center; γ_0 is given by

$$\cos\gamma_0 = \sin\vartheta_2 \frac{\cos\gamma_1}{\sin\alpha} + \cos\vartheta_2 \sqrt{1 - \left(\frac{\cos\gamma_1}{\sin\alpha}\right)^2}, \qquad (9.15)$$

$$\gamma_0 = \frac{\pi}{2} - \vartheta_2 - \vartheta_3, \quad \vartheta_3 = \arccos\left(\frac{\cos\varphi_1}{\sin\alpha}\right). \qquad (9.16)$$

The volume and area of the spherical cap are given by the elementary relations

$$V = \pi R^3 \left(\frac{2}{3} - \cos\gamma_0 + \frac{1}{3}\cos^3\gamma_0\right), \qquad (9.17)$$

$$A = 2\pi R^2 (1 - \cos\gamma_0). \qquad (9.18)$$

The liquid volumes cut off by the N faces can be calculated by means of (9.2)–(9.6). The relevant dihedral angle 2α is the angle between the faces and the plane through the sphere's intersection with the edges.

Finally, we have to determine the liquid volume in the corner up to the plane through the sphere's intersection with the edges, and the corresponding area of the solid–liquid interface. We obtain

$$V = \frac{N}{6}\sin^2\vartheta_2 \cos\vartheta_2 \sin\frac{2\pi}{N}, \qquad (9.19)$$

$$A = N\sin\zeta\cos\zeta. \qquad (9.20)$$

Adding up these volumes and areas with the proper signs and taking account of the solid–liquid interface with the factor $-\cos\gamma_1$ leads to

$$A = -\frac{3}{R}V. \qquad (9.21)$$

Figure 9.7 shows the curvature $\widehat{p}$ of a unit liquid volume $V = 1$ in a tripod as a function of the angle 2ζ between the edges and of the contact angle γ between the liquid and the container material. For large angles 2ζ and large contact angles α, the curvature of the liquid drop in the corner becomes positive. In that case a single drop in one corner of a regular polyhedron has a lower energy than does a set of drops of equal size in all corners; see Fig. 3.7a. With decreasing 2ζ and γ, the curvature becomes zero and then negative. In that case, the situation of drops of equal size in all corners of a regular polyhedron represents the minimum in energy. Eventually, if $\alpha + \gamma$ is smaller than $\pi/2$, the liquid penetrates into the wedges, such that the liquid volumes in the corners become connected by nearly cylindrical volumes in the wedges.

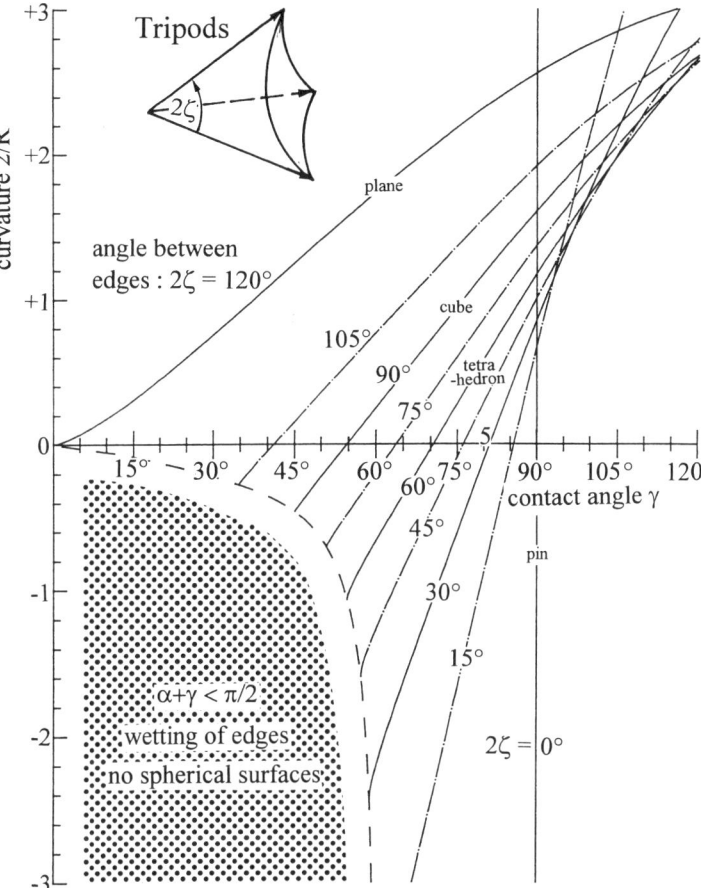

Fig. 9.7. The curvature $\widehat{p} = 2/R$ of a unit liquid volume $V = 1$ in a tripod as a function of the angle 2ζ between the edges and of the contact angle γ of the liquid with the container material

In the case of very small polar angles ϑ_1, ϑ_2 between the space diagonal and the face diagonals and edges, the N-pod becomes very long and thin. For calculating the curvature $\widehat{p}$ of the liquid surface, such N-pods may be approximated by a cylinder with N-gonal cross section. This allows one to apply the integral theorem introduced in Sect. 8.2. The error involved in the slight narrowing of the cross section is of order $O(\vartheta_1^2)$. If the side length of the N-pod is L in the vicinity of the liquid surface, one obtains the following for the volume V and the curvature $\widehat{p}$:

$$V = \frac{N}{3}\left(\frac{L}{2}\right)^3 \cot\left(\frac{\pi}{N}\right)\cot\vartheta_1, \quad \widehat{p}L = 4\cos\gamma\tan\left(\frac{\pi}{N}\right). \tag{9.22}$$

If a unit volume $V = 1$ is assumed as in Fig. 9.7, the curvature becomes proportional to $(\cot \vartheta_1)^{-1/3}$.

9.2 Transition Between the Corner and the Wedge

9.2.1 Liquid Volumes in Polyhedra

Let us now turn to the case of small dihedral angles and contact angles, when (9.9) is not satisfied. The liquid drop is no longer limited to the corner, but is sucked into the wedges because of the negative capillary pressure. Nevertheless, a surplus volume is left in the corners. Far from the corner, the liquid surface assumes a cylindrical shape. Constancy of the capillary pressure $\widehat{p}$ requires that the radius R_{crn} of the liquid surface at its intersection with the space diagonal equals twice the cylindrical radius R_{wdg} in the wedges, i.e.

$$\widehat{p} = \frac{2}{R_{\text{crn}}} = \frac{1}{R_{\text{wdg}}} . \tag{9.23}$$

When the total liquid volume in a polyhedron is reduced, the volume per unit length in the wedges decreases quadratically with the radius R_{wdg}, whereas the surplus volume in a corner decreases in proportion to the third power of the radius R_{crn}. If there is no effective overlap of the volumes piled up in different corners, we obtain

$$V = N_{\text{wdg}} L |\widehat{p}|^{-2} V_{\text{wdg}} + N_{\text{crn}} |\widehat{p}|^{-3} V_{\text{crn}} , \tag{9.24}$$

$$A = N_{\text{wdg}} L |\widehat{p}|^{-1} A_{\text{wdg}} + N_{\text{crn}} |\widehat{p}|^{-2} A_{\text{crn}} . \tag{9.25}$$

Here N_{wdg} and N_{crn} are the numbers of wedges and corners, respectively, of the polyhedron considered. For the dimensionless volume V_{wdg} and the surface area A_{wdg} of the cylindrical volume in a wedge (see Fig. 9.8), we obtain

$$V_{\text{wdg}} = \frac{\cos \gamma_1 [\cos \gamma_2 + \cos(2\alpha + \gamma_1)] + \cos \gamma_2 [\cos \gamma_1 + \cos(2\alpha + \gamma_2)]}{2 \sin(2\alpha)}$$
$$- \frac{\pi - 2\alpha - \gamma_1 - \gamma_2}{2} , \tag{9.26}$$

$$A_{\text{wdg}} = -2 V_{\text{wdg}} . \tag{9.27}$$

The surplus volume V_{crn} and the corresponding area A_{crn} in the corners have to be determined numerically; see Sect. 9.2.3.

The validity of (9.24) and (9.25) depends on the condition that the liquid volumes V_{crn} accumulated in adjacent corners are fully independent. Along each wedge, accumulation due to the left-hand corner must completely fade out before accumulation due to the right-hand corner starts. This is possible because of the exponential fading out of the corner volume far from the corner. A further correction to (9.24) and (9.25) has to be taken into account: the

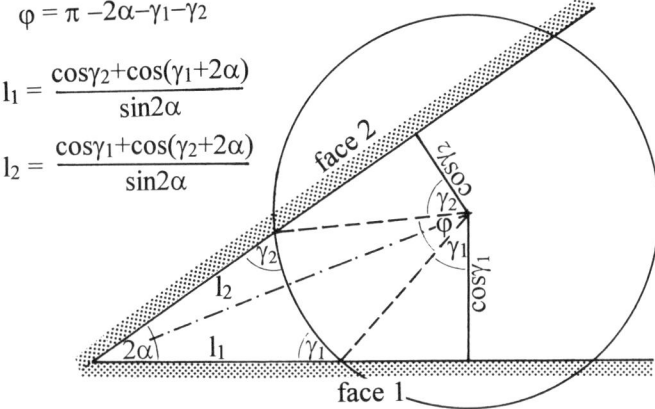

Fig. 9.8. The volume V_{wdg} and surface area A_{wdg} of a cylindrical liquid volume in a wedge. The contact angles with faces 1 and 2 are γ_1 and γ_2, respectively

wedge volumes V_{wdg} overlap in the corners. This overlap has to be eliminated by bisecting the wedge volumes along the symmetry planes with the other wedges.

9.2.2 Exponential Piling-Up in Corners

Figure 9.9 shows schematically the intersection of the liquid surface with the plane through an edge and the opposite face diagonal; z is the coordinate in the direction of the wedge.

The anticipated smooth asymptotic approach of the cylindrical surfaces in the wedges to the surfaces in the corners favors a one-dimensional approach. We assume that the liquid surface in the planes normal to the wedges may be described by circular sections with varying radius $R(z)$. We put

$$x(z,\varphi) = R(z)\sin\varphi, \quad y(z,\varphi) = R(z)\left(\frac{\cos\gamma}{\sin\alpha} - \cos\varphi\right). \quad (9.28)$$

If $R(z)$ smoothly approaches the value $R(\infty) = R_{\text{wdg}}$ at infinity, the boundary condition on the contact angle is satisfied for the same angle $\varphi = \pi/2 - \alpha - \gamma$ in all planes $z = $ const. The increase in $R(z)$ with decreasing z nevertheless entails a local increase of the effective contact angle. This requires an equivalent increase of the azimuth φ at the contact line. The relative rise of the contact line with decreasing z therefore is stronger than that of the surface along the symmetry plane.

The curvature in the x, y plane is $1/R(z)$, and that in the z direction equals $y''(z,0)/[1 + y'^2(z,0)]^3/2$. The capillary equation therefore takes the form

224   9. Liquid Surfaces in Polyhedral Containers

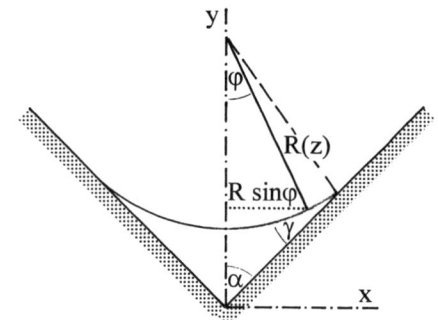

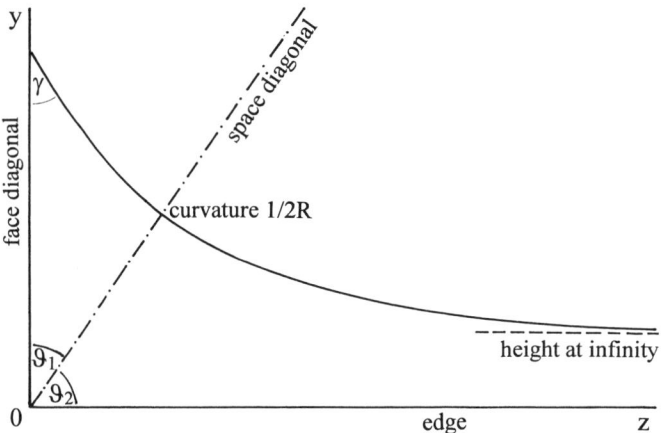

Fig. 9.9. Section of a liquid surface in a regular N-pod in the symmetry plane through an edge and a face diagonal (N odd)

$$\frac{1}{R(z)} + \frac{y''(z,0)}{\sqrt{1+y'(z,0)^2}^3} = \frac{1}{R_{\text{wdg}}}, \tag{9.29}$$

which is equivalent to

$$\frac{1}{R(z)} + \frac{\Gamma^2 R''(z)}{\sqrt{1+\Gamma^4 R'^2}^3} = \frac{1}{R_{\text{wdg}}}, \tag{9.30}$$

where

$$\Gamma = \sqrt{\frac{\cos\gamma}{\sin\alpha} - 1}. \tag{9.31}$$

Equation (9.30) may be integrated analytically. Multiplication by $R'(z)$ and adjusting the integration constant so that $R(z)$ approaches R_{wdg} smoothly yields

$$\log\left(\frac{R(z)}{R_{\text{wdg}}}\right) + \frac{1}{\Gamma^2}\left(1 - \frac{1}{\sqrt{1 + \Gamma^4 R'^2}}\right) = \frac{R(z) - R_{\text{wdg}}}{R_{\text{wdg}}}. \tag{9.32}$$

Solving (9.32) using a Taylor expansion for $[R(z) - R_{\text{wdg}}]/R_{\text{wdg}} \ll 1$ leads to

$$\frac{R(z) - R_{\text{wdg}}}{R_{\text{wdg}}} = c_1 \exp\left(-\frac{z}{\Gamma R_{\text{wdg}}}\right) - \frac{c_1^2}{3}\exp\left(-\frac{2z}{\Gamma R_{\text{wdg}}}\right) \pm \dots . \tag{9.33}$$

Thus, under the present assumptions, the surface approaches its cylindrical shape at infinity exponentially. If the length of decay ΓR_{wdg} is small compared with the length of the wedges, it is possible to split up the total liquid volume in a polyhedron into independent contributions, ascribed to the wedges and to the corners.

R_{wdg} depends mainly on the liquid volume per unit length of the wedge and on the contact angle γ. The decrement Γ given (9.31) decreases monotonically from a finite value at zero contact angle towards zero in the limiting case. From the numerical simulations presented below, it can be concluded that for a fixed liquid volume, R_{wdg} increases with γ and, in the product ΓR_{wdg}, counteracts the decrease of Γ for almost any γ except very close to $\alpha + \gamma = \pi/2$. In general, the exponential decay becomes slower with increasing contact angle, since more and more liquid is piled up in the corner and the volume in the wedge decreases.

The analytical considerations presented are valid far from the corner and therefore are applicable also to wedges, which join in the same corner but have different dihedral angles $2\alpha_k$ and different contact angles γ_i and γ_j, such that $\gamma_i + \gamma_j + 2\alpha_k \geq \pi$.

9.2.3 Numerical Calculation of Corner Volume

The exponential protrusion of the corner volume V_{crn} into the wedges and its quantitative integration has been numerically checked by de Lazzer [1998, de Lazzer and Langbein 1998]. Calculations of liquid surfaces in regular tripods were performed using SURFACE-EVOLVER, a numerical code developed by Brakke [1995]. The code solves the Laplace equation, starting with a given initial configuration and proceeding towards the nearest minimum of the system's energy. In contrast to other codes that remove the contact line by applying an augmented Laplace equation [Wong, Morris & Ratke 1992 a,b], contact lines are preserved in SURFACE-EVOLVER and contact angles arise owing to the energy terms $-\cos\gamma$ on the bounding solid faces. The same code has been used also to simulate a variety of different surfaces in unit cubes [Mittelmann & Hornung 1992, Mittelmann 1993a,b, 1995] and in complex-shaped surface tension tanks [Dominick & Tegart 1994].

For the numerical simulation, a tripod with wedges of fixed length was used. At the end of the wedges, a contact angle of $\pi/2$ between the surface and an imaginary end plane perpendicular to the wedge was imposed (i.e. the surface is assumed to have zero slope along the wedge). The surface shapes arising for different contact angles γ and hedral angles 2ζ between of the edges, or and for different dihedral angles 2α of the wedges were calculated. The results presented here mainly refer to the rectangular tripod, $2\zeta = \pi/2, 2\alpha = \pi/2$. Calculations for other wedge angles 2α yield similar results.

Considering the surfaces formed by identical liquid volumes with different contact angles γ in a given tripod, the results of the numerical calculations show that the central surface height approaches its minimum value y_{wdg} very fast; see Fig. 9.10. The decay of the central height apparently follows an exponential law. To judge whether the decay is indeed exponential, the logarithm of the slope of the central height $\log[-dy(0, z)/dz]$ is plotted in Fig. 9.11 for the rectangular tripod and different contact angles. Where the numerical data do not scatter, the decay definitely tends to become linear. The graphs for different contact angles also show that the exponential behavior develops much faster for small contact angles than for large ones. This is due to the fact that the surface shape becomes more and more spherical close to the corner with increasing contact angle until at $\alpha + \gamma = \pi/2$ a spherical cap is left. Furthermore, the exponential slope is steeper for contact angles close to zero. This can be described in terms of the decrement ΓR_{wdg}, as shown in Fig. 9.12, which illustrates the influence of varying curvature on the decrement. The increase of R_{wdg} for a fixed liquid volume counteracts the decrease of Γ. Thus the dimensional decrement ΓR_{wdg} has a finite value for zero contact angle, increases slowly with increasing γ and reaches a maximum value

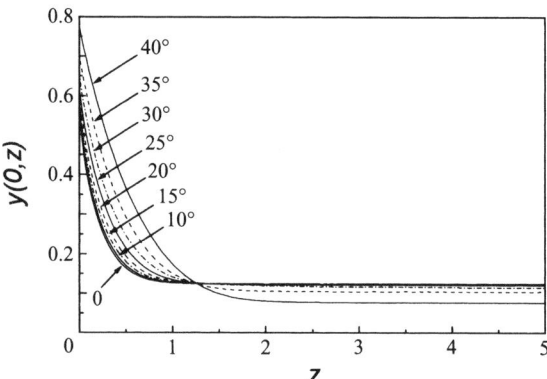

Fig. 9.10. Numerically obtained central meniscus height $y(0, z)$ versus the wedge coordinate z of a rectangular tripod for different contact angles γ of the liquid. The corner of the tripod is located at $z = 0$, and the y axis is in the direction of the face diagonal. At $z = 5$ the surface is required to show a contact angle $\gamma = \pi/2$ with the x, y plane

9.2 Transition Between the Corner and the Wedge   227

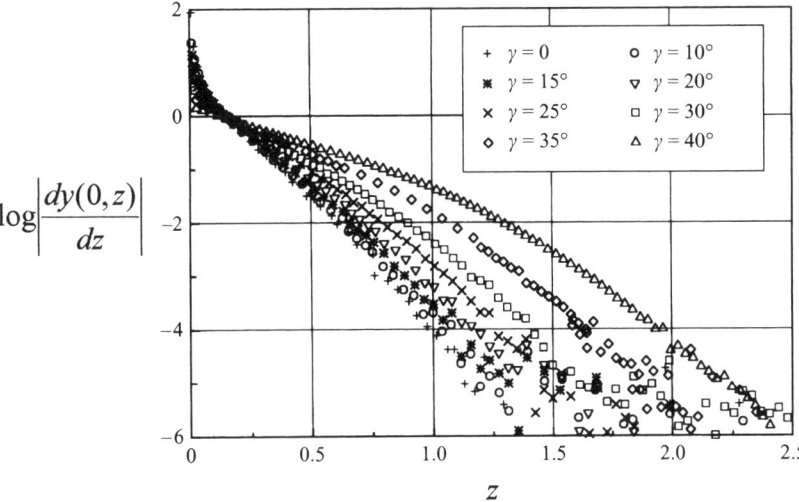

Fig. 9.11. Logarithmic decay of the central surface height $y(0,z)$ versus the wedge coordinate z of a rectangular tripod for different contact angles γ. The scatter of the results at logarithmic decays $\log[-\mathrm{d}y(0,z)/\mathrm{d}z] < -4$ is clearly due to the limited numerical accuracy

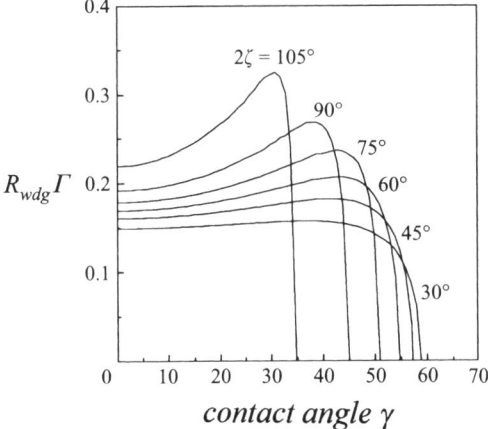

Fig. 9.12. Decrement ΓR_{wdg} of the surface shape versus the contact angle γ for different edge angles 2ζ. Plots for fixed liquid volume $V = 0.1\pi$ in tripods with wedge length $z = 5$

rather close to $\alpha + \gamma = \pi/2$. It tends to zero when $\alpha + \gamma$ approaches $\pi/2$, since in that limiting case the cylindrical surface in the wedge vanishes, whereas the spherical curvature in the corner remains finite.

9.2.4 Similarity of Corner Volumes

The numerical simulations by de Lazzer and Langbein [1998] clearly validate the suggestion (9.33) of an exponential protrusion of the corner volumes into the wedges. The corner volumes V_{crn} are integrable and the total liquid volume may be dissected into subvolumes as suggested by (9.24), (9.25). If the wedges are taken into account with their full length L, their overlap volumes have to be subtracted from V_{wdg}, or, more accurately, to be added to V_{crn}. To obtain the overlap volumes V_{bsc}, the cylindrical volumes in the wedges have to be bisected near the corner along the midplanes with the other wedges. Figure 9.13 depicts two of the three cylindrical liquid volumes meeting in the corner of a rectangular tripod; the edge towards the viewer is cutout and the cylindrical volumes are slightly pulled apart. This yields

$$V_{\text{bsc}} = \cot \zeta \left[\frac{\cos(a+\gamma)\cos\gamma \left[\cos(a+\gamma)+\cos\gamma\right]}{3\sin^2\alpha} + \frac{2}{3}(\cos\gamma - \sin\alpha) \right.$$
$$\left. - \left(\frac{\pi}{2} - \alpha - \gamma\right)\cos\gamma \cot\alpha \right]. \qquad (9.34)$$

Since the numerical simulations assume a fixed volume V and a fixed wedge length L and calculate the curvature $\widehat{p}$, the volume V_{crn} to be ascribed to the corner becomes

$$V_{\text{crn}} = V - N_{\text{wdg}} \left(L|\widehat{p}|^{-2} V_{\text{wdg}} - |\widehat{p}|^{-3} V_{\text{bsc}} \right). \qquad (9.35)$$

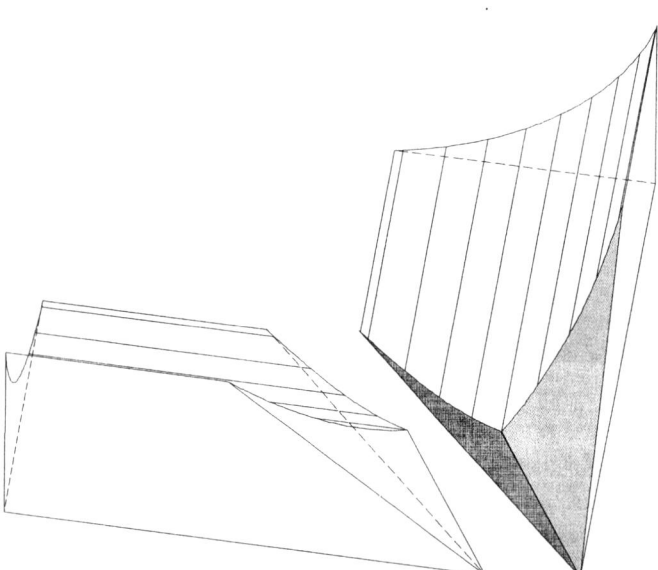

Fig. 9.13. Sketch of two of the three cylindrical wedge surfaces meeting in a tripod. The intersections are given by the planes through the space diagonal and the adjacent face diagonals

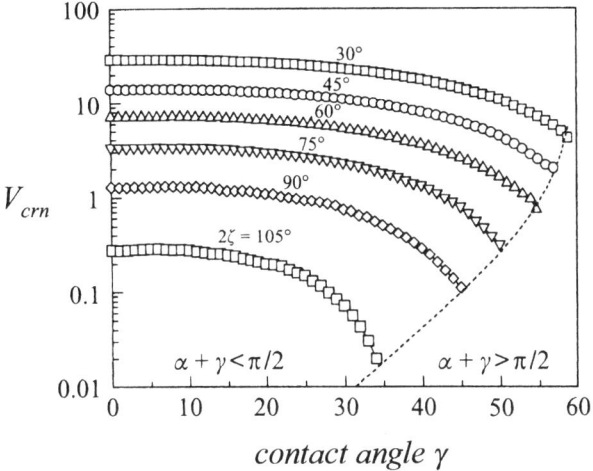

Fig. 9.14. Numerically obtained corner volumes V_{crn} for regular tripods with angles 2ζ between the edges versus contact angle γ

Figure 9.14 depicts numerically obtained corner volumes V_{crn} for regular tripods with angles 2ζ between the edges versus the contact angle γ. As γ approaches zero, the slopes of the plots of V_{crn} tend to zero. To obtain the curvature $\widehat{p}$ for the case where a unit volume is associated with the corner,

$$V_{crn}|\widehat{p}|^{-3} = 1 \tag{9.36}$$

has to be solved for $\widehat{p}$. Figure 9.15 shows the curvature $\widehat{p}$ for various tripods and contact angles γ, comprising wetted wedges and also nonwetted wedges. Since the surface assumes a concave shape, the mean curvature has been given a negative sign. It is obvious that the results obtained for $\alpha + \gamma < \pi/2$ provide a continuous extension of the analytical results for $\alpha + \gamma \geq \pi/2$, although the interface configuration behaves discontinuously. It is also worth noting that for all angles 2ζ between the edges, the curvature approaches its value for a contact angle $\gamma = 0$ with zero slope, as can be anticipated from the behavior of V_{crn} as given in Fig. 9.14.

9.2.5 Finite Wedge Length

The preceding considerations have been based on the assumption that the dimensionless extent of the wedge $L/\Gamma R_{wdg}$ considerably exceeds unity. The boundary condition set at the end of the wedges is that of zero axial slope. No surface deformation at the far end of the wedge has been considered, although wedges of finite length L may connect corners with different geometry but with equal wetting conditions. For what follows, let us recall that the one-dimensional model according to (9.29) accounts for the axial curvature simply

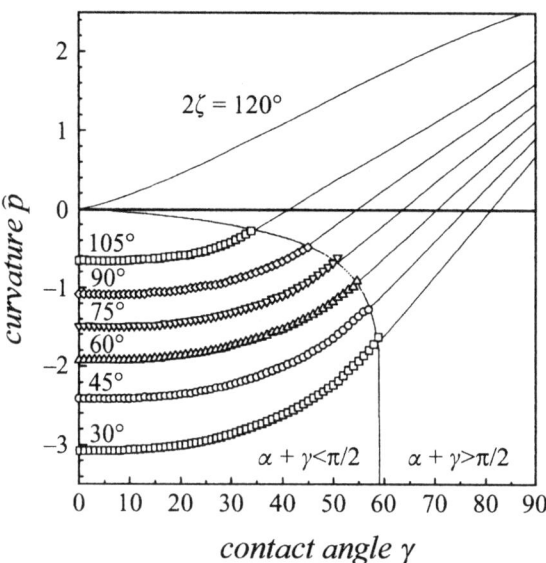

Fig. 9.15. Curvature $\hat{p}$ of a drop with unit corner volume V_{crn} in regular tripods with angles 2ζ between edges versus contact angle γ. The numerical simulations yield a continuous extension of the analytic results for $\alpha + \gamma \geq \pi/2$ into the region $\alpha + \gamma < \pi/2$

by adjusting the radius $R(z)$, i.e. the radius or curvature perpendicular to the wedge.

The range of applicability of the similarity solutions nevertheless allows for a certain overlap of the volumes in opposite corners. If a wedge is sufficiently long, the surface arising can be approximated by the underlying surface appropriate to the wedge plus a superposition of the two corner volumes. The one-dimensional model yields a favorable compensation of errors when the liquid volumes ascribed to opposite corners are cutoff in the middle of the wedge. Let us mention in this context the additivity of curvatures and slopes: the curvatures add up, whereas the slopes cancel owing to their opposite signs.

The similarity solution remains valid as long as the surface shape in the middle of the wedge is well into the regime of exponential decrease and the slopes of the isolated corner volumes are sufficiently small not to violate the one-dimensional approach of (9.29). The similarity solution breaks down when the slopes of the isolated corner volumes cease to follow the exponential decrease. In this case, the contributions of the tails of the corner volumes to the overall slope do not mutually cancel any more and the contact angle boundary condition becomes important.

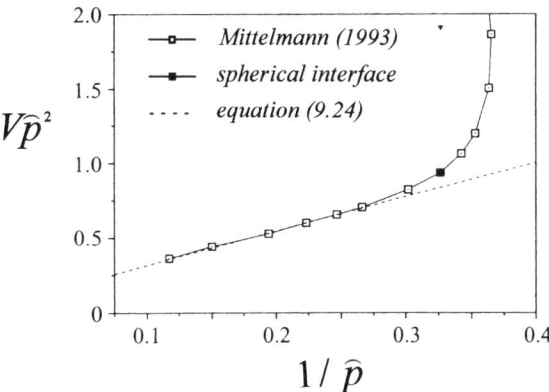

Fig. 9.16. Comparison of $V|\widehat{p}|^2$ obtained by the present approach and by the numerical calculations of Mittelmann [1993b] versus the inverse curvature $1/|\widehat{p}|$, for a liquid wetting the edges ($\gamma = 40°$) of a unit cube

9.2.6 Accuracy of the Present Approach

The capability of the present approach may be compared with numerical calculations by Mittelmann [1993b] of the surface of a liquid with a contact angle $\gamma = 40°$ in a unit cube. In the present approach, the edges of the cube are divided into two halves, each being associated with the adjacent corner. Now, on applying (9.24) to each of the eight corners, the total liquid volume within the cube may be obtained. An equivalent procedure can be applied to any polyhedral container.

Rebuilding the unit cube from eight rectangular tripods ($N = 3$) with length $L = 1/2$, one obtains from (9.24) and (9.35)

$$V = 8\left[NLV_{\text{wdg}}|\widehat{p}|^{-2} + (V_{\text{crn}} - NV_{\text{bsc}})|\widehat{p}|^{-3}\right]. \tag{9.37}$$

V_{wdg} can be calculated from (9.26) and V_{bsc} from (9.34), and V_{crn} can be obtained from numerical simulation as summarized in Fig. 9.14; the curvature $\widehat{p}$ of the concave surface is assumed to have negative sign. The result is shown in Fig. 9.16, depicting the dependence of the squared curvature times the liquid volume on the radius of curvature $-1/\widehat{p}$ for both of the approaches. For radii of curvature up to about 0.33 and filling levels up to 10 percent, the results obtained by the two approaches fit with a maximum deviation in the curvature $\widehat{p}$ of less than 4 percent. At radii of curvature below 0.27, the maximum deviation diminishes by one order of magnitude.

This agreement is better than expected, since the sum $\alpha + \gamma$ is rather large but not very close to $\pi/2$, and thus the decrement ΓR_{wdg} of the corner surfaces is close to its maximum value (compare Fig. 9.12). Furthermore, it has to be taken into account that the surface for $V = 0.1$ is almost spherical, i.e. it does not fulfill the requirements that $\Gamma R_{\text{wdg}} \ll L$ and that the axial curvature of the surface at the end of the wedge is negligible.

On the other hand, Mittelmann's results were obtained numerically, too. To judge their reliability, we consider the particular situation in which the liquid surface in a cube assumes a spherical shape. In this particular configuration, we obtain

$$R_{\mathrm{sph}} \cos \gamma = \frac{1}{2}, \quad \widehat{p} = -4 \cos \gamma \tag{9.38}$$

and

$$V = 1 - \frac{4\pi}{3} R_{\mathrm{sph}}^3 + 2\pi R_{\mathrm{sph}}^3 \left(2 - 3\cos \gamma + \cos^3 \gamma\right) \approx 0.099756 \tag{9.39}$$

for $\gamma = 40°$. This shows Mittelmann's result to agree well with the exact value. Since the present method yields an error of less than 4 percent for the spherical surface, it can be applied for rather high filling levels, as a view of the resulting surface shape (Fig. 3.8) shows. The corner volume has not faded away at half the length of the wedge, neither is the axial curvature of the surface negligible there. The error involved in the approach used is thus strikingly small even in the case of the spherical surface.

This provides additional validation of the superposition of the surfaces that arise in a wedge joining two corners of similar wetting conditions. In the case of the spherical surface with contact angle $\gamma = 40°$ in the unit cube, we obtain the following for the decrement $\varGamma R_{\mathrm{wdg}}$ of the corner surface:

$$\varGamma R_{\mathrm{wdg}} = \frac{1}{4\cos\gamma} \sqrt{\frac{\cos\gamma}{\sin\alpha} - 1} \approx 0.09422 \ . \tag{9.40}$$

This is about one fifth of the length $L/2$ attributed to each corner. It coincides well with the minimum wedge length required for the surface to show the exponential behavior. The applicability of the similarity solutions to the corner volumes thus extends to decrements $\varGamma R_{\mathrm{wdg}}$ of about an order of magnitude lower than the wedge length.

9.2.7 Prospects

The liquid surfaces in a polyhedron whose wedges $[i,j]$ fulfill the wetting condition $2\alpha_{ij} + \gamma_i + \gamma_j \leq \pi$ may be pasted together from cylindrical surfaces in the wedges and surplus volumes piled up in the corners. The corresponding similarity relation is based on decomposing the liquid volume into portions associated with the corner, the wedges and the mutual intersection of the wedge volumes at the corner. The specific coefficients depend merely on the container shape and contact angle(s). Although the specific corner volume requires numerical simulation, the present approach has the advantage that this volume has to be determined only once for a given geometry. For the determination of specific parameters of an interface configuration, such as the dependence of curvature on volume, the technique used thus effectively reduces the amount of numerical simulation in the limit of small volumes and likewise allows very convenient calculations of the pressure or curvature of surfaces in arrays of wedges and corners.

The curvature of a surface in a corner changes smoothly when the wetting limit $2\alpha_{ij} + \gamma_i + \gamma_j = \pi$ of the adjacent wedges is crossed, although the mathematical problem contains a discontinuity. Thus fluids of technological interest that are initially confined within a corner show a smooth change in pressure/curvature even if the wedges become wetted, for example owing to chemical processes within the fluid or the wall (surface aging), owing to electrical fields or owing to the addition of surfactants.

The approach presented can be extended to more general polyhedral configurations than the examples given above without major numerical effort. It can be applied to any scarcely filled closed container or capillary, thereby significantly extending existing relations for the shape, curvature and volume of liquid surfaces in long capillaries. Furthermore, it allows modeling of foams or porous structures as coupled clusters of polyhedra filled with low liquid volumes. Within a cluster, all coupled surfaces have identical internal pressures. Concerning porous media, simulation of materials containing pores of different shapes and volumes becomes possible.

References

1. Brakke K: *SURFACE-EVOLVER Manual*, version 1.98. The Geometry Center, Minneapolis (1995)
2. de Lazzer A, Langbein D: Liquid surface in regular N-pods. J. Fluid Mech. **358** (1998) 203–221
3. de Lazzer A: Zum Verhalten kapillarer Flüssigkeitsgrenzflächen in Ecken und Kanten. Dissertation, University of Bremen (1998) 1–120. dto: Progress Reports VDI (*Verein Deutscher Ingenieure*), Series 7, flow technology, No. 345
4. Dominick S, Tegart J: Orbital test results of a vaned liquid acquisition device. AIAA Paper (1994) 94-3027
5. Langbein D: Liquid surfaces in polyhedral containers. Proceedings of the International Conference on Advances in Geometric Analysis and Continuum Mechanics. P. Concus, K. Lancaster (eds.) (1994) 168–173
6. Langbein D: Liquid surfaces in polyhedral containers. Microgravity Sci. Technol. **8** (1995a) 148–154
7. Langbein D: Liquid surfaces in polyhedral containers. In: *Festschrift zum 70. Geburtstag von Prof. Julius Siekmann*. Universität-GH Essen (1995b) 301–312
8. Mittelmann HD, Hornung U: Symmetric capillary surfaces in a cube. Lawrence Berkeley Laboratory, LBL-31850 (1992)
9. Mittelmann HD: Symmetric capillary surfaces in a cube. Math. Comput. Simulat. **35** (1993a) 139–152
10. Mittelmann HD: Symmetric capillary surfaces in a cube; part 2: near the limit angle. Lect. Appl. Math. **29** (1993b) 339–361

11. Mittelmann HD: Symmetric capillary surfaces in a cube; part 3: more exotic surfaces, gravity. In: *Advances in Geometry Analysis and Continuum Mechanics*. P. Concus and K. Lancaster (eds.). International Press, Boston (1995) 199–208
12. Wong H, Morris S, Radtke CJ: Two-dimensional menisci in nonaxissymmetric capillaries. J. Colloid Interf. Sci. **148** (1992a) 284–287
13. Wong H, Morris S, Radtke CJ: Three-dimensional menisci in polygonal capillaries. J. Colloid Interf. Sci. **148** (1992b) 317–336

10. Playing with Stability

A liquid drop on a horizontal plane is in a metastable state. Its energy is independent of its position. In a microgravity environment, this is likewise true for liquid drops in a spherical container. Apart from these trivial cases, several container shapes with exotic properties are considered in this chapter. It is possible to construct cylindrical container shapes (so-called proboscides) in which liquid surfaces with equal curvatures but different energies and volumes are possible, so that $dp/dV = 0$. The wedges in these containers are wetted by a fluid only if the contact angle falls short of a critical contact angle. However, at infinitesimal fill levels, the capillary underpressure which causes the fluid to wet the wedges does not diverge towards infinity as for planar wedges.

It is also possible to construct exotic axisymmetric containers in which liquid surfaces with equal energies and volumes but different curvatures are possible, so that $dV/dp = 0$. This may be achieved either by adjusting the coating of a cylindrical tube to achieve a particular contact angle variation or by adjusting the container shape to achieve a constant contact angle. From the minimum-volume condition, it is clear that the resulting surfaces must be unstable.

10.1 Proboscides

10.1.1 Finite Rhombic Prisms

Let us consider a long rhombic cylinder with acute dihedral angles 2α and obtuse dihedral angles $\pi - 2\alpha$. Let the container be initially filled with fluid 2, and then an immiscible fluid 1 is introduced, displacing fluid 2. Let the contact angle of fluid 1 with the faces of the cylinder be γ and satisfy

$$\alpha + \gamma < \frac{1}{2}\pi, \quad \gamma < \alpha. \tag{10.1}$$

In this case, the acute wedges are wetted by fluid 1, whereas the obtuse wedges are not; see Fig. 10.1a. Assume further that the contact angle of fluid 1 with the end caps of the vessel is approximately $\pi/2$. Fill ports for fluid 1

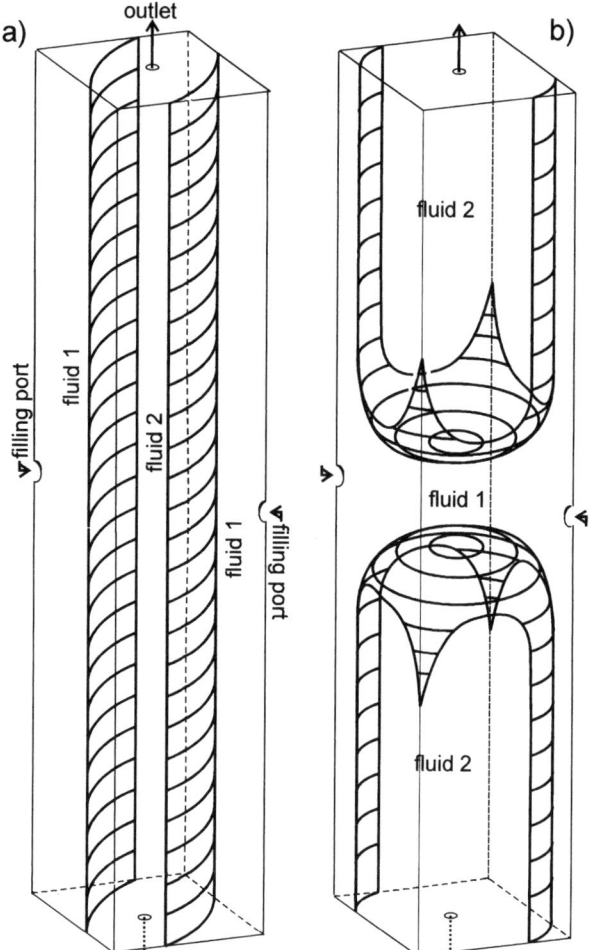

Fig. 10.1. Liquid interfaces in a rhombic cylinder. (**a**) At low filling ratios, fluid 1 forms cylindrical interfaces of radius R along the acute wedges. (**b**) At large filling ratios, a plug of fluid 1 arises in the middle of the container, whereas fluid 2 forms long drops extending from the ends to the middle of the container

are provided in the middle of the acute wedges, and outflow holes for fluid 2 are located in the middle of the end caps.

In the limit of low fill levels of fluid 1 one obtains cylindrical interfaces of radius R along the acute wedges of the vessel. The areas of the wetted wedge segment Σ_1, the free interface Σ_2 and the wetted cross section Ω_1 are given by (Fig. 8.7a)

$$\Sigma_1 = R\frac{\sin \delta}{\sin \alpha}, \quad \Sigma_2 = R\delta, \quad \Omega_1 = \frac{R^2}{2}\left(\frac{\cos \gamma}{\sin \alpha}\sin \delta - \delta\right). \quad (10.2)$$

In (10.2), all quantities refer to a quadrant of the cross section; $\delta = \pi/2 - \alpha - \gamma$ is the angle of the circular arc of the interface of fluid 1. Thus, the capillary pressure is given by $\widehat{p} = 1/R$. The absolute value of the capillary pressure decreases in inverse proportion to the square root of the filling level. In order to avoid hysteresis, the filling ratio has to be increased very slowly. From (10.2), it is possible to show that (see also (8.15))

$$2\widehat{p}\Omega_1 = \Sigma_1 \cos\gamma - \Sigma_2 . \tag{10.3}$$

Equations (10.2), (10.3) are valid until, during further filling, the contact lines of the opposing interfaces reach the obtuse wedges. One would then expect a circular cylindrical interface of the extruded fluid 2 to evolve, which, however, would be susceptible to the well-known Rayleigh instability. In this limit, because the container is long relative to its diameter, such a circular cylindrical column will break up into droplets. And since the fill ports for fluid 1 are located centrally in the acute wedges, it appears most likely that the opposing interfaces will meet in this region first. This means that a plug of fluid 1 in the middle of the container may be anticipated, whereas fluid 2 will form long drops extending from the ends to the middle of the container; see Fig. 10.1b. If a different filling procedure was applied, fluid 1 might cover both of the container ends, whereas fluid 2 might form a long drop in the middle.

In both cases, the pressure and curvature can be obtained from the integral theorem, which states that the cross-sectional integral of the curvature equals the contact line integral of the normal; see Sect. 8.2. If the side length Σ of the rhombic cylinder equals L, one has the following for the cross section Ω (again related to one quadrant; see Fig. 8.7b):

$$\Sigma = L , \quad \Omega = \frac{1}{2}L^2 \sin\alpha \cos\alpha , \tag{10.4}$$

$$\widehat{p}(\Omega - \Omega_1) = (\Sigma - \Sigma_1)\cos\gamma + \Sigma_2 . \tag{10.5}$$

Using (10.3), (10.4) can be simplified to

$$\widehat{p}(\Omega + \Omega_1) = \Sigma \cos\gamma , \tag{10.6}$$

$$(\widehat{p}L)^2 \sin\alpha \cos\alpha - 2(\widehat{p}L)\cos\gamma + \left(\frac{\cos\gamma}{\sin\alpha}\sin\delta - \delta\right) = 0 . \tag{10.7}$$

The length of the plug of fluid 1 in the middle of the container may be arbitrary, whereas the capillary pressure is independent of the fill level.

If, subsequently, the fluid 1 fill level is reduced, the length of the plug of fluid 1 will shorten and the long drops of fluid 2 will lengthen until the point of coalescence is reached. At this point the interface will return to its initial state, with opposing interfaces in both acute wedges. This effect shows strong hysteresis: the volumes of fluid 1 in the acute wedges meet in the obtuse wedges during filling at much higher filling levels than the filling levels at which the drops of fluid 2 meet in the middle of the container during volume reduction; see Fig. 10.2.

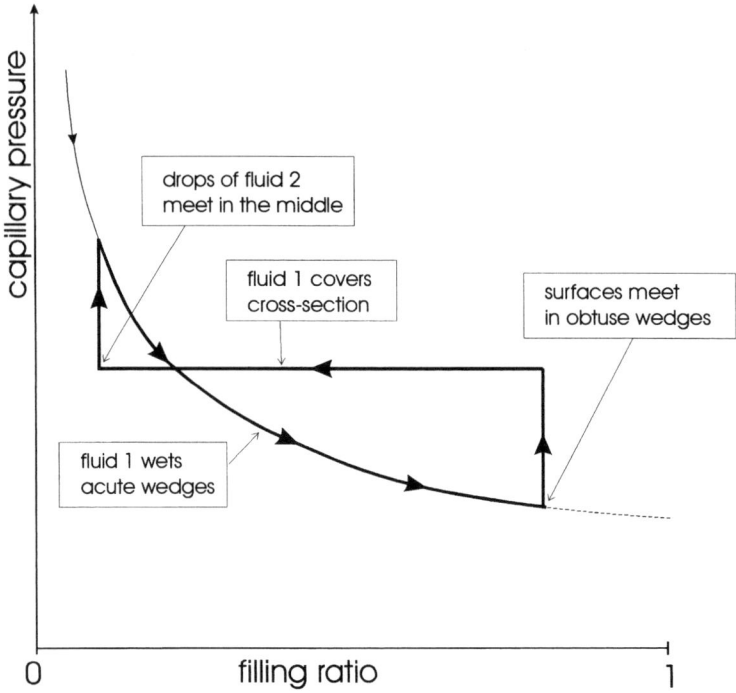

Fig. 10.2. When the filling ratio is increased, the interfaces of fluid 1 in the acute wedges meet in the obtuse wedges. A plug of fluid 1 arises, whereas fluid 2 forms long drops extending from the ends to the middle of the container. When the filling ratio is lowered, the drops of fluid 2 meet in the middle, the plug of fluid 1 vanishes and, as at the beginning of the procedure, separate interfaces appear in the two acute wedges

10.1.2 Canonical Proboscides

Fischer and Finn [1993] have proposed the study of wedges that are not formed by planes, but are shaped in such a manner that the capillary pressure becomes independent of the volume of liquid in the wedge (see also Finn and Leise [1994], Finn and Matek [1996]). These authors have termed such wedges *canonical proboscides*. These wedges are wetted by a fluid only if the contact angle falls short of a critical contact angle γ_0. However, at infinitesimal fill levels, the capillary underpressure which causes the fluid to wet the wedges does not diverge towards infinity as for the planar wedge, but remains constant. This considerably impedes filling on the basis of capillary forces alone.

The shape of these wedges maintains the curvature $\widehat{p} = 1/R$ and the critical contact angle γ_0 of the cylindrical interfaces. This results in an increase of the azimuth δ with increasing fill level. If the shape is described parametrically by $X(\delta), Y(\delta)$, one obtains the requirement (Fig. 10.3a)

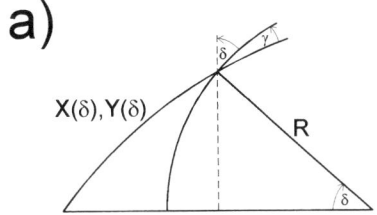

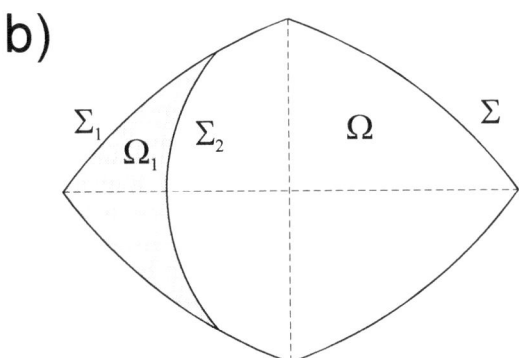

Fig. 10.3. Canonical proboscides are shaped in such manner that the pressure of a cylindrical interface becomes independent of the filling ratio. (**a**) Parametric description of the shape by $X(\delta), Y(\delta)$. (**b**) Two diametrically opposite proboscides used to construct a closed container

$$\frac{\mathrm{d}Y}{\mathrm{d}X} = \cot(\delta + \gamma_0) \,. \tag{10.8}$$

(The cylindrical surfaces in wedges formed by planes have a constant azimuth δ.) Equation (10.8) is solved by

$$Y(\delta) = R \sin \delta \,,$$
$$X(\delta) - X_0 = R \sin \gamma_0 \log\left(\frac{1 + \sin(\delta + \gamma_0)}{\cos(\delta + \gamma_0)}\right) - R \cos \delta \,. \tag{10.9}$$

Equation (10.9) is conveniently proved by differentiation with respect to δ.

Let us construct a container from two such proboscides pointing in opposite directions. If the same filling procedure is applied as for the rhombic prism in the preceding section, fluid 1 will initially form cylindrical interfaces of radius R along the wedges. During that filling process, the radius and curvature of the surfaces do not change. The wetted side length Σ_1, the length of the free interface Σ_2 and the wetted cross-sectional area Ω_1 are given by (Fig. 10.3b)

$$\Sigma_1 = \int dy \sqrt{1 + \left(\frac{dX}{dY}\right)^2} = R \int_0^\delta d\varphi \frac{\cos\varphi}{\cos(\varphi+\gamma_0)}, \quad \Sigma_2 = R\delta, \quad (10.10)$$

$$\Omega_1 = \int dx\, Y - \frac{1}{2}R^2(\delta - \sin\delta\cos\delta) = R^2 \int_0^\delta d\varphi \frac{\sin\varphi \sin\gamma_0}{\cos(\varphi+\gamma_0)}. \quad (10.11)$$

The integrals in (10.10) and (10.11) are easily derived from the general integral

$$\int d\varphi \frac{\cos\varphi}{\cos(\varphi+\gamma)} = \varphi\cos\gamma - \sin\gamma \log[\cos(\varphi+\gamma)]. \quad (10.12)$$

During filling, the interfaces grow in the two proboscides while facing each other, until they meet in the obtuse wedges. And, as above, one may suggest that a circular cylindrical interface of the extruded fluid 2 would evolve if not for the well-known Rayleigh instability. Since the container is assumed to be long relative to its diameter, such a cylindrical interface must break into droplets. And since the filling ports for fluid 1 are in the middle of the acute wedges, it appears most likely that the interfaces facing one another will meet there first. This means that, in the middle of the container, a plug of fluid 1 may be anticipated, whereas fluid 2 will form long drops extending from the end caps to the middle of the container.

This situation, however, does not satisfy the integral theorem. According to the latter, the curvature of an interface covering the cross section is given by (10.5), where Ω is the cross-sectional area of the proboscides and Σ the wetted side length (again related to one quadrant), see Fig. 3.5. However, instead of (10.3) for planar wedges, we now obtain the following relation from (10.10) and (10.11)

$$\widehat{p}\Omega_1 = \Sigma_1 \cos\gamma - \Sigma_2. \quad (10.13)$$

In the case of proboscides, the additional terms due to the wetting of the wedges just balance each other. The integral theorem clearly takes account of the fact that the curvature is independent of the fill level. Independent of the fill level, the integral theorem leads to

$$\widehat{p}\Omega = \Sigma \cos\gamma, \quad (10.14)$$

$$\frac{1}{2}\widehat{p}R(\delta_0 - \sin\delta_0 \cos\delta_0) = \delta_0, \quad (10.15)$$

where δ_0 is the azimuth of the circular arcs when they meet in the obtuse wedges. The curvature obtained from (10.14) and (10.15) is much larger than the curvature required for construction of the proboscides. Thus, the integral theorem contradicts the construction of the proboscides, and it cannot be applied to such a container.

We found in Sect. 9.2.2 that in planar wedges, the cylindrical interface shape is approached exponentially. This exponential approach requires that the radius of curvature depends linearly on the depth of the liquid in the wedge. This condition is clearly violated in the case of the proboscides, where the curvature is independent of the fluid depth. The fluid wets the wedges, but does not approach the cylindrical form asymptotically. Rather, the shape of the interface depends explicitly on the length of the container and the fluid volume. In consequence, the integral theorem is not applicable in such wedges.

There is just one possible way to realize any of the possible surfaces contained in the class of proboscides: one has to mutually adjust the capillary pressures required by the integral theorem, on one hand, and the capillary pressures required for construction of the proboscides, on the other hand. In order to achieve that, one may construct a container from a proboscis and a circular (or polygonal) cylinder with dimensions such that the ratio of the surface integral with weight $\cos\gamma$ to the area of the cross section equals the curvature of the liquid surface in the proboscis.

The above considerations require that the contact angle γ is exactly equal to γ_0. Any deviation in the contact angle causes a nonlinear dependence of the radius of curvature on the depth of the liquid volume in the wedge. This means that, in the middle of the container, a plug of fluid 1 may be anticipated, whereas fluid 2 will form long drops extending from the end caps to the middle of the container, as suggested earlier.

10.1.3 Interface Configuration Experiment

Let us recall some essential points:

- If a wedge is formed by planes, increasing the amount of wetting lowers the curvature.
- For proboscides, increasing the amount of wetting does not change the curvature.
- Likewise, one may construct containers such that increasing the volume increases the curvature.
- $\gamma > \gamma_0$: too low a pressure in the proboscis, no fluid in wedge.
- $\gamma < \gamma_0$: too high a pressure in the proboscis, wedge is completely filled.

The configurations of interfaces in proboscides have been investigated by Concus et al. [2000] and Chen et al. [1995, 1997] in a Space Shuttle glovebox experiment during the USML-2 mission. Each of the three acrylic containers used was designed from two proboscides with different critical angles γ_1 and γ_2, which were oriented opposite to each other and connected to a central, circular cylindrical core. The values of γ_1 and γ_2 selected for the three containers were based on the approximate value of the equilibrium contact angle of 32° measured in a terrestrial environment. The spread of values of contact angle covered by the three containers was intended to allow observation of possible effects of contact angle hysteresis, which is not included in the classical theory. The coordinates were computed from (10.9), all dimensions being scaled to correspond to a circular boundary arc of radius

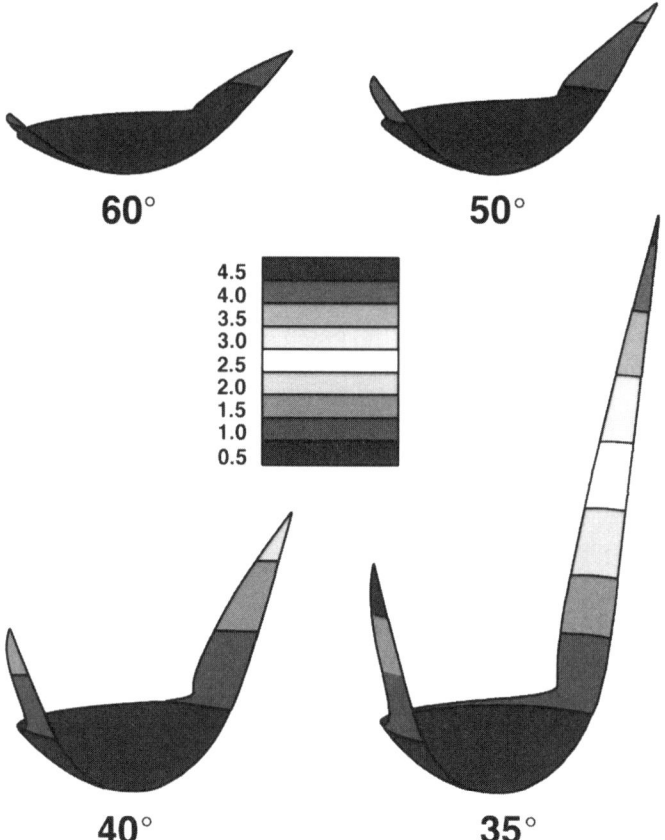

Fig. 10.4. Computed equilibrium interface for a 30°/34° double proboscis section for contact angles of 60°, 50°, 40° and 35°, $\gamma_0 = 34°$ [from Concus et al. 2000]

1.5 cm (Fig. 10.4). The test liquid for all the double-proboscis vessels was an aqueous-ethanol solution, 50% by volume. (The long-time contact angle of the aqueous-ethanol/machined-acrylic system was measured by the sessile-drop method to be 32° ± 2°; receding, 20°; advancing, 41°.)

When γ is larger than γ_0 there exists an equilibrium configuration that covers only the base of the cylindrical container, while for contact angles smaller than γ_0 no such equilibrium configuration is possible. In the latter case the liquid moves to the walls and can rise arbitrarily high along a part of the wall, uncovering a portion of the base if the container is tall enough. A practical challenge in this connection is to design cross sections for which a large enough portion of the liquid will rise up the walls for ease of observation as the critical value of contact angle is crossed, without the containers being unrealistically tall, and so that the change will be abrupt enough to allow accurate determination of the critical contact angle value.

Fig. 10.5. Static interface shapes for ICE-P1 ($20°/26°$) vessel. (**a**) After completion of fill; (**b**) after disturbance by payload specialist [from Concus et al. 2000]

For $\gamma_0 \leq \gamma < \pi/2$ and for a liquid volume sufficient to cover the base, the height of the free surface S can be given in closed form by considering a portion of a lower hemisphere meeting the walls with the prescribed contact angle γ. Thus, for a given volume of liquid, the height is bounded uniformly for values of γ throughout this range. For $0 \leq \gamma < \gamma_0$, the liquid will move to the corner and rise arbitrarily high, uncovering the base if the container is tall enough.

ICE-P1 ($20°/26°$) *vessel.* Both proboscides are subcritical for a liquid with $\gamma = 32°$. Figure 10.5a was recorded shortly after the fill procedure, and Fig. 10.5b was recorded after significant disturbances had been imparted to the vessel. Very little change in the interface can be distinguished between the initial and final states.

ICE-P2 ($30°/34°$) *vessel.* The left-hand proboscis is subcritical, and the right-hand proboscis is supercritical. As the disturbances were increased in magnitude, the liquid, instead of returning to the initial state of Fig. 10.6a, rose noticeably and roughly equally in the proboscides (Fig. 10.6b). As seen in Figs. 10.6c,d, subsequent larger disturbances led to an increased rise only in the 34° proboscis on the right, which was supercritical.

ICE-P3 ($38°/44°$) *vessel.* Both proboscides are supercritical here. Disturbances to this vessel caused large shifts of the liquid up both proboscides, with larger shifts up the right-hand proboscis. These results are in accordance with the predictions, except that the liquid did not move spontaneously. Significant disturbances ($Bo \leq 0.3$) were necessary to bring about "equilibrium

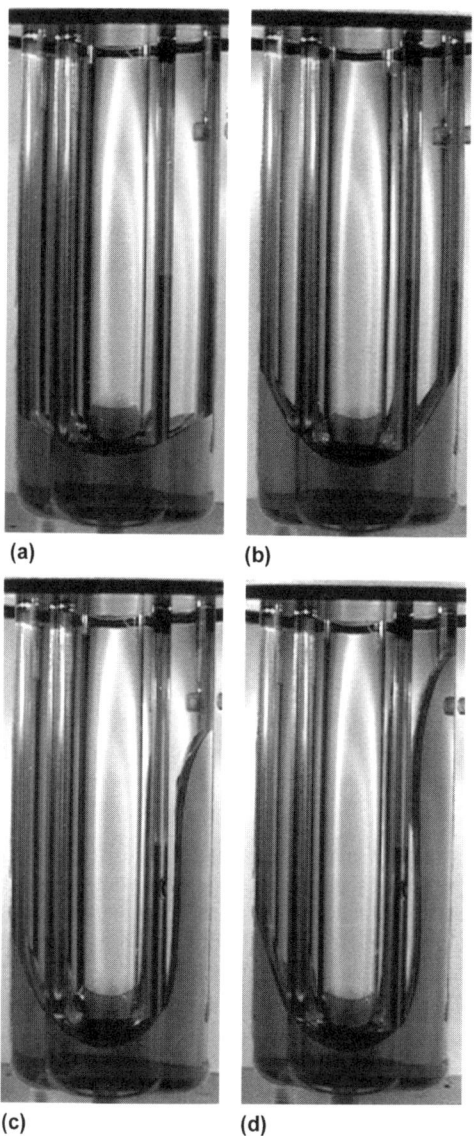

Fig. 10.6. Static interface shapes for ICE-P2 (30°/34°) vessel. (**a**) After completion of fill; (**b**)–(**d**) after successive disturbances by payload specialist

behavior". After completion of the the ICE-P3 procedures, the crew placed the vessel in the aft end cone of the Spacelab module, where it was allowed to remain for seven days. During that time it was observed that the liquid continued to creep toward the end state configuration of Fig. 10.7. The liquid is seen to have risen further in the left-hand proboscis, while the liquid in the

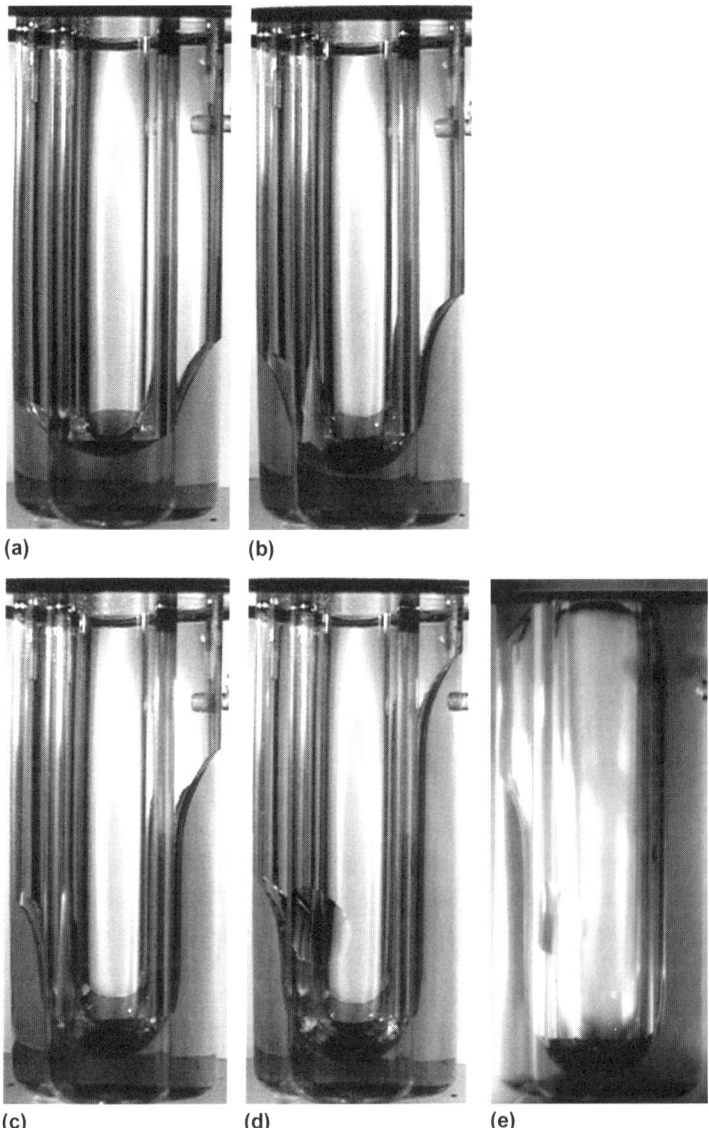

Fig. 10.7. Static interface shapes for ICE-P3 (38°/44°) vessel. (**a**) After completion of fill; (**b**)–(**d**) after successive disturbances by payload specialist; (**e**) after one week

right-hand proboscis rose to the lid, covered it at least partially, and then started advancing down the left-hand proboscis.

We conclude that contact angle hysteresis does not prevent the predicted behavior, but only impedes it noticeably.

10.2 Exotic Containers

The proboscides are characterized by the fact that the capillary pressure is independent of the liquid volume in the wedge, i.e. $dp/dV = 0$. An alternative interesting situation arises if the liquid volume becomes independent of the capillary pressure, i.e. $dV/dp = 0$.

A liquid drop on a horizontal plane rests in a metastable position. It may be shifted sideways without a change in energy. Work has to be done against viscous friction only. This is also true for a liquid drop in a spherical container in a microgravity environment. Spherical geometry under microgravity is equivalent to a horizontal plane under normal gravity. In addition to these trivial cases, several container shapes with exotic properties have been reported. It is possible to construct axisymmetric container shapes in which liquid surfaces with equal energies and volumes but different curvatures (capillary pressure) are possible. These surfaces, in contrast to those described above, require well-defined contact angles between the liquid and the container material. They become unstable otherwise. A second cause of instability may be residual gravity.

10.2.1 Circular Tubes with Unusual Properties

A liquid surface in a circular tube adopts a spherical shape that depends on the contact angle γ. If the tube has radius R, the radius r and pressure p of the liquid surface are given by

$$r \cos \gamma = R, \quad \widehat{p} = \frac{2}{r} = \frac{2 \cos \gamma}{R}. \tag{10.16}$$

Let us consider first a tube with contact angle $\gamma_1 < \pi/2$ in the upper part $z > 0$ and the complementary contact angle $\gamma_2 = \pi - \gamma_1$ in the lower part $z < 0$, which is filled up to the height $z = 0$, as depicted in Fig. 10.8. In this case there exist three types of fluid interfaces with the same volume:

1. the surface wets the upper part and is concave
2. the surface wets the lower part and is convex
3. the surface sticks to the wetting barrier and is plane.

The first and the second configuration are stable, whereas configuration 3 is metastable. An accidental wetting of the upper part will draw more liquid into that section, and configuration 1 will arise. Similarly, an accidental dewetting of the lower section will bring about configuration 2. However, under certain conditions, a family of nonaxisymmetric configurations, including canthotaxis, may arise.

Let us now ask whether it might be possible to coat the tube, i.e. to vary the contact angle γ in such a manner that all the surfaces shown in Fig. 10.9a have the same volume and energy. The concave liquid surface 1 has a large

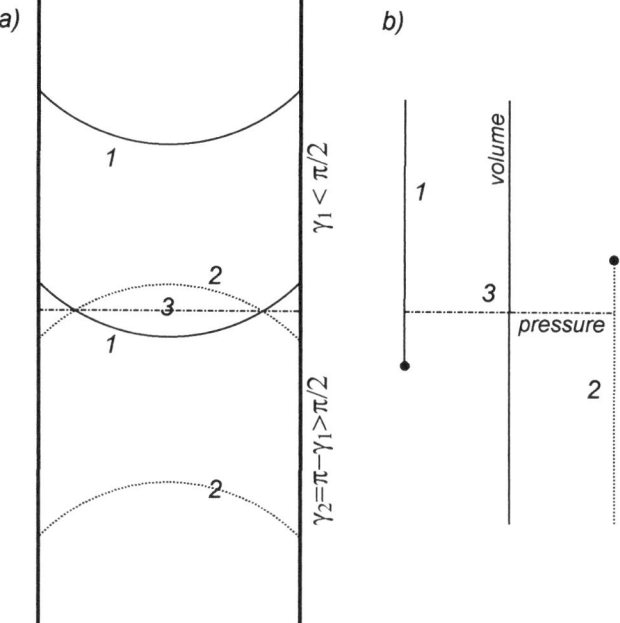

Fig. 10.8. (a) Spherical interfaces in a circular tube which is wetting in its upper part and nonwetting in its lower part. (b) Sketch of the corresponding volume-versus-pressure diagram

surface area. This disadvantage, however, is balanced by good wetting, by a large wetted tube area. The convex liquid surface 5 has a large surface area also, but this disadvantage is balanced a small wetted tube area. The plane liquid surface 3 corresponds to the average, to a contact angle of $\pi/2$. Is it possible that the contact angle γ can be adjusted in such a manner that all these surfaces have the same volume and energy? If the height z of the planar surface is denoted by $z = 0$, we obtain the following for the liquid volume enclosed by any of the surfaces:

$$V = \pi R^2 z - \frac{\pi r^3}{3}\left[2 - 3\cos\left(\frac{\pi}{2} - \gamma\right) + \cos^3\left(\frac{\pi}{2} - \gamma\right)\right] \tag{10.17}$$

$$V = \pi R^2 z - \pi R^3 \frac{\sin(\pi/4 - \gamma/2)\left[2 + \cos(\pi/2 - \gamma)\right]}{6\cos^3(\pi/4 - \gamma/2)}. \tag{10.18}$$

Hence, the condition that the volume is independent of z leads to

$$z = R\frac{\sin(\pi/4 - \gamma/2)\left[1 + 2\cos^2(\pi/4 - \gamma/2)\right]}{6\cos^3(\pi/4 - \gamma/2)}; \tag{10.19}$$

see Fig. 10.10. If $\gamma(z)$ satisfies (10.19), the family of surfaces shown in Fig. 10.9a has constant liquid volume, i.e. $dV/dp = 0$; $\gamma(z)$ must decrease

248   10. Playing with Stability

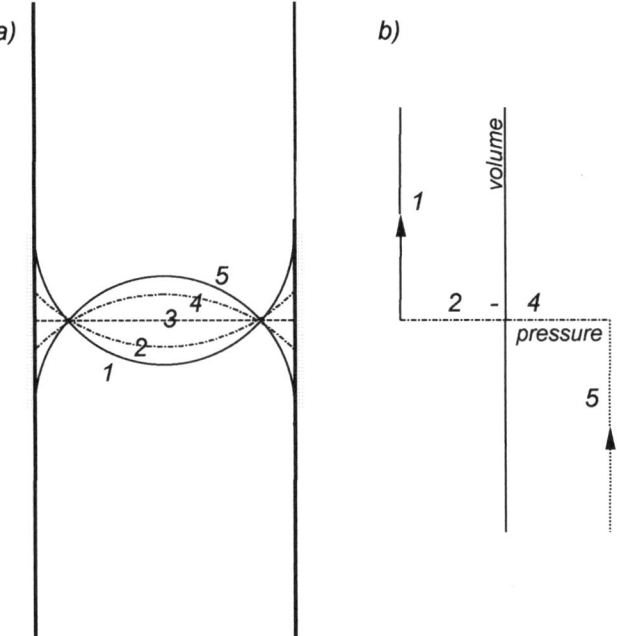

Fig. 10.9. (a) Spherical interfaces in a circular tube with the same volume and energy. (b) Sketch of the volume-versus-pressure diagram

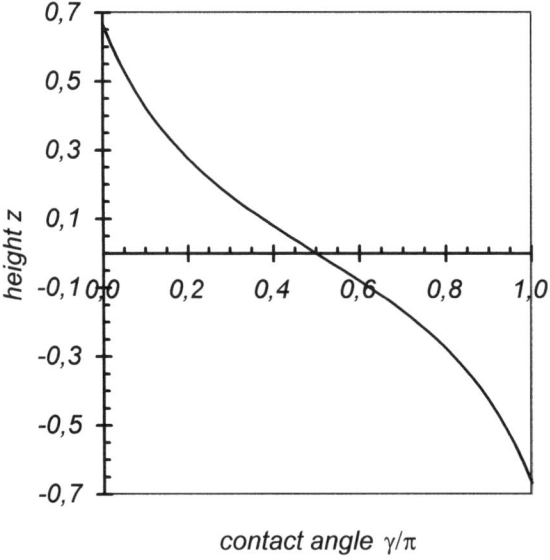

Fig. 10.10. The variation of the contact angle with height required to achieve constant volume and energy in a circular tube

from π to 0 within the range $-2/3 \leq z/R \leq +2/3$. Now we consider the energy of the surfaces shown. We obtain the following for the liquid surface area A_l and the wetted solid area A_s (weighted with $\cos\gamma$):

$$A_l = 2\pi r^2 (1 - \sin\gamma) = 2\pi R^2 \frac{1 - \sin\gamma}{\sin^2(\pi/2 - \gamma)}$$

$$= \frac{\pi R^2}{\cos^2(\pi/4 - \gamma/2)}, \tag{10.20}$$

$$A_s = 2\pi R \int_0^z d\zeta \cos\gamma. \tag{10.21}$$

All surfaces of the family considered have the same energy $A_l - A_s = \pi R^2$ if

$$R\tan^2\left(\frac{\pi}{4} - \frac{\gamma}{2}\right) = 2\int_0^z d\zeta \cos\gamma. \tag{10.22}$$

Differentiation of the energy condition (10.22) with respect to γ yields

$$-R\frac{\sin(\pi/4 - \gamma/2)}{\cos^3(\pi/4 - \gamma/2)} = 2\cos\gamma\frac{dz}{d\gamma}, \quad -R\frac{d\gamma}{dz} = 4\cos^4\left(\frac{\pi}{4} - \frac{\gamma}{2}\right). \tag{10.23}$$

Differentiation of condition (10.19) with respect to γ, holding the volume fixed, also yields (10.23), which proves that all the surfaces of the family considered have the same volume and energy.

Which of these configurations is stable? The volume-versus-pressure diagram sketched in Fig. 10.9b indicates that all of them are metastable. Any deviation of the contact angle from the intended value changes the volume-versus-pressure curve, such that the surface becomes caught in an unintended position, or even may switch to a nonaxisymmetric configuration. The latter suggestion is strongly supported by the investigations of Concus and Finn [1991] on exotic containers, where the container shape is varied rather than the contact angle to obtain families of surfaces showing constant volume and energy.

10.2.2 Adjustment of Container Shape

Let us now consider the stability of liquid surfaces in exotic containers which, for a constant contact angle γ, allow spherical capillary surfaces with different radii r and pressures $\widehat{p} = 2/r$ but equal liquid volumes V and effective surface areas $A_l - A_s \cos\gamma$. Special attention will be paid to deviations of the actual contact angle γ from the contact angle γ_0 used for the design and to residual accelerations (variations of gravity).

There exist axisymmetric containers that permit a continuum of distinct, axisymmetric free surfaces, all enclosing the same liquid volume and having the same energy and contact angle. The special case of zero gravity and

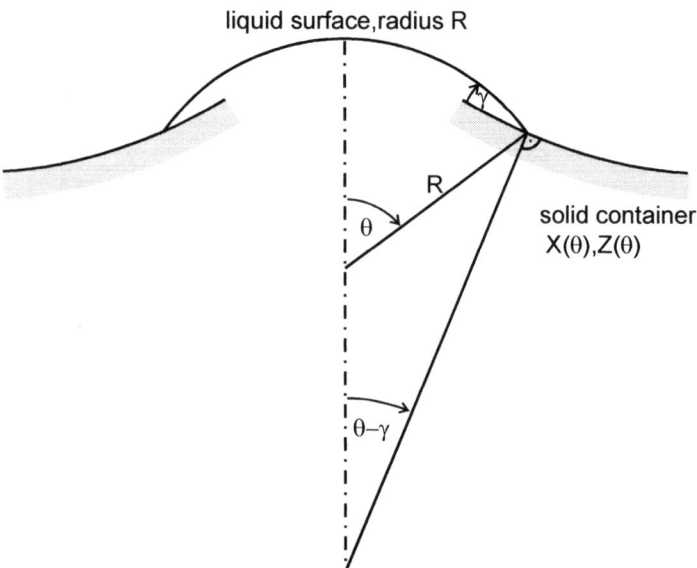

Fig. 10.11. Sketch of the calculation of the shape $X(\theta), Z(\theta)$ of an exotic container. This shape allows spherical fluid surfaces that differ in polar angle $\vartheta = \theta$ and radius $R(\theta)$ but are equal in volume and energy

contact angle $\pi/2$ has been studied by Gulliver and Hildebrandt [1986], where the authors derived a closed-form solution; the general case has been studied by Finn [1988]. It has been shown further, by Finn [1988] and Concus and Finn [1989, 1991], that the families of symmetric surfaces are unstable, in that certain asymmetric deformations yield surfaces with lower energy.

Assuming zero gravity, $Bo = 0$, let us consider spherical liquid surfaces whose radius $R(\theta)$ depends on the polar angle $\vartheta = \theta$ at which they contact an axisymmetric container with shape $X(\theta), Z(\theta)$; see Fig. 10.11. We obtain

$$X(\theta) = R(\theta) \sin \theta, \quad \frac{\mathrm{d}Z}{\mathrm{d}X} = -\tan(\theta - \gamma). \tag{10.24}$$

The liquid volume is given by

$$V(\theta) = \frac{\pi R(\theta)^3}{3} \left(2 - 3\cos\theta + \cos^3\theta\right) + \int^\theta \mathrm{d}\vartheta\, \pi X(\vartheta)^2 \frac{\mathrm{d}Z}{\mathrm{d}\vartheta}. \tag{10.25}$$

The condition that the liquid volume is independent of θ and R, $\mathrm{d}V/\mathrm{d}\theta = 0$, leads to

$$R(\theta)^2 \frac{\mathrm{d}R}{\mathrm{d}\theta} \left(2 - 3\cos\theta + \cos^3\theta\right) + R(\theta)^3 \sin^3\theta$$

$$= X(\theta)^2 \frac{\mathrm{d}X}{\mathrm{d}\theta} \tan(\theta - \gamma), \tag{10.26}$$

$$\frac{\mathrm{d}R}{R\mathrm{d}\theta} = \frac{\sin\gamma(1+\cos\theta)}{\cos\gamma(1-\cos\theta) - 2\sin\gamma\sin\theta}$$

$$= \frac{1}{\tan(\theta/2)\left(\cot\gamma\tan(\theta/2) - 2\right)} \, . \tag{10.27}$$

Integration of (10.27) yields

$$R(\theta) = \frac{R_0}{\sin(\theta/2)} [2\cos(\theta/2)\tan\gamma - \sin(\theta/2)]^{1/(1+4\tan^2\gamma)}$$

$$\times \exp\left(-\frac{\tan\gamma}{1+4\tan^2\gamma}\theta\right) \, , \tag{10.28}$$

where R_0 is the integration constant and is taken as the unit radius in the following.

As in Sect. 10.2.1 we have to make sure that the family of surfaces obtained has constant energy. The spherical liquid surface area A_l and the contact area A_s with the solid are given by

$$A_l = 2\pi R^2(1-\cos\theta) = 4\pi \left[R\sin\left(\frac{1}{2}\theta\right)\right]^2 \, , \tag{10.29}$$

$$A_s = \int^{\theta} \mathrm{d}\vartheta \, 2\pi X(\vartheta)\sqrt{\left(\frac{\mathrm{d}X}{\mathrm{d}\vartheta}\right)^2 + \left(\frac{\mathrm{d}Z}{\mathrm{d}\vartheta}\right)^2} \, . \tag{10.30}$$

The lower integration limit may be chosen arbitrarily. For the energy to be independent of θ, we require $\mathrm{d}E/\mathrm{d}\theta = 0$, yielding

$$\frac{\mathrm{d}}{\mathrm{d}\theta}R^2(1-\cos\theta) - \frac{\cos\gamma}{\cos(\theta-\gamma)}R\sin\theta\frac{\mathrm{d}}{\mathrm{d}\theta}(R\sin\theta) = 0 \, . \tag{10.31}$$

Equation (10.31) is equivalent to (10.27), i.e. the spherical liquid surfaces considered here indeed have equal volume and equal energy. It does not actually matter whether the contact angle or the container shape is adjusted. Obviously, the latter is easier experimentally.

10.2.3 Integration of Container Shape

From (10.24) and (10.28), we obtain for the container shape

$$X(\theta) = 2R_0 \cos\left(\frac{\theta}{2}\right)\left[2\cos\left(\frac{\theta}{2}\right)\tan\gamma - \sin\left(\frac{\theta}{2}\right)\right]^{1/(1+4\tan^2\gamma)}$$

$$\times \exp\left(-\frac{\tan\gamma}{1+4\tan^2\gamma}\theta\right) \tag{10.32}$$

$$\frac{dZ}{d\theta} = R_0 \frac{\sin(\theta-\gamma)}{\cos\gamma} \left[2\cos\left(\frac{\theta}{2}\right)\tan\gamma - \sin\left(\frac{\theta}{2}\right)\right]^{-4\tan^2\gamma/(1+4\tan^2\gamma)}$$
$$\times \exp\left(-\frac{\tan\gamma}{1+4\tan^2\gamma}\theta\right). \tag{10.33}$$

Explicit integration of (10.33) is possible for $\gamma \approx \pi/2$. Putting

$$Z(\theta) = R_0 \zeta(\theta) \left[2\cos\left(\frac{\theta}{2}\right)\tan\gamma - \sin\left(\frac{\theta}{2}\right)\right]^{1/(1+4\tan^2\gamma)}$$
$$\times \exp\left(-\frac{\tan\gamma}{1+4\tan^2\gamma}\theta\right) \tag{10.34}$$

leads to a power series in $\cot\gamma$ for $\zeta(\theta)$, which for $\gamma \approx \pi/2$ reads

$$\zeta(\theta) = -\left[2\sin\left(\frac{\theta}{2}\right) + \log\left(\tan\left[\frac{\pi-\theta}{4}\right]\right)\right] \tag{10.35}$$
$$+ \frac{\cot\gamma}{4}\int_0^\theta d\vartheta \left\{\frac{\sin(\vartheta/2)}{\cos^2(\vartheta/2)} - \log\left[\tan\left(\frac{(\pi-\vartheta)}{4}\right)\right]\right\} + O\left(\cot^2\gamma\right).$$

The container shape obtained for $\gamma \approx \pi/2$ is shown in Fig. 10.12c. The corresponding shapes for $\gamma = \pi/6$ and $\gamma = \pi/3$, depicted in Figs. 10.12a,b, were obtained by numerical integration. Figure 10.13a shows the family of liquid surfaces in the container designed for $\gamma = \pi/2$. In Fig. 10.13b, the container has been segmented, and supplemented by two spherical sections with polar angle $5\pi/12$.

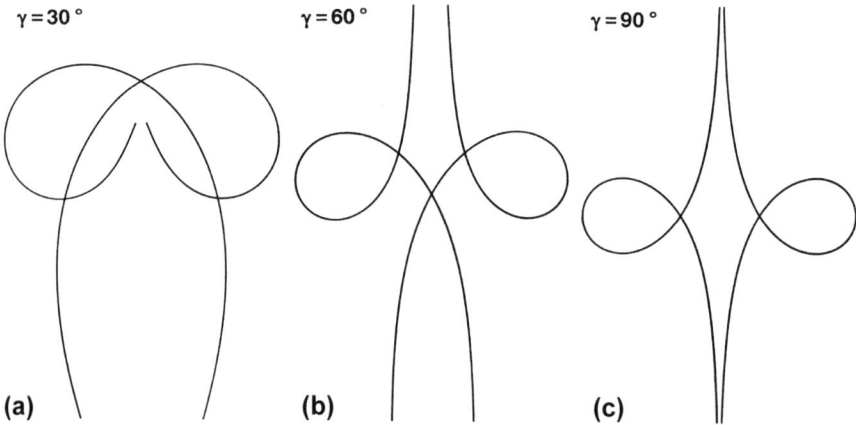

Fig. 10.12. The container shapes obtained for $\gamma = \pi/6, \gamma = \pi/3$ and $\gamma \approx \pi/2$. The first two shapes ((a) and (b)) were obtained by numerical integration, and the last one (c) by analytical integration

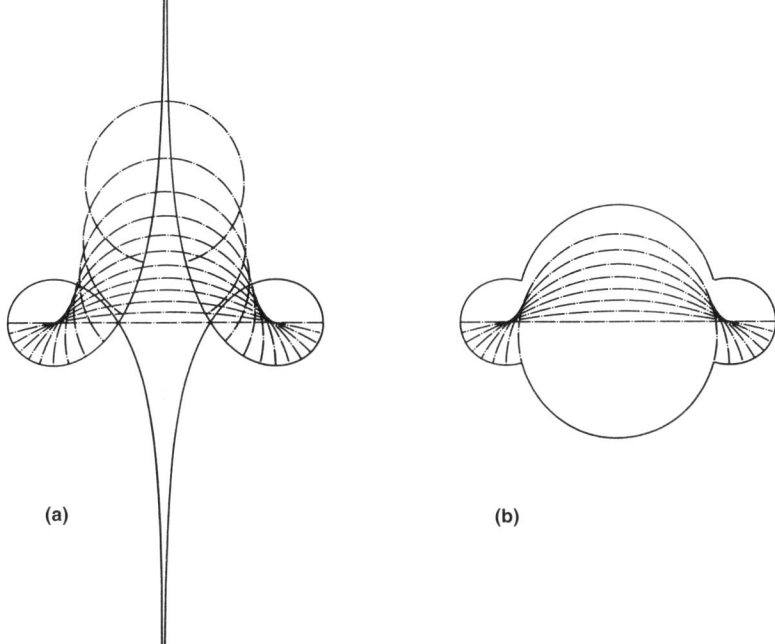

Fig. 10.13. (a) The family of liquid surfaces in the container designed for $\gamma = \pi/2$. (b) The container has been segmented, and supplemented by two spherical sections with polar angle $5\pi/12$

The design of exotic container shapes is by no means limited to zero gravity or to density matched fluids. Equivalent shapes also exist for finite Bond numbers [Concus and Finn 1990].

10.2.4 Mismatch of Volume and/or Contact Angle

The relation $dV/d\theta = 0$ means carrying the stability criterion to its extreme. All the spherical capillary surfaces in this case are metastable. On the other hand, metastability (in contrast to the metastable state in spherical containers) depends on filling the container with the exact liquid volume and on the accuracy of the contact angle γ. Any difference $\delta\gamma$ between the true contact angle and that used for the container design has the potential to strongly affect the stability. The liquid surface contacting the container at the contact line $X(\theta), Z(\theta)$ has a different radius $R + \delta R$ and pressure $p + \delta p$ and encloses a different volume $V + \delta V$.

In the following we consider the stability of liquid surfaces in containers designed for microgravity conditions when the actual contact angle γ differs from the contact angle γ_0 used for the design or if there is residual gravity. Let us check whether the capillary surfaces are stabilized or destabilized by a mismatch in contact angle and/or volume. If the true contact angle is larger than anticipated by $\delta\gamma$, this implies a corresponding increase of θ, $\theta \to \theta + \delta\gamma$, along the contact line. For the resulting change in radius $R \to R + \delta R$ of the spherical surface and in its pressure and volume, we obtain

$$X(\theta) = (R + \delta R)\sin(\theta + \delta\gamma), \tag{10.36}$$

$$\delta R = -R\cot\theta\,\delta\gamma, \quad \widehat{\delta p} = \frac{2}{R}\cot\theta\,\delta\gamma, \quad \delta V = \pi\frac{X^3(\theta)}{(1+\cos\theta)^2}\delta\gamma. \tag{10.37}$$

Thus, families of spherical capillary surfaces exist if $\delta\gamma$ and δV have the same sign. In that case we obtain

$$\operatorname{sgn}\left(\frac{dV}{d\theta}\right) = \operatorname{sgn}(\theta\,\delta\gamma), \quad \frac{dp}{d\theta} > 0, \quad \operatorname{sgn}\left(\frac{\partial V}{\partial p}\right) = \operatorname{sgn}(p\,\delta\gamma). \tag{10.38}$$

The stability condition $dV/dp < 0$ is satisfied only if

- the liquid surface is convex and both the contact angle and the liquid volume are smaller than anticipated, or
- the liquid surface is concave and both the contact angle and the liquid volume are larger than anticipated.

If the contact line moves to a position $X(\theta), Z(\theta)$ where the container has been segmented, it will stick to the edge owing to canthotaxis.

The above rules are also valid if the empty and filled fluid volumes, together with their contact angles γ and $\pi - \gamma$, are interchanged. Figure 10.14 gives the stable regions in a quantitative plot of volume versus pressure for the container shown in Fig. 10.13b.

10.2.5 Residual Gravity

A further important condition for the stability of the liquid surfaces considered here is zero gravity. A nonzero Bond number causes a flattening of a convex spherical surface at its pole and thus reduces the liquid volume enclosed. This volume reduction becomes larger the stronger the curvature and the capillary pressure of the surface are. It may therefore be expected that the minimum-volume condition $dV/dp < 0$ is satisfied, i.e. that the flattened liquid surfaces are stable.

In the limit of low Bond numbers it can be shown that for the family of flattened liquid surfaces to contact the container $X(\theta), Z(\theta)$ of the exact angle θ, their polar radius must be larger than $R(\theta)$ by

$$R(\theta, Bo) = R(\theta, 0)\left[1 + \frac{1}{6}Bo(1 + 2\cos\theta)\sin\left(\frac{\theta}{2}\right)\tan\left(\frac{\theta}{2}\right)\right]. \tag{10.39}$$

The corresponding decrease δV in liquid volume is given by

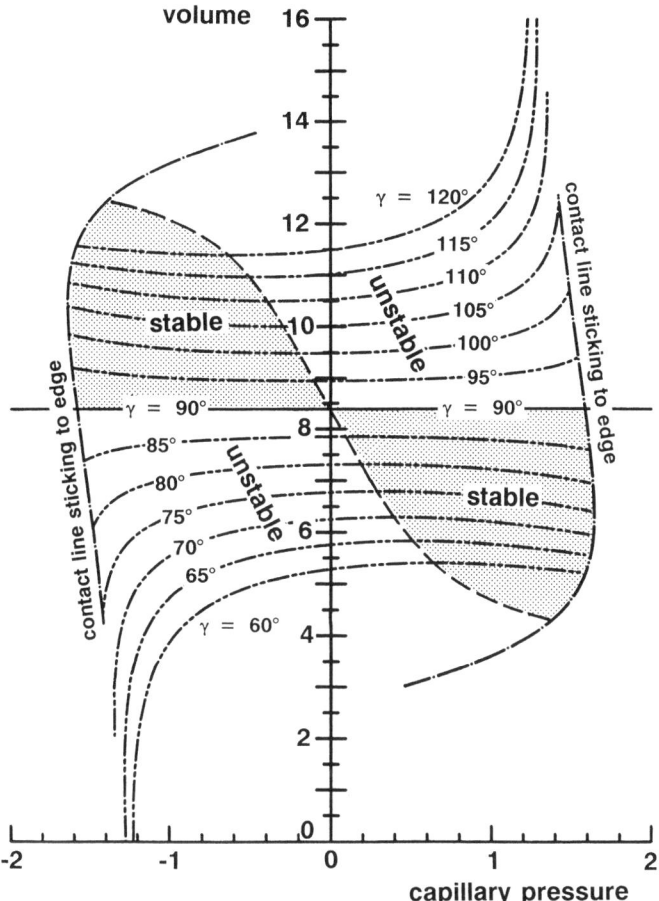

Fig. 10.14. Quantitative plot of volume versus pressure for the container shown in Fig. 10.13b, with the stable regions indicated

$$\delta V(\theta) \approx -\pi R^3(\theta)\, Bo \left[2\sin^6\left(\frac{\theta}{2}\right) + \frac{4}{3}\sin^8\left(\frac{\theta}{2}\right) - \frac{41}{36}\sin^{10}\left(\frac{\theta}{2}\right) \pm \ldots \right]. \tag{10.40}$$

Here δV is always negative and increases with increasing polar angle at the contact line, for both positive and negative values of θ. The volume increase of the concave surface is equal to the volume decrease of the convex surface. This is conveniently proved by interchanging the empty and filled regions and the sign of the Bond number.

A positive Bond number therefore tilts the volume-versus-pressure diagram shown in Fig. 10.14 downward at the right, such that a broad region of stable surfaces arises. It even allows some mismatch of the volume and the contact angle. A negative Bond number, on the other hand, tilts Fig. 10.14

downward at the left, such that the stable regions become fully separated. A positive Bond number strongly stabilizes the liquid surfaces considered here, and a negative Bond number strongly destabilizes them.

10.2.6 Drop Tower Tests

Drop tower investigations of water in acrylic containers have indicated that the liquid surface with the lowest energy may be nonaxisymmetric [Concus 1990, Concus et al. 1992]. Also, Finn [1988] has shown that a particular planar surface can be embedded into a one-parameter family with decreasing energy. This surface must undergo a nonaxisymmetric deformation.

The planar surface is perhaps best adapted to drop tower tests, where the transition from $1g$ to $0g$ is quite sudden. In preparatory work, equilibrium contact angle values of $80°$ for distilled water on acrylic were measured and were shown to be repeatable to $\pm 2°$.

Several experiments were performed, observing the interface over the 5 s period of free fall. The experiment of central interest was one in which the vessel was filled with the prescribed liquid volume with an initially flat interface. The flat interface is an equilibrium solution for any gravity level, but for zero gravity this interface is unstable in this vessel according to the idealized theory, as are all members of the symmetric equilibrium family. This experiment was carried out twice. In one case the surface remained essentially flat, in its initial configuration. In the other case the surface reoriented to a nonsymmetric configuration, as shown in Fig. 10.15. Which situation occurred probably depended on the nature of the small perturbation imparted to the fluid at the initiation of free fall. It is significant, however, that in one case the fluid did move in the 5 s of free fall from its initial flat equilibrium configuration to an obviously nonsymmetric one, in accordance with the mathematical theory.

Two experiments were carried out with the vessel tilted by a small amount from vertical, in an attempt to overcome surface friction (resistance at the contact line) that might be preventing reorientation. This procedure biased the surface toward a particular nonsymmetric situation from the outset. For an experiment with an initial tilt of $2.5°$, reorientation did not occur, whereas for a $5°$ initial tilt the shape depicted in Fig. 10.15 resulted. Thus, in the total of four experiments performed with initially horizontal or tilted planar interfaces, two interfaces reoriented to an asymmetric configuration as depicted in Fig. 10.15.

Additional experiments were carried out to investigate the effect of varying the fluid volume from that corresponding to a planar interface. The fill volume for a horizontal planar interface was $314\,\text{cm}^3$, of which the volume of the portion of the fluid in the bulge was approximately $48\,\text{cm}^3$. The amount of fluid less than or in excess of the planar-interface fill level was generally about $10\,\text{cm}^3$. For an initially vertical vessel, the one underfilled case resulted in reorientation during free fall to a symmetric surface that domed upward in

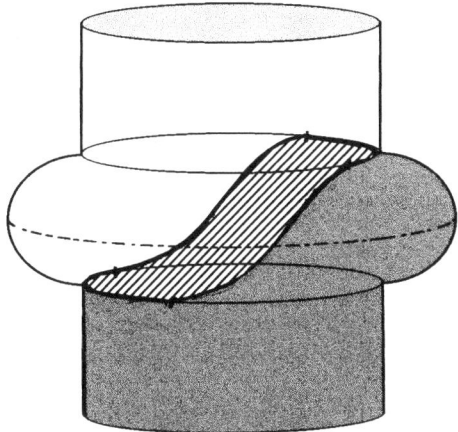

Fig. 10.15. Of the total of four drop experiments performed with initially horizontal or tilted planar interfaces, two interfaces reoriented to an asymmetric configuration

the center. The overfilled cases were tested on initially tilted vessels; of these, two with initial tilts of 2.5° and 5° reoriented to a surface similar to the one shown in Fig. 10.15. In the other overfilled case, with a 1° tilt and only about 3 cm^3 overfill, although an apparently stable domed-down meniscus was formed, the surface remained off-axis, maintaining the initial tilt with respect to the container.

The off-axis sticking phenomenon was investigated in similar experiments performed with right circular cylinders and some fluid/solid pairs with a larger contact angle, i.e. with wetting angles substantially closer to 90° than to 0°, such as water and acrylic. (Typically, contact angle hysteresis is more prominent at larger contact angles.) The tests were carried out with 2 cm diameter cylinders in a smaller, less-controlled, but more readily available 2.2 s drop tower. The containers were tilted initially by amounts between 5° and 10°. For the smaller tilts, the initial surface inclination with respect to the container persisted after the container was put into free fall. For the larger initial tilts, the ensuing motion during free fall resulted in the contact line becoming unstuck, and the surface reoriented toward the axially symmetric configuration.

For the materials used, the motion-inhibiting effects of surface friction can apparently be overcome by the forces resulting from a sufficiently large initial tilt.

References

1. Chen A, Concus P, Finn R: On cylindrical container sections for a capillary free-surface experiment. AIAA Paper 95-0271 (1995) 1–8
2. Chen A, Concus P, Finn R, Weislogel M: On cylindrical container sections for a capillary free surface experiment. Microgravity Sci. Technol. **9** (1997) 169–174
3. Concus P, Finn R: Instability of certain capillary surfaces. Manuscr. Math. **63** (1989) 209–213
4. Concus P, Finn R: Capillary surfaces in microgravity. In: *Progress in Low-Gravity Fluid Dynamics and Transport Phenomena*. J.N. Koster, R.L. Sani (eds.). Progress in Astronautics and Aeronautics **130**. AIAA Washington (1990) 183–205
5. Concus P, Finn R: Exotic containers for capillary surfaces. J. Fluid Mech. **224** (1991) 383–394
6. Concus P, Finn R, Weislogel MM: Drop-tower experiments for capillary surfaces in an exotic container. AIAA J. **55** (1992) 134–137
7. Concus P, Finn R, Weislogel M: Measurement of critical contact angle in a microgravity space experiment. Exp. Fluids **28** (2000) 197–205
8. Finn R, Leise TL: On the canonical proboscis. Zeit. Anal. Anwend. **13** (1994) 443–462
9. Finn R, Matek JM: The modified canonical proboscis. Zeit. Anal. Anwend. **15** (1996) 95–108
10. Finn R: Nonuniqueness and uniqueness of capillary surfaces. Manuscr. Math. **61** (1988) 347–372
11. Fischer B, Finn R: Non-existence theorems and measurement of capillary contact angle. Zeit. Anal. Anwend. **12** (1993) 405–423
12. Gulliver R, Hildebrandt S: Boundary configurations spanning continua of minimal surfaces. Manuscr. Math. **54** (1986) 323–347

11. Liquid Penetration into Tubes and Wedges

Following an outline of the main features of fluid motion, the penetration of liquids into tubes and wedges is considered. There are two cases of particular interest: the primary intrusion of the liquid into a cylindrical flow channel, and the pressure gradient required for maintaining the flow in a long cylindrical channel.

The main difference between liquid intrusion into wedges from penetration into smooth flow channels is the extremely high (indeed infinite) capillary underpressure that arises in the tip, of the intruding liquid. Though the no-slip condition is imposed at the tip it nevertheless moves. The liquid reaches the tip and accumulates there by flowing along the surface and within the bulk. These flow characteristics need to be artificially required in models other than the one considered here.

The volume flow in a long channel is proportional to the square of the cross-sectional area times a geometric friction coefficient Φ, which for the well-known Hagen–Poiseuille flow in a circular tube equals $1/8\pi$. The flow in rectangular cylinders and parallelograms is conveniently treated by using an antimetric combination to form a periodic lattice, and application of a Fourier expansion.

11.1 About the Momentum, or Navier–Stokes, Equation

Let us outline the main features of fluid flow, as they appear in the momentum, or Navier–Stokes, equation:

- Liquids flow from regions of high pressure to regions of low pressure. In order for a liquid to flow through a tube, a pressure difference is required, which may be provided by a piston or by some form of suction.
- Liquids are usually viscous. If two adjacent liquid strings flow in parallel but with differing velocity, the faster string causes the slower one to speed up, which in turn slows down the faster string. The coefficient describing this viscous friction is the dynamic shear viscosity η.
- In order to flow, the liquid mass has to be accelerated. This requires an inertial force.

260   11. Liquid Penetration into Tubes and Wedges

- The ratio of the dynamic viscosity η to the density ρ is termed the kinematic viscosity ν.
- Additional external forces such as gravity, centrifugal forces, van der Waals forces and electrodynamic forces may act on the fluid. They drive flows as well.

These effects are collated in the momentum equation, or Navier–Stokes equation, for the flow velocity $\boldsymbol{u}(\boldsymbol{r})$:

$$\rho \frac{d\boldsymbol{u}}{dt} = \rho \left(\frac{\partial \boldsymbol{u}}{\partial t} + \boldsymbol{u} \cdot \nabla \boldsymbol{u} \right) = \boldsymbol{f} - \nabla p - \eta \nabla \times \nabla \times \boldsymbol{u}, \tag{11.1}$$

$$\frac{\text{mass}}{\text{volume}} \text{acceleration} = \frac{\text{force}}{\text{volume}} - \text{pressure gradient} - \text{viscous friction}. \tag{11.2}$$

The momentum equation is basically Newton's law applied to a moving volume element. The transformation from the corresponding moving coordinate system to a fixed coordinate system means replacing the total derivative $d\boldsymbol{u}/dt$ of the flow velocity with respect to time by the local derivative $\partial \boldsymbol{u}/\partial t$ and the momentum transport by the fluid $\boldsymbol{u} \cdot \nabla \boldsymbol{u}$.

The momentum equation has to be solved subject to the continuity equation

$$\nabla \cdot \boldsymbol{u}(\boldsymbol{r}) = 0. \tag{11.3}$$

The continuity equation is identically satisfied by introducing the vector potential $\boldsymbol{U}(\boldsymbol{r})$ such that

$$\boldsymbol{u}(\boldsymbol{r}) = \nabla \times \boldsymbol{U}(\boldsymbol{r}). \tag{11.4}$$

In axisymmetric problems $\boldsymbol{U}(\boldsymbol{r})$ has an azimuthal component only, which is referred to as the stream function. By applying the curl operator, the pressure gradient may be eliminated from the momentum equation (11.1). One obtains

$$\rho \left(\frac{\partial}{\partial t} \nabla \times \nabla \times \boldsymbol{U}(\boldsymbol{r}) + \nabla \times (\boldsymbol{u} \cdot \nabla \boldsymbol{u}) \right) = \nabla \times \boldsymbol{f} - \eta \nabla \times \nabla \times \nabla \times \nabla \times \boldsymbol{U}(\boldsymbol{r}). \tag{11.5}$$

By introducing the vorticity

$$\boldsymbol{V}(\boldsymbol{r}) = \nabla \times \boldsymbol{u}(\boldsymbol{r}) = \nabla \times \nabla \times \boldsymbol{U}(\boldsymbol{r}), \tag{11.6}$$

one can reduce (11.5) to

$$\rho \left(\frac{\partial}{\partial t} \boldsymbol{V}(\boldsymbol{r}) + \nabla \times (\boldsymbol{u} \cdot \nabla \boldsymbol{u}) \right) = \nabla \times \boldsymbol{f} - \eta \nabla \times \nabla \times \boldsymbol{V}(\boldsymbol{r}). \tag{11.7}$$

If the external force is conservative, i.e. if a potential exists, and the creeping-flow approximation is used, one is left with

$$\frac{\partial}{\partial(\nu t)}\boldsymbol{V}(\boldsymbol{r}) = -\nabla \times \nabla \times \boldsymbol{V}(\boldsymbol{r}), \quad \boldsymbol{V}(\boldsymbol{r}) = \nabla \times \boldsymbol{u}(\boldsymbol{r}) = \nabla \times \nabla \times \boldsymbol{U}(\boldsymbol{r}).$$
(11.8)

One is left with two differential equations of second order: the differential equation for the vorticity $\boldsymbol{V}(\boldsymbol{r})$ is homogeneous and time dependent, whereas in the corresponding equation for the vector potential $\boldsymbol{U}(\boldsymbol{r})$, the vorticity $\boldsymbol{V}(\boldsymbol{r})$ appears as an inhomogeneous term.

11.2 Penetration into Capillaries

11.2.1 Cylindrical Vessels

An exact treatment of liquid penetration into tubes and wedges should deal with

- sticking and/or moving contact lines
- advancing and/or receding contact angles.

The dynamics of moving contact lines, however, are not known. The inherent problems are discussed in detail in Sect. 2.4.

The common method of overcoming this situation is to restrict oneself to the resulting energy balance:

1. there is an energy gain due to wetting
2. the energy of wetting decreases, if the advancing contact angle exceeds the stationary contact angle
3. there is an energy loss due to viscous friction of the intruding liquid string
4. the contribution of the moving contact line to viscous friction is considered small.

Neglecting contributions 2 and 4, let us just balance the energy of wetting (1) against the viscous friction (3) of the intruding string. The simplest case is the circular tube, which leads to the well-known Hagen–Poiseuille relation. This relation is based on the flow of a quasi-infinite liquid string. In the absence of acceleration and external forces, the momentum equation (11.1) reduces to

$$(\nabla p)_z = -\eta (\nabla \times \nabla \times \boldsymbol{u})_z = \eta \nabla . \nabla u_z(r),$$
(11.9)

where z is the coordinate in the axial direction. Equation (11.9) is not a momentum equation any more, it is Stokes' law. Requiring zero flow velocity at the wall $r = R$ (Newton's boundary condition) leads to a quadratic velocity profile

$$u_z(r) = u_0(R^2 - r^2), \quad \frac{dp}{dz} = \eta \nabla \cdot \nabla u_z(r) = -4\eta u_0. \tag{11.10}$$

The volume flow is given by

$$\frac{\text{vol}}{\text{time}} = \int_0^R dr\, 2\pi r u_z(r) = \pi u_0 \frac{R^4}{2} = \frac{(\pi R^2)^2}{8\pi\eta} \frac{dp}{dz}. \tag{11.11}$$

Introducing the mean flow velocity $\bar{u}$ and generalizing to cylindrical vessels with arbitrary cross section Ω and boundary Σ (see Fig. 3.5), one obtains the general relations

$$\bar{u} = \frac{u_0 R^2}{2}, \quad \frac{\text{vol}}{\text{time}} = \Omega \bar{u} = \Phi \frac{\Omega^2}{\eta} \frac{dp}{dz}, \tag{11.12}$$

where Φ is a geometric friction coefficient, or hydraulic factor ($\Phi = 1/(8\pi)$ for a circular tube). Thus, for the intrusion of a liquid string of length l, the basic differential equation reads

$$\frac{dl}{dt} = \Phi \frac{\Omega}{\eta} \frac{p(l)}{l}. \tag{11.13}$$

The driving energy of wetting, E_wet, is given by

$$\frac{d}{dt} E_\text{wet} = -\bar{u}\sigma \Sigma \cos\gamma \tag{11.14}$$

(compare the derivation of the integral theorem in Sect. 8.2). If gravity acts, work has to be done against the potential energy E_pot.

$$\frac{d}{dt} E_\text{pot} = \bar{u} g \Delta\rho\, \Omega l. \tag{11.15}$$

Thus, in a flow channel of cross-sectional area Ω and length l, the balance of energy and forces leads to

$$l \frac{dl}{dt} = \frac{\Phi}{\eta} (\sigma\Sigma \cos\gamma - g\Delta\rho\, \Omega l). \tag{11.16}$$

Finally, taking into account the kinetic energy of the moving liquid string, we find

$$E_\text{kin} = \frac{2\rho\Omega l \bar{u}^2}{3}, \quad \frac{d}{dt} E_\text{kin} = \frac{2\rho\Omega}{3} \frac{dl}{dt} \left(\left(\frac{dl}{dt}\right)^2 + 2l \frac{d^2 l}{dt^2} \right), \tag{11.17}$$

$$l \frac{dl}{dt} = \frac{\Phi}{\eta} \left[\sigma\Sigma \cos\gamma - g\Delta\rho\, \Omega l - \frac{2\rho\Omega}{3} \left(\left(\frac{dl}{dt}\right)^2 + 2l \frac{d^2 l}{dt^2} \right) \right]. \tag{11.18}$$

The factor $(2/3)\bar{u}^2$ in E_kin, rather than $(1/2)\bar{u}^2$, results from the quadratic flow profile (11.10), according to the general relation

$$\int_0^R dr\, 2\pi r u_z^n = \pi \int_0^{R^2} dr^2\, u_0^n \left(R^2 - r^2\right)^n = \pi R^2 \frac{u_0^n R^{2n}}{n+1} = \pi R^2 \frac{(2\bar{u})^n}{n+1}. \tag{11.19}$$

11.2.2 Liquid Rise in Capillaries

The differential equation (11.17) for the penetration of a liquid into a cylindrical capillary with cross-section Ω and boundary Σ has a very preliminary character. It is based not on exact kinematics, but on energetic reasoning. It is not only the friction coefficient Φ which is unknown. (This coefficient may be determined numerically.) Other, more questionable parameters are the effective lengths l that arise in the potential energy and kinetic energy according to (11.15) and (11.17). And there are numerous problems related to the advancing contact angle.

Looking for solutions of (11.16) neglecting liquid acceleration, we note that the energy gain driving the process vanishes if the equilibrium length $l = l_0$ between capillary pressure and gravity is reached. Putting

$$l_0 = \frac{\sigma \Sigma \cos \gamma}{g \Delta \rho \Omega} \tag{11.20}$$

yields

$$\frac{dl}{dt} \frac{l_0 l}{l_0 - l} = \frac{\Phi}{\eta} \sigma \Sigma \cos \gamma \tag{11.21}$$

and

$$-l_0^2 \log\left(1 - \frac{l}{l_0}\right) - l_0 l = \frac{\Phi \sigma \Sigma \cos \gamma}{\eta} t . \tag{11.22}$$

This result was first reported by Lucas [1917]. In the limit $l \ll l_0$, i.e. under conditions of zero gravity, (11.22) reduces to a liquid rise proportional to $\sqrt{t}$, the so-called Washburn equation [Washburn 1921],

$$\frac{l^2}{2} = \frac{\Phi \sigma \Sigma \cos \gamma}{\eta} t . \tag{11.23}$$

The inertial terms in (11.18) usually become important at the beginning of liquid rise, for very short liquid strings, when viscous friction is still negligible. In that case one finds

$$\sigma \Sigma \cos \gamma = \frac{2\rho \Omega}{3}\left(\left(\frac{dl}{dt}\right)^2 + 2l \frac{d^2 l}{dt^2}\right) \tag{11.24}$$

and

$$\frac{3\sigma \Sigma \cos \gamma}{2\rho \Omega} = \left(\frac{dl}{dt}\right)^2 . \tag{11.25}$$

11.2.3 Liquid Penetration into Wedges

The relative importance of the different contributions to wetting changes drastically if cylindrical vessels containing sharp wedges are considered. The distinctions made in Sect. 8.1 regarding the stationary wetting of wedges also govern the dynamics. A planar wedge is wetted if the sum of the dihedral angle 2α and the contact angles γ_1 and γ_2 with faces 1 and 2 is smaller than the sum of angles π in a triangle; see (8.1). In the presence of gravity the hyperbolic rise reported by Taylor [1712] results; see Sect. 3.

The main change compared with liquid penetration into capillaries with smooth surfaces is the extremely high (even infinite) capillary underpressure; that arises in the tip of the intruding liquid string. On the one hand, the fluid surface in the wedge causes a high capillary underpressure, on the other hand, the no-slip boundary condition in the wedge strongly hinders liquid motion. It is the specific properties of the wedge, not the overall properties of the capillary, which govern penetration.

Figure 11.1 depicts the wetting of the acute wedges of rhombic prisms with acute angles of 60° (top) and 65° (bottom) during drop tower experiments at ZARM, Bremen. The contact angle γ of the Cargille 50350 fluid used here with the uncoated quartz cells is 19°. The intruding liquid volumes exhibit a "sword-like" shape; see Sect. 11.2.4 for more details.

In the following, the basic assumption is made that in the planes perpendicular to the wedge, the liquid surface is given by a circular section of radius $R(z)$, z being the coordinate in the wedge direction. This is the same principle as used in Sect. 9.2.2 for calculating the static liquid surfaces in polyhedral containers. The differential equation for liquid penetration into the wedge is derived from the continuity condition that the change in liquid volume on the side to be filled, that is for $\zeta \geq z$, must equal the flow through the cross section at z. This volume flow is generally proportional to the squared area of the cross section times the pressure gradient, yielding

$$\frac{\text{vol}}{\text{time}} = -\frac{\Phi}{\eta}(\text{cross section})^2 \frac{\partial p}{\partial z} \tag{11.26}$$

and

$$\frac{\partial}{\partial t}\int_z^\infty d\zeta\, \Omega_1 R^2 = -\frac{\Phi}{\eta}(\Omega_1 R^2)^2 \frac{\partial p}{\partial z}, \tag{11.27}$$

where $\Omega_1 R^2$ is the area of the local cross section, η is the dynamic viscosity and Φ is the friction coefficient introduced above [Langbein et al. 1994, 1996, 1998]. The normalized cross section Ω_1 is given by

$$\Omega_1 = \frac{\cos\gamma}{\sin\alpha}\sin\delta - \delta, \tag{11.28}$$

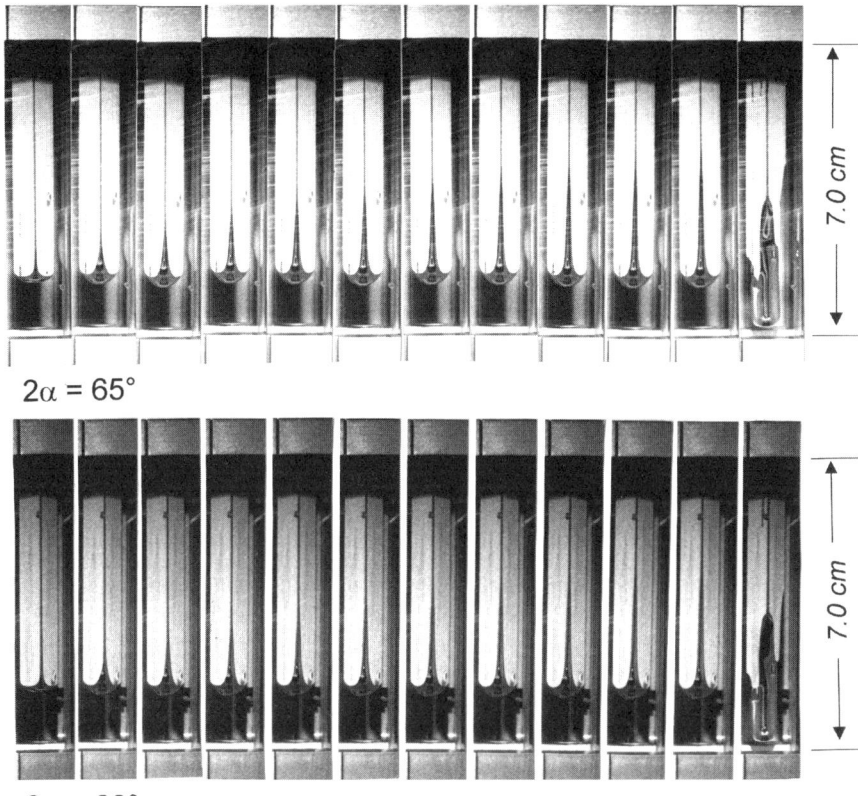

Fig. 11.1. Drop tower test with uncoated quartz cells, using Cargille 50350 fluid at ambient temperature. Rhombic angles 60° (*top*), 65° (*bottom*); the time interval between successive frames is 0.4 s

where $\delta = \pi/2 - \alpha - \gamma$. Differentiation with respect to z on both sides of (11.27) leads to

$$\frac{\partial \Omega_1 R^2}{\partial t} = +\frac{\Phi}{\eta}\frac{\partial}{\partial z}(\Omega_1 R^2)^2 \frac{\partial p}{\partial z} \ . \tag{11.29}$$

The capillary pressure, under the above assumption of circular sections of radius $R(z)$, is given by

$$\widehat{p} \equiv \frac{p}{\sigma} = -\left(\frac{1}{R} + \frac{\Gamma^2 R''}{\sqrt{1 + (\Gamma^2 R')^2}^3}\right) \ . \tag{11.30}$$

The first and second terms in the large parentheses in (11.30) represent the curvatures perpendicular and parallel, respectively, to the wedge direction. Γ^2 is the normalized depth of the liquid volume in the wedge,

$$\Gamma^2 = \frac{\cos\gamma}{\sin\alpha} - 1 \; ; \tag{11.31}$$

see also (9.30) and (9.31). Hence we obtain

$$\frac{\partial R^2}{\partial t} = -\Phi \frac{\sigma \Omega_1}{\eta} \frac{\partial}{\partial z} R^4 \frac{\partial}{\partial z} \left(\frac{1}{R(z)} + \frac{\Gamma^2 R''}{\sqrt{1+(\Gamma^2 R')^2}^3} \right) . \tag{11.32}$$

This is a fourth order partial differential equation for $R(z,t)$. In general, (11.32) has to be solved numerically. However, it is similar in character to a diffusion equation with the modification that different powers of $R(z,t)$ appear on its two sides. Equation (11.32) is an equation in which the effective diffusion coefficient decreases with decreasing cross section.

Scaling the time with the velocity σ/η makes it into a length. Scaling of lengths should be based on the pressure (curvature) $\hat{p} = 1/R_{\text{wdg}}$ in the container base. Hence, we scale lengths and time as

$$\tilde{R} = \frac{R}{R_{\text{wdg}}}, \quad \tilde{t} = t \frac{\Phi \sigma \Omega_1}{2\eta R_{\text{wdg}}} . \tag{11.33}$$

This scaling of time is similar to that obtained for the capillary rise in a circular tube. The essential difference is the strong dependence on curvature via the contact angle and the dihedral angle, which enter the normalized cross section Ω_1. If the critical wetting condition $\alpha + \gamma = \pi/2$ is approached, Ω_1 approaches zero in proportion to $\delta^2 = (\pi/2 - \alpha - \gamma)^2$ while R_{wdg} remains finite. This makes penetration very slow as γ approaches $\pi/2 - \alpha$ from below.

In the stationary limit, when the liquid surface has reached its final shape in the wedge, we reobtain (9.30). The liquid volume in the base approaches the cylindrical volume in the wedge exponentially. Γ is the decrement of that exponential protrusion (see Sect. 9.2.2).

11.2.4 Similarity Solutions for Long Times

To study the time dependence of the solutions of (11.32), let us first assume that the curvature of the surface in the axial direction is negligible in comparison with the curvature perpendicular to the wedge direction. This assumption is valid if ample time is allowed for the fluid to extend over a sufficient length such that it may be considered a slender column. The drop tower experiments, as depicted in Fig. 11.1, strongly support this assumption regarding the distinct "sword-like" shape of the advancing surface. In this case (11.32) reduces to

$$\frac{3}{2} \frac{\partial R^2}{\partial t} = \frac{\partial^2 R^3}{\partial z^2} , \tag{11.34}$$

where R, t and z are dimensionless (the tilde symbols ~ are dropped for convenience). Equation (11.34) yields similarity solutions under the general substitution

$$R(z,t) = \frac{1}{z^n} f(\zeta), \quad \zeta = \frac{z}{(2t)^{1/(n+2)}}. \tag{11.35}$$

Substitution of (11.35) into (11.34) leads to

$$\frac{3}{n+2} \frac{1}{\zeta^{2n-1}} \frac{df^2}{d\zeta} + \frac{d^2}{d\zeta^2}\left(\frac{f}{\zeta^n}\right)^3 = 0. \tag{11.36}$$

For arbitrary n, (11.36) has the particular solution $f(\zeta) = -\zeta^{n+2}/5$. Other exact solutions of (11.36) can be obtained: for $n = 1$, $f(\zeta) = (\zeta^{4/3} - \zeta^3)/5$, and for $n = -1$, $f(\zeta) = (1-\zeta)/\zeta$.

The solution best adapted to the motion of the surface in a wedge, however, arises for $n = 0$. Numerical integration of (11.36) for $n = 0$ shows excellent convergence of $f(\zeta)$ as $\zeta \to -\infty$. Using the fact that (11.36) is invariant under the substitution $f \to f/a^{n+2}, \zeta \to \zeta/a$, we choose $a = \zeta_{\text{tip}}$, yielding $\zeta_{\text{tip}} = 1$, $(df/d\zeta)_{\text{tip}} = -0.5$ and $f(0) = 0.345163$, $f(-\infty) = 0.465847$.

With the assumption $\partial^2 f/\partial \zeta^2, \partial f/\partial \zeta \ll f(\zeta)$, we obtain the following from (11.36) for $n = 0$:

$$\frac{\partial^2 f/\partial \zeta^2}{\partial f/\partial \zeta} + 2\frac{\partial f/\partial \zeta}{f} = -\frac{\zeta}{f(-\infty)}. \tag{11.37}$$

Integration over ζ yields

$$f(\zeta) \approx f(-\infty) + f_1 \int_{-\infty}^{\zeta} d\xi \exp\left(-\frac{\xi^2}{2f(-\infty)}\right), \tag{11.38}$$

where $f_1 = -0.100563$. This is the expected behavior of a diffusion equation for R^3 in which the diffusion coefficient decreases with decreasing R. Solutions for $n = -1, -0.5, 0, 0.5$ and 1 are shown graphically in Fig. 11.2. Figure 11.3 shows the specific case of $n = 0$, including the continuation of the graph along the negative ζ axis. The solution for $n = 0$ leads to a tip motion proportional to $\sqrt{t}$, whereas the solution for $n = -1$ would lead to a tip motion proportional to t. Furthermore, the solution for $n = 0$ shows satisfactorily the experimentally observed sword-like shape of the advancing surface.

Noting that $\zeta_{\text{tip}} = \text{const}$, we may express the asymptotic tip location in closed form as

$$z_{\text{tip}} + z_0 = \sqrt{C(t+t_0)}, \tag{11.39}$$

the general form of which was successfully used in correlating the drop tower results of Fig. 11.1. The above may be redimensionalized, producing

$$z_{\text{tip}} + z_0 = \sqrt{3 \frac{\Phi \sigma \Omega_1 R_{\text{wdg}}}{\eta}(t+t_0)}, \tag{11.40}$$

268   11. Liquid Penetration into Tubes and Wedges

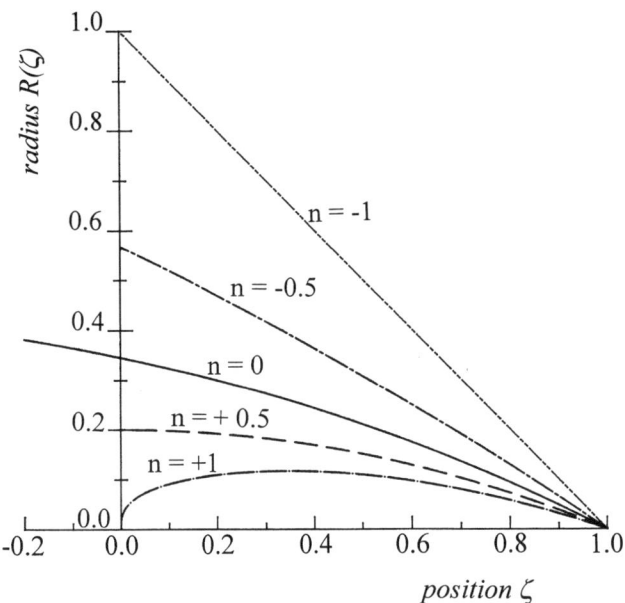

Fig. 11.2. The similarity solutions for a fixed value of ζ of 1 and $-1 \le n \le +1$. The tip position varies in proportion to $(2t)^{1/(n+2)}$. In particular, $f(\zeta) = (1-\zeta)/\zeta$ for $n = -1$ and $f(\zeta) = 0.2(\zeta^{4/3} - \zeta^3)$ for $n = +1$

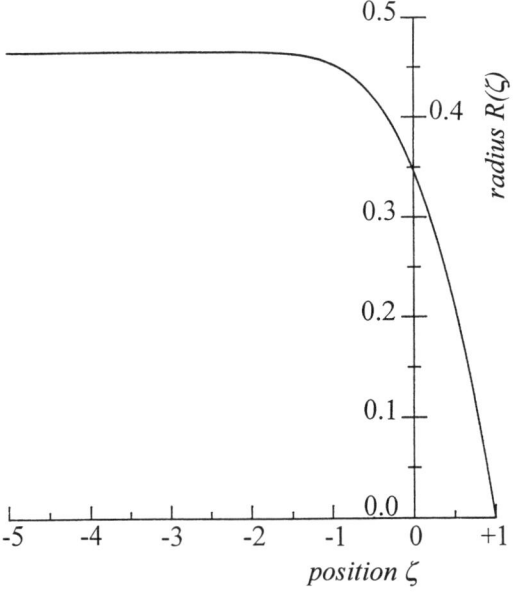

Fig. 11.3. The similarity solution for $n = 0$, leading to a constant surface height at minus infinity

from which the tip velocity may be determined as

$$v_{\text{tip}} = \sqrt{\frac{3\Phi\sigma\Omega_1 R_{\text{wdg}}}{4\eta(t+t_0)}}, \qquad (11.41)$$

where Φ is the friction coefficient, which can be computed numerically or measured experimentally. The prediction $v_{\text{tip}} \propto (\sigma/\eta)^{1/2}$ obtained from (11.41) corresponds to the drop-tower finding that the tip velocity increases with increasing temperature. Since $d\sigma/dT$ (= driving force) is smaller than $d\eta/dT$ (= retarding force), this aspect of the flow comes as no surprise.

11.2.5 Numerical Solution

A scheme for numerical solution of the basic flow equation (11.32) based on finite differences has been developed. After division by R, (11.32) reads

$$\frac{\partial R}{\partial t} = 2\left(\frac{\partial R}{\partial z}\right)^2 + R\frac{\partial^2 R}{\partial z^2} - \left(4R^2 \frac{\partial R}{\partial z}\frac{\partial}{\partial z} + R^3 \frac{\partial^2}{\partial z^2}\right) \frac{\Gamma^2 R''(z)}{\sqrt{1+(\Gamma^2 R')^2}^3}. \qquad (11.42)$$

Three basic initial surface shapes have been tested:

- a step function
- a linear function
- a parabolic function.

In each case (11.42) has been solved both by neglecting and by accounting for the axial curvature. In the latter case Γ^2 has been equated to 1.

When the numerical solution is started with an initial step function and the axial curvature is neglected, the similarity solution for $n = 0$ appears immediately; see Fig. 11.4a. If the strong positive curvature in the axial direction at the step is taken into account, an the axial pressure maximum arises, i.e. the liquid flows to both sides of the step, thus causing a maximum in height on the positive side of the step (Fig. 11.4b). Eventually, with considerable retardation, the same surface shape as that obtained when the axial curvature is neglected appears.

Figure 11.5 shows the result of the numerical integration when the initial surface shape is either linear or parabolic. At $z = 0$, a constant surface height and slope have been assumed. The height and slope have been chosen such that the surface approaches the tip position linearly in Fig. 11.5a and slightly parabolically in Fig. 11.5b. The surface penetrates faster into the wedge in the former case than in the latter. This is mainly due to the smaller slope of the surface in the linear case, which leads to a higher final surface. The results in the two cases are $R_{\text{wdg}} = 0.7006$ and 0.5894, respectively.

Furthermore, in all physically relevant cases, the solution for $n = 0$ dominates the solution right from the beginning, since the tip of the surfaces

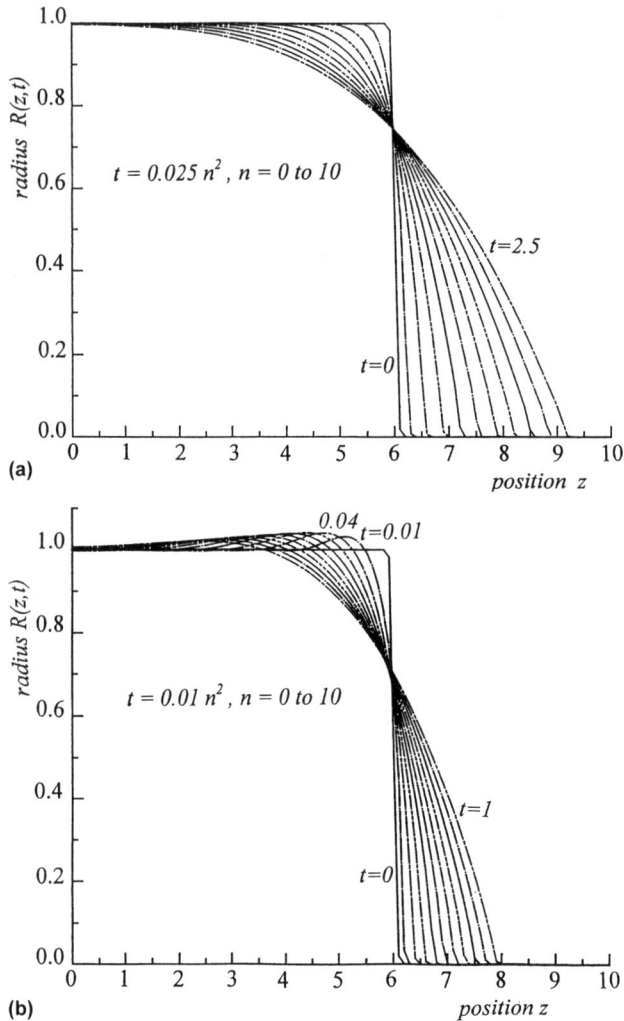

Fig. 11.4. Wetting of a wedge if the initial surface shape is a step function. (**a**) Neglect of axial curvature (yields $n = 0$ solution). (**b**) Inclusion of axial curvature. The sharp pressure maximum at the step causes the liquid to flow to both sides initially

advances in proportion to the square root of the time. This is demonstrated in Fig. 11.6 for the linear and parabolic initial surface profiles. (Note that if, instead of putting $\zeta_{\text{tip}} = 1$, we scale so that $f(-\infty) = 1$, we obtain $f(0) = 0.740937$ and $\zeta_{\text{tip}} = 1.465138$.)

Both the numerical solution and the similarity analysis yield solutions of particular interest: though no slip condition has been imposed at the tip, it nevertheless moves, and the slope of the interface has a nonzero value at the

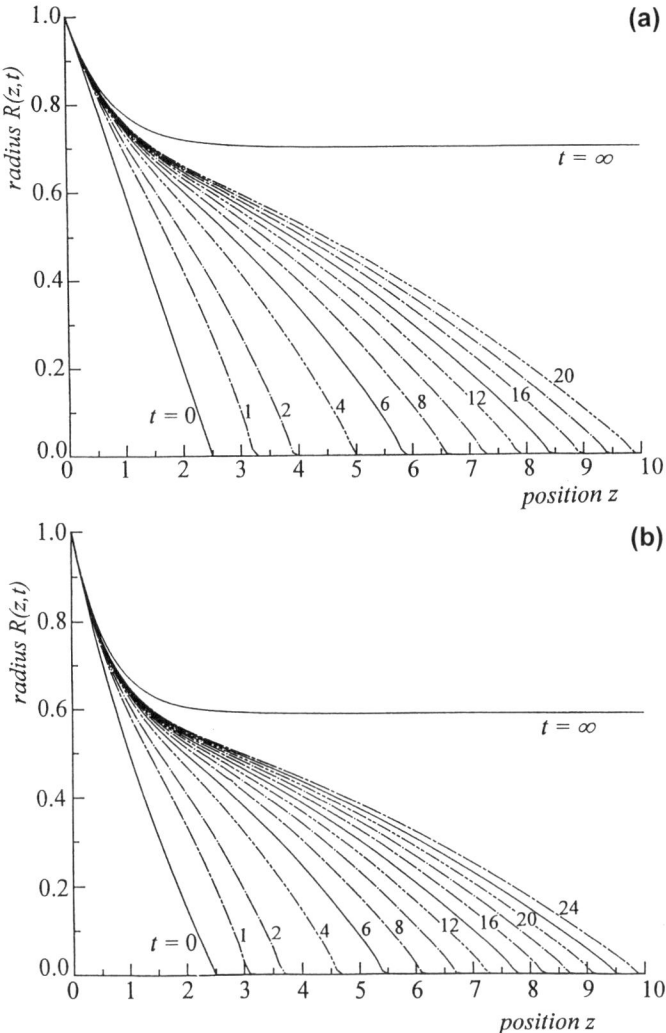

Fig. 11.5. Wetting of a wedge if the initial surface shape is either (**a**) linear or (**b**) parabolic. The surface penetrates faster into the wedge in the former case than in the latter. This is mainly due to the smaller negative slope of the surface in the linear case, which causes a higher final surface

tip irrespective of the value of the contact angle. The liquid reaches the tip and accumulates there by flowing along the surface and within the bulk. At the tip, the flow has a sink. These flow characteristics need to be required artificially in other models, but they arise naturally from (11.32). The reason is the higher power of R on the left-hand side of (11.32) than on the right-hand side.

272   11. Liquid Penetration into Tubes and Wedges

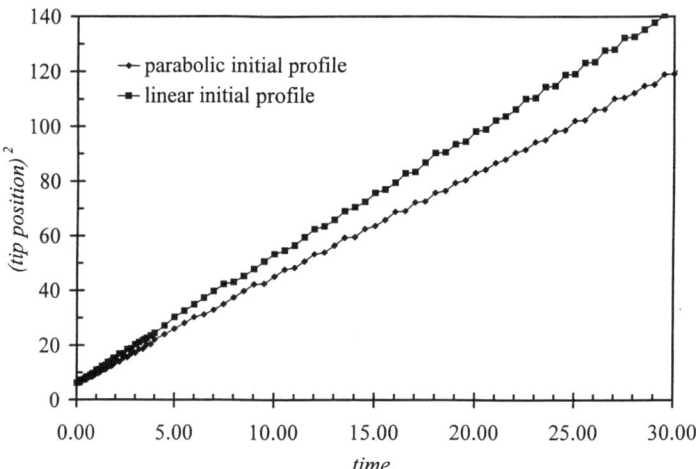

Fig. 11.6. Squared tip position of the surface shown in Fig. 11.6 versus time. The similarity solution for $n = 0$, tip position $\sim \sqrt{t}$, holds right from the beginning. The *solid squares* denote the results for a linear initial profile, and the *solid diamonds* denote the results for a parabolic initial profile

The $\sqrt{t}$ dependence obtained here for the length of liquid penetration into a circular tube and also into a wedge (see (11.23) and (11.39)) is, after all, a consequence of the balance between the energy gain due to wetting and the energy loss due to viscous friction.

11.3 Dynamics of Liquids in Edges and Corners

11.3.1 The DYLCO Experimental Module

The Dynamics of Liquids in Edges and Corners experiment (DYLCO, Sect. 13.16) on the second International Microgravity Laboratory (IML-2) Shuttle flight STS-65 was proposed as a simple experiment to probe the particular behavior of capillary surfaces in containers of rhombic cross section in the limit of near-critical wetting, $\alpha + \gamma \approx \pi/2$. The objective was to utilize temperature control to vary the fluid–solid contact angle γ. Its change in a system where all else is fixed provides a unique ability to observe the variety of surfaces which can form in a single vessel. Not only may changes in local shape be anticipated, but gross reorientation of the fluid is also possible in some cases.

While the IML-2 experiment focused on comparatively slow changes in fluid orientation in the limit of bad wetting of the wedges, rapid changes of orientation occur, for example, in the course of the spreading of a well-wetting liquid along a wedge. This situation was considered in drop tower experiments using the IML-2 experimental cells.

11.3 Dynamics of Liquids in Edges and Corners   273

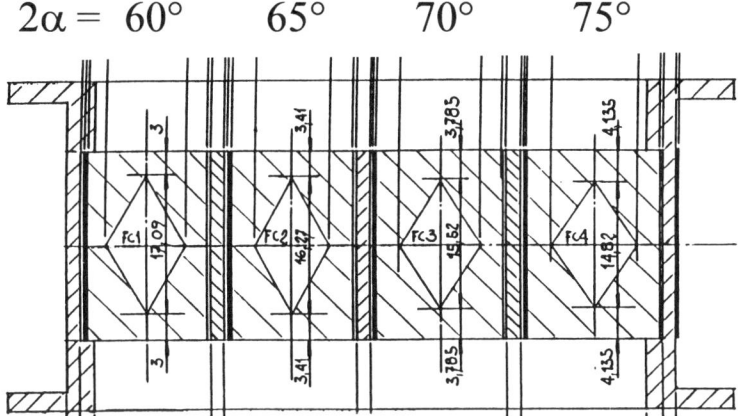

Fig. 11.7. Schematic of the test cell cross sections for the DYLCO IML-2 experiment and the drop tower tests

For the experiments, four cylindrical vessels of rhombic cross section with dihedral angles $2\alpha = 60°, 65°, 70°$ and $75°$ and identical inner cross-sectional areas made of fused silica (quartz) were used (denoted TC-60, TC-65, TC-70 and TC-75). Figure 11.7 shows the cross sections. The cells were made as large as the Bubble, Drop and Particle Unit (BDPU) would allow. This was to minimize the effects of surface roughness, contamination, etc. on the anticipated surface motion. The test liquid was Cargille 50350 fluid, which had an index of refraction matched to minimize optical distortions. The interior surfaces of the test cells were coated with the surface modifier FC-724. The coating was used to establish the desired wetting conditions in the cells. The equilibrium contact angles in the FC-724-coated cells TC-70 and TC-65 at 20°C and 100°C, respectively, guaranteed that the interfaces would be spherical and not wet any of the wedges. The uncoated cells, as used in the Bremen drop tower for the experiments on capillary spreading in wedges, showed a contact angle of approximately 19° with the liquid.

11.3.2 Drop Towers Tests for DYLCO

For the drop tower tests the cells were left uncoated, and thus the contact angle was approximately 19°, and $\alpha + \gamma < \pi/2$ applied for all the cells. The tests simply required a step reduction of gravity with a vessel partially filled with the test fluid. The resulting flows were captured on high-speed cine camera film. Observation of the film stimulated the development of the theory which now represents well the capillary-driven wedge flow that results when $\alpha + \gamma < \pi/2$. The effects of wedge angle and contact angle could also be confirmed; the results reveal clearly the decay of the wedge flow velocity to zero as γ approaches $\pi/2 - \alpha$.

A photographic sequence of sample data using the flight cells in the drop tower at Bremen is shown in Fig. 11.1. The rise of the fluid in the wedge can be seen by tracking the tip of a "sword-like" sword-like shape dark region along the center line (principal dihedral-angle axis) of the cell. Since microgravity was limited to the 4.74 s of free fall, the final two images of each sequence reveal the final retardation of the drop capsule during impact. Figure 11.8 displays the digitized tip location data for all four cells for the system at ambient temperature (Fig. 11.8a), 40°C (Fig. 11.8b) and 60°C (Fig. 11.8c). Two important observations can be readily made from these results:

- the rise rate increases with increasing temperature, since with increasing temperature the decrease in viscosity exceeds the decrease of surface tension
- the surface advances in proportion to $\sqrt{t}$.

In these results, a scaling law for the tip position following $z = \sqrt{C(t+t_0)}$ is observed, showing a strong dependence on temperature and a rather weak dependence on the actual dihedral angle of the prism; see Fig. 11.9. This latter observation agrees with the theoretical analysis in Sect. 11.2.4.

11.3.3 Conduct of the IML-2 Experiment

During the IML-2 flight, after insertion into the BDPU, the cells were partially filled via four independent fill ports. The filling of the cells was free of bubbles and no liquid fountain arose. During filling, the surfaces advanced smoothly, and thus the wetting conditions on the interfaces can be assumed to be smooth, too. Obviously, the interface in TC-60 was rather close to the wetting limit, as can be deduced from the sword-like surface tip extending along part of the wedge.

The temperature was then cycled from 20°C to 80°C in increments of 20 K, which required approximately 20 to 30 minutes per increment. Though the originally planned peak temperature was 80°C, since ample experimental time was available, the system temperature was increased further to 95°C. Cooling steps followed similarly, with the surface shapes being recorded on cine film after each step. Sample photographs of the fluid interfaces during thermal cycling are shown in Fig. 11.10. The change in the curvature of the interface from the initial ambient condition to the final ambient condition is easily observed and is attributable to the effects of contact angle hysteresis. Even in reprints, the interfaces can be clearly distinguished, leaving no need for postflight image enhancement. Furthermore, comparison of the advancing interfaces during the filling sequence as depicted in Fig. 11.10 with the almost-equilibrium shape prior to heating reveals clearly the effect of the dynamic contact angle on the curvature of the interface, and the subsequent contact angle relaxation after the end of the filling process.

11.3 Dynamics of Liquids in Edges and Corners 275

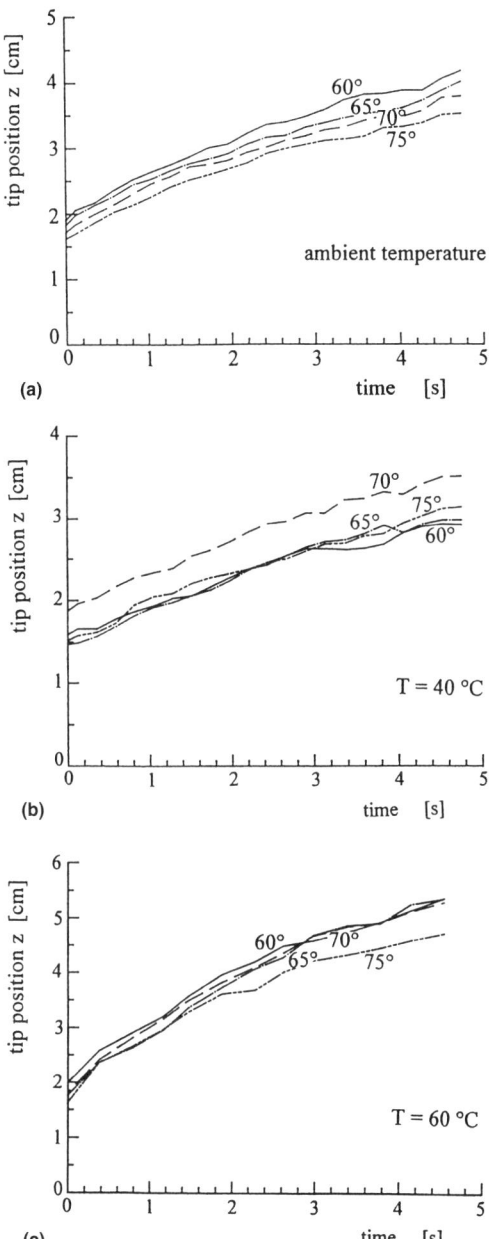

Fig. 11.8. Tip position z versus time t in the drop tower tests at (**a**) ambient temperature, (**b**) 40°C, (**c**) 60°C. The squared tip position increases in an approximately linear manner in all cases. Owing to the decrease in viscosity, the spreading velocity clearly increases with temperature

a) $z = 2.09 \sqrt{t+0.2}$ T = 60 °C
b) $z = 1.36 \sqrt{t+0.6}$ T = 40 °C
c) $z = 1.14 \sqrt{t+1.2}$ ambient temperature

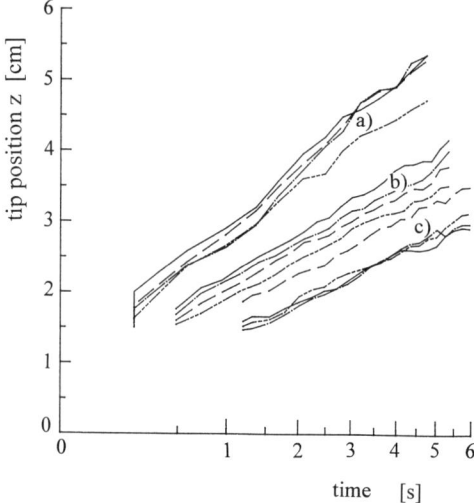

Fig. 11.9. Tip position versus $\sqrt{t}$ for drop tower tests. In the relation $z = C(t + t_0)^{0.5}$, the spreading coefficient C increases and the delay time t_0 decreases with increasing temperature

11.3.4 Results of the DYLCO IML-2 Experiment

As has already been stated, when $\pi/2-\alpha \leq \gamma \leq \pi/2+\alpha$, the resulting surfaces are portions of spherical surfaces. It is this feature of the surfaces achieved in orbit which allowed for their easy characterization. Once the radius of curvature of the surfaces had been measured from the film records, the angle of intersection of the surface with the wall in the plane of the photographs could be transformed to give the contact angle for the particular cell, at each temperature.

The spherical radius was determined by curve-fitting the digitized surface to the equation of a circle. The degree of precision of the fit was a direct indication of the degree of sphericity of the interface. A sample of values for γ determined in this way is provided in Table 11.1 for surfaces digitized at four selected time increments.

In all but increment 4, it was possible to fit the curves with a portion of a circle and to determine the contact angle. Since the surface is not spherical in increment 4, the local slope of the interface at the wall was measured and the local contact angle was extrapolated. The contact angle values shown in Table 11.1 for increment 4 are the results of this procedure. The results agree closely with the measured value for the receding contact angle. These contact angle values are well below the critical angle yet no flow, even locally, was

11.3 Dynamics of Liquids in Edges and Corners

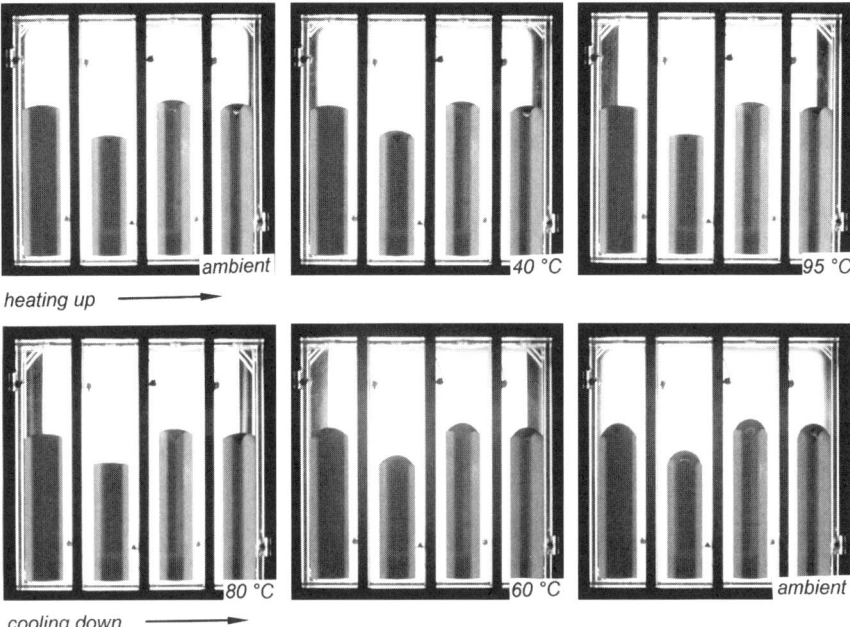

Fig. 11.10. IML-2 experimental data: the liquid surfaces during heating from ambient temperature to 40°C and 95°C and subsequent cooling down to 80°C, 60°C and ambient temperature. The dihedral angles 2α of the cells are 75°, 70°, 65° and 60° from *left* to *right*

Table 11.1. Contact angle γ (degrees) obtained by digitization of the interface in the plane perpendicular to the line of sight containing the vertices of the wedges with the larger angle. For increment 4, local contact angles measured near to the vertices of the vessels are tabulated

Increment	Effective contact angle	TC-60	TC-65	TC-70	TC-75
1	After filling, prior to heating	51.7	50.3	70.0	60.7
2	During the heating steps	54.6	52.3	44.9	65.7
3	At the peak temperature	65.2	64.8	80.8	72.8
4	At end of experiment (no spherical surfaces)	42.7	42.5	41.1	50.7
	Critical contact angle γ_{cr}	60.0	57.5	55.0	52.5

observed. The data selected in Table 11.1 for cells TC-70 and TC-75 appear to exhibit a somewhat anomalous behavior as compared with cells TC-60 and TC-65, a contributor to which could be a sporadic "stick-slip" nature of the fluid motion at the contact line.

The changes in interface shape between one heating step and the next were governed by the fact that the fluid expanded or contracted with an increase or decrease, respectively, in temperature. Owing to the contact angle hysteresis, relatively minor motion of the contact line was detected during each temperature step, with the change in fluid volume being accommodated primarily by the change in interface curvature. Surprisingly, the hysteretic angles were constant around the contact line in the majority of the tests. This allowed measurement of the effective contact angles in these instances, i.e. increments 1, 2 and 3 in Table 11.1. As pointed out in Table 11.1, however, a nonuniform contact angle distribution about the contact line was observed in increment 4. No theory developed to date is able to predict such surfaces, let alone their stability, as information on the history of the fluid is necessary, in addition to quasi-equilibrium constitutive data on the hysteretic behavior of the fluid–solid pair. Models incorporating variable contact angle distributions at the contact line may serve well in this regard. Local contact angle measurements in the vertices of the cells in increment 4 show that the receding limit is reached in the wedge, since γ_{rec} is approximately 40° at room temperature.

What is most interesting about the results in Table 11.1, for cells TC-60 and TC-65 in particular, is that the critical wetting condition was exceeded by the hysteretic surface curvature. What resulted were stable interfaces that did not follow the classical capillary equation. Indeed, the extreme situation of a receding surface that theoretically satisfied the condition $\alpha + \gamma_{rec} < \pi/2$ but nevertheless was unable to wet the wedge occurs here. The receding contact angle falls short of the wetting limit of the wedge but at the same time the liquid is unable to wet the wedge, because the stationary contact angle exceeds the wetting limit, i.e. $\alpha + \gamma_{sta} > \pi/2$. It is exactly this situation that is observed in increment 4.

11.4 The Geometric Friction Coefficient Φ

11.4.1 Flow in Rectangular Tubes

In Sect. 11.2.3 we introduced the friction coefficient Φ. This is a geometric measure of the flow resistance, which takes account of the shape of the flow channel considered. In the case of the well-known Hagen–Poiseuille flow in a circular tube, one obtains $\Phi = 1/(8\pi)$ from the quadratic flow profile. In the case of rectangular tubes, Φ is conveniently obtained by putting together tubes with opposite flow directions to create an infinite lattice, as shown in Fig. 11.11. This works even for parallelograms. However, while for rectangular tubes it follows from symmetry that the flow velocity vanishes on the tube boundaries, in the case of parallelograms quite different flows in opposite directions must be imposed.

Let us consider a periodic lattice of cylindrical flow channels, where the cross section of a channel is a parallelogram with sides $\boldsymbol{a}_1, \boldsymbol{a}_2$. From the translational invariance of the assumed flow in the axial direction z, we conclude that the pressure gradient is constant over the cross section. In neighboring

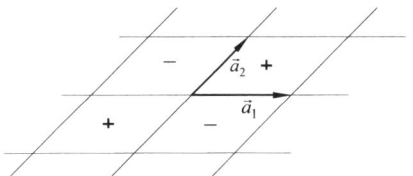

Fig. 11.11. Periodic, antimetric extension of a tube with parallelogram cross section; this can be expanded as a Fourier series

tubes, the flow velocity u_z and thus the pressure gradient $\nabla_z p$ have opposite directions. The period of the lattice is given by $2\boldsymbol{a}_1$ and $2\boldsymbol{a}_2$, where

$$\boldsymbol{a}_1 = a_1(1,0), \quad \boldsymbol{a}_2 = a_2(\cos\alpha, \sin\alpha). \tag{11.43}$$

The reciprocal vectors of this lattice are defined by

$$\boldsymbol{a}_i \cdot \boldsymbol{b}_j = \pi \delta_{ij}, \tag{11.44}$$

yielding

$$\boldsymbol{b}_1 = \frac{\pi}{a_1 \sin\alpha}(\sin\alpha, -\cos\alpha), \quad \boldsymbol{b}_2 = \frac{\pi}{a_2 \sin\alpha}(0,1). \tag{11.45}$$

We expand $\nabla_z p$ and the flow velocity u_z into Fourier series. Requiring

$$\nabla_z p = \begin{cases} -1 & \text{for } 0 < y < a_2 \sin\alpha \text{ and } y\cot\alpha < x < a_1 + y\cot\alpha \\ +1 & \text{for } -a_2 \sin\alpha < y < 0 \text{ and } y\cot\alpha < x < a_1 + y\cot\alpha \\ -1 & \text{for } -a_2 \sin\alpha < y < 0 \text{ and } -a_1+y\cot\alpha < x < y\cot\alpha \\ +1 & \text{for } 0 < y < a_2 \sin\alpha \text{ and } -a_1+y\cot\alpha < x < y\cot\alpha \end{cases}, \tag{11.46}$$

we obtain the direct product

$$\nabla_z p = \mp 1 = -\sum_{m \text{ odd}}^{\infty} \frac{4}{m\pi} \sin(m\boldsymbol{b}_1 \cdot \boldsymbol{x}) \sum_{n \text{ odd}}^{\infty} \frac{4}{n\pi} \sin(n\boldsymbol{b}_2 \cdot \boldsymbol{x}), \tag{11.47}$$

where m and n run over odd integers only, $m, n = 1, 3, 5, \ldots$ Now putting

$$u_z(x,y) \tag{11.48}$$
$$= \sum_{m \text{ odd}}^{\infty} \sum_{n \text{ odd}}^{\infty} [A_{mn} \cos(m\boldsymbol{b}_1 \cdot \boldsymbol{x})\cos(n\boldsymbol{b}_2 \cdot \boldsymbol{x}) + B_{mn} \sin(m\boldsymbol{b}_1 \cdot \boldsymbol{x})\sin(n\boldsymbol{b}_2 \cdot \boldsymbol{x})],$$

we obtain

$$\nabla \cdot \nabla u(x,y) = -\sum_{m \text{ odd}}^{\infty} \sum_{n \text{ odd}}^{\infty} \tag{11.49}$$

$$\left\{ \left[A_{mn}\left(m^2 \boldsymbol{b}_1 \cdot \boldsymbol{b}_1 + n^2 \boldsymbol{b}_2 \cdot \boldsymbol{b}_2\right) + 2B_{mn} mn \boldsymbol{b}_1 \cdot \boldsymbol{b}_2 \right] \cos(m\boldsymbol{b}_1 \cdot \boldsymbol{x})\cos(n\boldsymbol{b}_2 \cdot \boldsymbol{x}) \right.$$
$$\left. + \left[B_{mn}\left(m^2 \boldsymbol{b}_1 \cdot \boldsymbol{b}_1 + n^2 \boldsymbol{b}_2 \cdot \boldsymbol{b}_2\right) + 2A_{mn} mn \boldsymbol{b}_1 \cdot \boldsymbol{b}_2 \right] \sin(m\boldsymbol{b}_1 \cdot \boldsymbol{x})\sin(n\boldsymbol{b}_2 \cdot \boldsymbol{x}) \right\}.$$

Comparison of (11.49) with (11.47) leads to

$$B_{mn} = \frac{4}{m\pi}\frac{4}{n\pi}\frac{m^2\mathbf{b}_1\cdot\mathbf{b}_1 + n^2\mathbf{b}_2\cdot\mathbf{b}_2}{(m^2\mathbf{b}_1\cdot\mathbf{b}_1 + n^2\mathbf{b}_2\cdot\mathbf{b}_2)^2 - (2mn\mathbf{b}_1\cdot\mathbf{b}_2)^2},$$

$$A_{mn} = -\frac{4}{m\pi}\frac{4}{n\pi}\frac{2mn\mathbf{b}_1\cdot\mathbf{b}_2}{(m^2\mathbf{b}_1\cdot\mathbf{b}_1 + n^2\mathbf{b}_2\cdot\mathbf{b}_2)^2 - (2mn\mathbf{b}_1\cdot\mathbf{b}_2)^2}. \tag{11.50}$$

The denominator in (11.50) equals the product of the distances between the reciprocal lattice points. The cosine terms A_{mn} in (11.50) imply that the flow velocity $u_z(x,y)$ does not vanish on the tube boundaries except when $\mathbf{b}_1\mathbf{b}_2 \propto \cos\alpha = 0$, i.e. if the tube has a rectangular cross section. Integration over the cross section Ω, i.e. over $0 \le y \le a_2\sin\alpha$, $y\cot\alpha \le x \le a_1 + y\cot\alpha$, yields

$$\int_0^{a_2\sin\alpha} dy\,\sin(n\mathbf{b}_2\cdot\mathbf{x}) \int_{y\cot\alpha}^{a_1+y\cot\alpha} dx\,\sin(m\mathbf{b}_1\cdot\mathbf{x}) = \frac{2a_1}{m\pi}\frac{2a_2}{n\pi}\sin\alpha. \tag{11.51}$$

The mean axial velocity is

$$\frac{1}{\Omega}\int d\Omega\,u_z(x,y) \tag{11.52}$$

$$= \sum_{m\,\text{odd}}^{\infty}\frac{8}{(m\pi)^2}\sum_{n\,\text{odd}}^{\infty}\frac{8}{(n\pi)^2}\frac{m^2\mathbf{b}_1\cdot\mathbf{b}_1 + n^2\mathbf{b}_2\cdot\mathbf{b}_2}{(m^2\mathbf{b}_1\cdot\mathbf{b}_1 + n^2\mathbf{b}_2\cdot\mathbf{b}_2)^2 - (2mn\mathbf{b}_1\cdot\mathbf{b}_2)^2}.$$

The double infinite summation in (11.52) with m odd, n odd may be reduced to a single infinite summation by using the representation of the hyperbolic tangent by its poles and residues. The hyperbolic tangent has poles at $x = \pm in\pi/2$, yielding

$$\tanh(x) = \sum_{n\,\text{odd}}^{\infty}\left(\frac{1}{x - in\pi/2} + \frac{1}{x + in\pi/2}\right) = \sum_{n\,\text{odd}}^{\infty}\frac{2x}{x^2 + (n\pi/2)^2}, \tag{11.53}$$

$$\frac{\tanh(x)}{x^2} - \frac{1}{x} = \sum_{n\,\text{odd}}^{\infty}\frac{1}{(in\pi/2)^2}\left(\frac{1}{x - in\pi/2} + \frac{1}{x + in\pi/2}\right). \tag{11.54}$$

The coefficients B_{mn} given by (11.50) may be rewritten as

$$B_{mn} = -\frac{4}{mn\pi^2}\frac{1}{im b_1 \sin\alpha}\left(\frac{1}{mb_1 e^{-i\alpha} + nb_2} - \frac{1}{mb_1 e^{+i\alpha} + nb_2}\right.$$
$$\left. + \frac{1}{mb_1 e^{-i\alpha} - nb_2} - \frac{1}{mb_1 e^{+i\alpha} - nb_2}\right), \tag{11.55}$$

so that we obtain

11.4 The Geometric Friction Coefficient Φ

$$\frac{1}{\Omega}\int d\Omega\, u_z(x,y) = \sum_{m=1}^{\infty}\frac{2\pi}{(m\pi)^2}\frac{1}{mb_1b_2\sin\alpha}$$

$$\times\left(\frac{\tanh\left[m(b_1/b_2)e^{-i\alpha}i(\pi/2)\right]}{[m(b_1/b_2)e^{-i\alpha}i(\pi/2)]^2} - \frac{1}{m(b_1/b_2)e^{-i\alpha}i(\pi/2)}\right.$$

$$\left. -\frac{\tanh\left[m(b_1/b_2)e^{+i\alpha}i(\pi/2)\right]}{[m(b_1/b_2)e^{+i\alpha}i(\pi/2)]^2} + \frac{1}{m(b_1/b_2)e^{+i\alpha}i(\pi/2)}\right)\quad(11.56)$$

Applying the addition theorem of the tangent

$$ie^{+2i\alpha}\tan(x-iy) - ie^{-2i\alpha}\tan(x+iy)$$
$$= 2\frac{\cos(2\alpha)\sinh(2y) - \sin(2\alpha)\sin(2x)}{\cos(2x) + \cosh(2y)}\quad(11.57)$$

eventually leads to

$$\frac{1}{\Omega^2}\int d\Omega\, u_z = \frac{8a_1\sin\alpha}{a_2}\sum_{m\text{ odd}}^{\infty}\frac{1}{(m\pi)^4}\quad(11.58)$$

$$\left[1 + \frac{\cos(2\alpha)\sinh[(m(a_2/a_1)\pi\sin\alpha] - \sin(2\alpha)\sin[m(a_2/a_1)\pi\cos\alpha]}{m(a_2/a_1)(\pi/2)\sin\alpha\,\{\cos[m(a_2/a_1)\pi\cos\alpha] + \cosh[m(a_2/a_1)\pi\sin\alpha]\}}\right],$$

In the case of rectangular tubes, $\alpha = \pi/2$ and one is left with (see Fig. 11.12)

$$\Phi = \frac{1}{\Omega^2}\int d\Omega\, u_z = \frac{8a_1}{a_2}\sum_{m\text{ odd}}^{\infty}\frac{1}{(m\pi)^4}\left(1 - \frac{\tanh[m(a_2/a_1)(\pi/2)]}{m(a_2/a_1)(\pi/2)}\right).$$
(11.59)

The summation over m in (11.59) is conveniently performed by exploiting the fact that at large arguments $a_2/a_1 \geq 1$ the hyperbolic tangent exponentially approaches 1. Using

$$\sum_{m\text{ odd}}^{\infty}\frac{1}{(m\pi)^2} = \frac{1}{8},\quad \sum_{m\text{ odd}}^{\infty}\frac{8}{(m\pi)^4} = \frac{1}{12},$$

$$\sum_{m\text{ odd}}^{\infty}\frac{16}{(m\pi)^5} = 0.052520739690,\quad(11.60)$$

one obtains

$$\frac{1}{\Omega^2}\int d\Omega\,|u_z(x,y)|\quad(11.61)$$

$$= \frac{a_1}{12a_2} - \left(0.052521 - \sum_{m\text{ odd}}^{\pi}\left(\frac{2}{m\pi}\right)^5\frac{1}{1+\exp(m\pi a_2/a_1)}\right)\left(\frac{a_1}{a_2}\right)^2$$

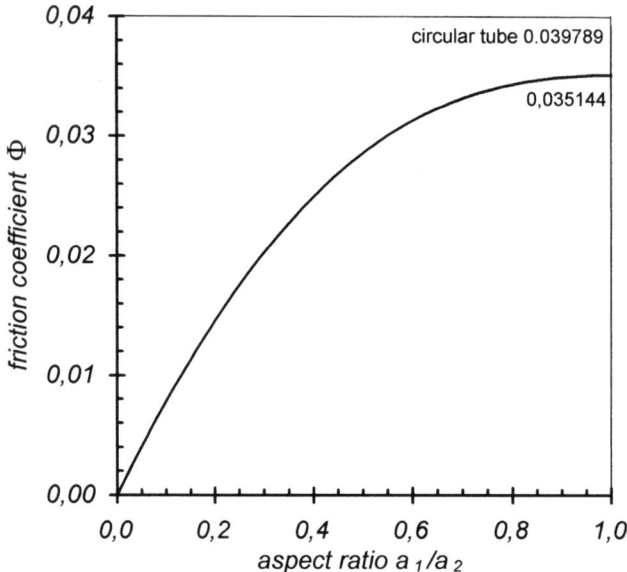

Fig. 11.12. The friction coefficient Φ in rectangular tubes

(see Fig. 11.12). For a tube with square cross section, $a_1 = a_2$, (11.61) gives a friction coefficient $c = 0.03514425$, to be compared with the value $\Phi = 1/(8\pi) = 0.03978874$ for a circular tube.

In the limit $a_1 \ll a_2$ one is left with the flow between two parallel planes a distance a_1 apart. In that case a quadratic velocity profile arises in the a_1 direction. One obtains

$$u_z(x) = u_0 \left[\left(\frac{a_1}{2}\right)^2 - \left(x - \frac{a_1}{2}\right)^2\right], \quad \nabla \cdot \nabla u_z(x) = 2u_0, \qquad (11.62)$$

$$\int d\Omega\, u_z(y) = a_2 \int_0^{a_1} dx\, u_0 \left[\left(\frac{a_1}{2}\right)^2 - \left(x - \frac{a_1}{2}\right)^2\right] = \frac{a_1^3 a_2}{12},$$

$$\frac{1}{\Omega^2} \int d\Omega\, |u_z(x)| = \frac{a_1}{12 a_2}. \qquad (11.63)$$

Comparison of (11.63) and (11.61) shows that the term $0.052521(a_1/a_2)^2$ accounts for the reduced flow at both ends of the a_2 interval. The deviation from the quadratic profile is finite and is integrable. Also, the decreases of the flow profile at the ends of the interval overlap exponentially. This result compares well with the exponential overlap of the liquid volumes piled up in opposite corners of a polyhedron, as found in Sects. 9.2.2–9.2.5.

11.4.2 Flow in Parallelograms

If the cross section of the tube is a parallelogram rather than a rectangle, (11.48) and (11.50) do not satisfy the condition of vanishing flow velocity $u_z(x, y) = 0$ along the tube boundaries. Fulfillment of this condition could be deduced from symmetry arguments in the case of rectangular tubes. In order to achieve $u_z(x, y) = 0$, the symmetry between parallelograms 1 and 3, on the one hand, and parallelograms 2 and 4, on the other hand, has to be given up. Since in the principal parallelogram 1, i.e. for $0 \leq y \leq a_2 \sin \alpha$, $y \cot \alpha \leq x \leq a_1 + y \cot \alpha$, and also in parallelogram 3, constancy of the pressure gradient $\nabla_z p$ has to be maintained, one has to allow for more general functions $\nabla_z p$ in parallelograms 2 and 4. There actually is no reason why these functions should represent pressure gradients.

A function $f(x, y)$ defined in parallelograms 2 and 4 only, which is given by the Fourier expansion

$$f(x, y) \tag{11.64}$$
$$= \sum_{m=0}^{\infty} \sum_{n=0}^{\infty} [a_{mn} \cos(m\boldsymbol{b}_1 \cdot \boldsymbol{x}) \cos(n\boldsymbol{b}_2 \cdot \boldsymbol{x}) + b_{mn} \sin(m\boldsymbol{b}_1 \cdot \boldsymbol{x}) \sin(n\boldsymbol{b}_2 \cdot \boldsymbol{x})]$$

in these parallelograms, is equal to the following Fourier series in the enlarged unit cell $2\boldsymbol{a}_1, 2\boldsymbol{a}_2$ covering all four parallelograms:

$$f(x, y) \tag{11.65}$$
$$= \sum_{m=0}^{\infty} \sum_{n=0}^{\infty} \left[\hat{a}_{mn} \cos(m\boldsymbol{b}_1 \cdot \boldsymbol{x}) \cos(n\boldsymbol{b}_2 \cdot \boldsymbol{x}) + \hat{b}_{mn} \sin(m\boldsymbol{b}_1 \cdot \boldsymbol{x}) \sin(n\boldsymbol{b}_2 \cdot \boldsymbol{x}) \right],$$

where

$$\hat{a}_{kl} = \frac{1}{2} a_{kl} - 2 \sum_{m=0}^{\infty} \sum_{n=0}^{\infty} b_{mn} \frac{m\left[1-(-1)^{k-m}\right]}{(m^2-k^2)\pi} \frac{n\left[1-(-1)^{l-n}\right]}{(n^2-l^2)\pi}, \tag{11.66}$$

$$\hat{b}_{kl} = \frac{1}{2} b_{kl} - 2 \sum_{m=0}^{\infty} \sum_{n=0}^{\infty} a_{mn} \frac{k\left[1-(-1)^{k-m}\right]}{(k^2-m^2)\pi} \frac{l\left[1-(-1)^{l-n}\right]}{(l^2-n^2)\pi}. \tag{11.67}$$

For the flow velocity $u_z(x, y)$ to vanish on the boundaries of a tube with cross section $\boldsymbol{a}_1, \boldsymbol{a}_2$, the coefficients A_{mn} in (11.48) have to be equated to zero, yielding

$$u_z(x, y) = \sum_{m \text{ odd}}^{\infty} \sum_{n \text{ odd}}^{\infty} B_{mn} \sin(m\boldsymbol{b}_1 \cdot \boldsymbol{x}) \sin(n\boldsymbol{b}_2 \cdot \boldsymbol{x}). \tag{11.68}$$

In that case, we obtain

$$-\nabla \cdot \nabla u(x, y) = \sum_{m \text{ odd}}^{\infty} \sum_{n \text{ odd}}^{\infty} B_{mn} \left[2mn\boldsymbol{b}_1 \cdot \boldsymbol{b}_2 \cos(m\boldsymbol{b}_1 \cdot \boldsymbol{x}) \cos(n\boldsymbol{b}_2 \cdot \boldsymbol{x}) \right.$$
$$\left. + \left(m^2 \boldsymbol{b}_1 \cdot \boldsymbol{b}_1 + n^2 \boldsymbol{b}_2 \cdot \boldsymbol{b}_2\right) \sin(m\boldsymbol{b}_1 \cdot \boldsymbol{x}) \sin(n\boldsymbol{b}_2 \cdot \boldsymbol{x}) \right]. \tag{11.69}$$

Comparison with (11.47) leads to

$$B_{mn}^{(1)} = \frac{4}{m\pi} \frac{4}{n\pi} \frac{1}{m^2 \boldsymbol{b}_1 \cdot \boldsymbol{b}_1 + n^2 \boldsymbol{b}_2 \cdot \boldsymbol{b}_2}, \qquad (11.70)$$

where the superscript (1) means that $B_{mn}^{(1)}$ represents the first term in the subsequent complete induction of $B_{mn}^{(i)}$. $B_{mn}^{(i)}$ gives rise to the cosine terms

$$\hat{a}_{mn}^{(i)} = 2mn\boldsymbol{b}_1 \cdot \boldsymbol{b}_2 B_{mn}^{(i)}, \qquad (11.71)$$

which contribute to $f(x,y)$ in parallelograms 2 and 4 as in (11.65) but with double weight. In the unit cell $2\boldsymbol{a}_1, 2\boldsymbol{a}_2$, in compensation, we have to add

$$\hat{b}_{kl}^{(i+1)} = -\left(\frac{4}{\pi}\right)^2 \sum_{m=1}^{\infty} \sum_{n=1}^{\infty} \frac{2\boldsymbol{b}_1 \cdot \boldsymbol{b}_2 klmn}{(k^2 - m^2)(l^2 - n^2)} B_{mn}^{(i)} \qquad (11.72)$$

according to (11.67), and

$$B_{kl}^{(i+1)} = -\left(\frac{4}{\pi}\right)^2 \frac{2\boldsymbol{b}_1 \cdot \boldsymbol{b}_2 kl}{k^2 \boldsymbol{b}_1 \cdot \boldsymbol{b}_1 + l^2 \boldsymbol{b}_2 \cdot \boldsymbol{b}_2} \sum_{m=1}^{\infty} \sum_{n=1}^{\infty} \frac{mn B_{mn}^{(i)}}{(k^2 - m^2)(l^2 - n^2)}, \qquad (11.73)$$

with m, n running over odd and even integers for i odd and even, respectively. Contributions to the mean velocity arise only from the contributions $B_{mn}^{(i)}$ with i odd, yielding the friction coefficient

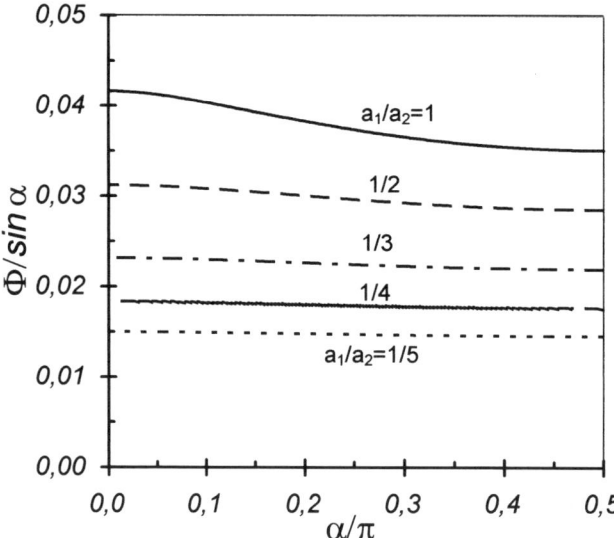

Fig. 11.13. The friction coefficient Φ in rhombic prisms

$$\Phi = \frac{1}{\Omega^2} \int d\Omega\, u_z(x,y) = \frac{1}{\Omega} \sum_{\substack{m \\ \text{odd}}}^{\infty} \frac{2}{m\pi} \sum_{\substack{n \\ \text{odd}}}^{\infty} \frac{2}{n\pi} \sum_{\substack{i \\ \text{odd}}}^{\infty} B_{mn}^{(i)}\,. \tag{11.74}$$

$B_{mn}^{(i)}$ is proportional to $(\cos\alpha)^{i-1}$, i.e. (11.74) represents a Taylor expansion with respect to $(\cos\alpha)^2$. Convergence is rapid up to $\cos\alpha = 1, \alpha = 1$. Φ is a dimensionless function of a_1/a_2 and $\cos\alpha$. Figure 11.13 depicts the resulting friction coefficient Φ for rhombic prisms. Φ allows a convergent Taylor expansion according to

$$\Phi = \sin\alpha \sum_{k=1}^{\infty} C_k \left(\frac{a_1}{a_2}\right) \frac{(2k-2)!}{2^{2k-2}(k-1)!^2} \frac{(\cos\alpha)^{2k}}{(2k-1)(2k)}\,, \tag{11.75}$$

where $C_k(a_1/a_2)$ approaches a constant value with increasing k. Summation of (11.75) leads to the inverse circular sine (arcsin) function. For small dihedral angles, $\sin\alpha \ll 1$, we obtain $\Phi \approx \sin\alpha\,(a_1/a_2)(2 - a_1/a_2)/24$.

Table 11.2. The coefficients $C_k(a_1/a_2)$ arising in the series expansion (11.75)

k	$C_k(1)$	k	$C_k(1)$	k	$C_k(1)$	k	$C_k(1)$	k	$C_k(1)$
1	0.070288	7	0.098556	13	0.099734	19	0.100013	25	0.100063
2	0.089181	8	0.098907	14	0.099810	20	0.100031	26	0.100060
3	0.094032	9	0.099167	15	0.099870	21	0.100045	27	0.100056
4	0.096158	10	0.099364	16	0.099919	22	0.100054	28	0.100050
5	0.097326	11	0.099518	17	0.099958	23	0.100060	29	0.100042
6	0.098061	12	0.099638	18	0.099989	24	0.100063	30	0.100033

References

1. Langbein D, Concus P, Finn R, Weislogel MM: DYLCO: Dynamics of liquids in edges and corners. In: *Forschung am ZARM*. H.J. Rath (ed.) ZARM, Bremen (1994) 149–153
2. Langbein D, Weislogel MM: Dynamics of liquids in edges and corners (DYLCO). Proceedings of the Second European Symposium on Fluids in Space. A. Viviani (ed.). Naples (1996) 122–129
3. Langbein D, Weislogel M: Dynamics of liquids in edges and corners (DYLCO): IML-2 experiment for the BDPU. NASA/TM-1998-207916 (1998)
4. Lucas R: Über das Zeitgesetz des kapillaren Aufstiegs von Flüssigkeiten. Kolloid-Zeitschrift **23** (1917) 15–22
5. Washburn EW: The dynamics of capillary flow. Phys. Rev. **17** (1921) 273
6. Taylor B: *Concerning the Ascent of Water between two Glas Planes*. Phil. Trans. Roy. Soc. London **27** (1712) 538

12. Oscillations of Liquid Columns

The resonance frequencies of liquid columns are a result of the balance between the kinetic energy of motion and the potential energy of the resulting surface enlargement.

A procedure for calculating these frequencies is to start with the oscillations of an infinite liquid column in terms of cylindrical Bessel functions. From these oscillations, one selects those that satisfy the correct boundary conditions on the flow velocity at the periphery of the supporting disks. The oscillations that are obtained for real frequencies and complex axial wavenumbers increase or decrease axially. From these, linear superpositions are formed which satisfy the boundary conditions on the supporting disks. The infinite number of such superpositions is truncated.

There are just two oscillations whose damping length is longer than the column's diameter. Of these, the first oscillation either has a wavelength larger than the stable length, equal to the circumference of the bridge, or is strongly damped, whereas the second oscillation has a shorter wavelength. Resonances, therefore, are primarily caused by the latter oscillation.

Regarding lateral oscillations of liquid columns, a one-dimensional approach is applied. These oscillations must induce neither a change in cross section of the column nor a flow in the axial direction. They are assumed to be independently superimposed on any simultaneous axial oscillation.

12.1 Introduction

In addition to the numerous experiments on the shape and stability of liquid bridges between coaxial circular disks, the resonance frequencies of liquid bridges were investigated in the D2 Spacelab mission in April 1993. Liquid bridges are very well suited for studies of their dynamics. All the longitudinal normal deformations $D\{m,0\}$ and the lateral normal deformations $D\{0,n\}$ discussed in Chaps. 5 and 6 may be excited resonantly by appropriate vibrations of the supporting disks. Various additional sensors for resonance detection may be integrated into the disks. At the stability limits, the resonance frequencies of the deformations vanish.

The theoretical models of the dynamics of liquid columns that have been developed include analytical calculations of infinite columns, simplified one-dimensional analytical investigations of finite bridges and several numerical models [Bauer 1984, Meseguer et al. 1985, 1992, Sanz 1985, Schilling et al. 1989, Schulkes 1990]. For evaluation of the D2 experiment LICOR, a theoretical model which exactly describes the oscillations of finite cylindrical columns between coaxial circular disks has been used. The oscillations are obtained as linear superpositions of axially decreasing and increasing waves, which originate at the supporting disks.

A nearly cylindrical column between circular disks requires a low Bond number Bo, where

$$Bo = \frac{g \Delta\rho R^2}{\sigma} \qquad (12.1)$$

(compare (1.3) and (6.18)). This condition may be achieved by

- theoretical modeling on the basis of the momentum equation, continuity equation, energy equation and capillary equation
- investigations of small liquid columns (R^2 small)
- investigations with density-adjusted liquids ($\Delta\rho$ small)
- experiments under low-gravity conditions (g small).

12.2 Theory

The resonance frequencies of liquid columns are determined by the balance between the kinetic energy of the liquid motion and the potential energy of the resulting surface enlargement. This means that it is appropriate to scale the resonance frequencies by

$$\frac{\text{kinetic energy}}{\text{surface energy}} = \frac{\rho R^2 L (\omega R)^2}{\sigma R L} = \frac{\rho \omega^2 R^3}{\sigma}. \qquad (12.2)$$

The sharpness of the resonance frequencies is determined by friction in the liquid. This friction can be scaled by the ratio of the transient time $\eta R/\sigma$ of motions of the liquid due to surface deformations to the damping time of these motions, R^2/ν. The square root of this ratio is the dimensionless Ohnesorge number

$$Oh = \sqrt{\frac{\eta R/\sigma}{R^2/\nu}} = \sqrt{\frac{\rho \nu^2}{\sigma R}}. \qquad (12.3)$$

The Ohnesorge number is also referred to as the capillary number or modified Reynolds number (ρ and ν are the density and kinematic viscosity, $\eta = \rho\nu$ is the dynamic viscosity and σ is the surface tension.)

In the one-dimensional slice model of oscillation, only the axial component of the flow velocity is taken into account in the kinetic energy [Lee

1974, Meseguer 1983]. The resonance frequencies are obtained by comparing the resulting kinetic energy with the potential energy associated with the surface deformation. An estimate of the kinetic energy that is too low results in resonance frequencies that are too high. The radial flow velocity, which primarily serves the purpose of satisfying the continuity equation, is found to depend linearly on the radial coordinate. The slice model is applicable to liquid bridges that have a large aspect ratio of length to diameter. In that case the axial flow indeed dominates, whereas for short liquid bridges the flow shows a strong radial component. An upper bound to the kinetic energy is obtained if one adds the kinetic energy corresponding to the radial flow velocity obtained from the continuity equation. This results in resonance frequencies that are too low, particularly if the oscillations have several nodes.

An alternative to the slice model is to start with the oscillations of an infinite liquid column in terms of cylindrical Bessel functions [Bauer 1984, 1986, 1992]. From these oscillations, one may select those that satisfy the correct boundary conditions on the flow velocity at the periphery of the supporting disks, while the corresponding boundary conditions at smaller radii are disregarded. Since the flow velocity in that model is less restricted than it is experimentally, resonance frequencies that are generally too low are obtained.

This procedure may be extended on the basis of the following steps [Langbein 1992a, 1992b]:

- The solutions of the momentum equation plus the continuity equation are determined in cylindrical coordinates.
- These solutions contain three independent sets of solutions in terms of cylindrical Bessel functions with respect to the radius, multiplied by exponential (periodic) functions $\exp(im\varphi + iqz)$ in the azimuth φ and in the radial coordinate z.
- These solutions are used to satisfy the boundary conditions on the radial component and on the two tangential components of the stress tensor at the surface of an infinite liquid column.
- Satisfying three boundary conditions by means of three linearly independent solutions leads to a 3×3 secular determinant.
- The solutions of this secular determinant for real axial wavenumbers q and complex frequencies ω are the natural frequencies of an inifinitely long liquid column.
- Instead of these natural oscillations, the forced oscillations for real frequencies ω and complex axial wavenumber q can be determined.
- These are axially increasing or decreasing oscillations of the liquid column.
- From these increasing or decreasing oscillations, linear superpositions are formed which satisfy the boundary conditions on the flow velocity at the supporting disks.
- Exact satisfaction of the boundary conditions at each radius would require one to take into account an infinite set of solutions.

- Satisfaction of the boundary conditions is therefore reduced to N radii on the supporting disks, which requires $2N$ Bessel functions.
- If $N = 12$ such radii are taken into account, the error is reduced to less than two percent even if their spacing is not optimized.

Among the most important results of this procedure, let us mention:

- The surface shape obtained at low frequencies is the unduloid, which fits with respect to length and volume.
- The pressure on the supporting disks is proportional to the amplitude a of the excitation and is conveniently scaled by $pR^2/a\sigma$.
- Resonant oscillations arise if the axially decreasing waves reach the disks at the end opposite to their origin and are reflected there with the correct phase.
- The latter requirement, generally, is satisfied by the first two pairs of solutions only, so that the resonance frequencies basically follow from a pair of these solutions that do not interfere destructively.
- The first resonance frequency approaches zero if the length L of the column becomes equal to the circumference $2\pi R$ (= Rayleigh instability $D\{1,0\}$).
- The height of the pressure maxima at the resonance frequencies varies in proportion to the Ohnesorge number.
- The sharpness of the resonances decreases with increasing number n of nodes, i.e. only a finite number of resonance frequencies can be measured experimentally.
- Decreasing the Ohnesorge number by one order of magnitude enables the detection of two of three further resonances.
- At high frequencies surface waves arise, i.e. the liquid motion vanishes along the axis.

This procedure is described in detail in the following sections.

12.2.1 Infinite Liquid Columns

Infinite cylindrical columns have the advantage that boundary conditions arise along their surface only. They have the disadvantage that they are unstable. Nevertheless, they provide a convenient basis for treating finite liquid bridges. The first step is to solve the linearized momentum equation together with the continuity equation to obtain the flow velocity $\boldsymbol{u}(r,t)$ and the pressure $p(r,t)$ in an infinite liquid medium. Using cylindrical coordinates r, φ, z and looking for periodic solutions in space and time

$$\boldsymbol{u}(\boldsymbol{r},t) = \{u_r(r), u_\varphi(r), u_z(r)\} \exp\left[\mathrm{i}(m\varphi + qz + \omega t)\right], \tag{12.4}$$

$$p(\boldsymbol{r},t) = p(r) \exp\left[\mathrm{i}(m\varphi + qz + \omega t)\right], \tag{12.5}$$

one finds three linear independent solutions in terms of cylindrical Bessel functions $I_m(qr), I_m(\hat{q}r)$:

$$u_r(r) = +c_1 i I'_m(qr) + c_2 i I'_m(\hat{q}r) + c_3 i m(\hat{q}r)^{-1} I_m(\hat{q}r) , \qquad (12.6)$$
$$u_\varphi(r) = -c_1 m(qr)^{-1} I_m(qr) - c_2 m(\hat{q}r)^{-1} I_m(\hat{q}r) - c_3 I'_m(\hat{q}r) , \qquad (12.7)$$
$$u_z(r) = -c_1 I_m(qr) - c_2 q^{-1} \hat{q} I_m(\hat{q}r) , \qquad (12.8)$$
$$p(r) = +c_1 \rho \omega q^{-1} I_m(qr) . \qquad (12.9)$$

Here q is the axial wavenumber; $\hat{q}$ includes the inertial term in the momentum equation,

$$\hat{q}^2 = q^2 + \frac{i\omega}{\nu} . \qquad (12.10)$$

12.2.2 The Free Fluid Surface

The second step is to satisfy the boundary conditions on the stress tensor along the free surface of the liquid column. The normal component must satisfy the capillary equation relating the pressure and surface tension, and the tangential components must vanish. By substitution of the local curvature caused by the surface deformation $u_r(r)/i\omega$ into the normal component of the stress tensor we obtain

$$\hat{p} = -\frac{1}{i\omega}\left(\frac{u_r}{r^2} + \frac{\partial^2 u_r}{(r\,\partial\varphi)^2} + \frac{\partial^2 u_r}{\partial z^2}\right) + 2\frac{\eta}{\sigma}\frac{\partial u_r}{\partial r} . \qquad (12.11)$$

For the tangential components, we have as usual

$$r\frac{\partial}{\partial r}\frac{u_\varphi}{r} + \frac{\partial u_r}{r\,\partial\varphi} = 0 , \quad \frac{\partial u_z}{\partial r} + \frac{\partial u_r}{\partial z} = 0 . \qquad (12.12)$$

Equations (12.11) and (12.12) constitute three boundary conditions for the flow velocity $\boldsymbol{u}(\boldsymbol{r},t)$ and the pressure $p(\boldsymbol{r},t)$, which are of the form (12.6)–(12.9). Since we have three independent solutions distinguished by the three coefficients c_1, c_2, c_3, we obtain three linear equations for these coefficients. A nontrivial solution exists only if the secular determinant vanishes. The structure of (12.11) reveals that the resulting dispersion relation may be conveniently written in the form

$$2i\frac{\omega\eta}{\sigma}\det(A_{ij}) + (q^2 + m^2 - 1)\det(B_{ij}) = 0 , \qquad (12.13)$$

where the elements of the 3×3 determinants A_{ij}, B_{ij} are given by

$$A_{1j}: \quad \frac{\hat{q}^2 - q^2}{2q^2} I_m(q) + q I''_m(q) \qquad \hat{q} I''_m(\hat{q}) \qquad m\left[I'_m(\hat{q}) - \hat{q}^{-1} I_m(\hat{q})\right]$$

$$B_{1j}: \quad I'_m(q) \qquad I'_m(\hat{q}) \qquad m\hat{q}^{-1} I_m(\hat{q})$$

$$A_{2j}, B_{2j}: 2m\left[I'_m(q) - q^{-1} I_m(q)\right] \quad 2m\left[I'_m(\hat{q}) - \hat{q}^{-1} I_m(\hat{q})\right] \quad \hat{q}\left[2I''_m(\hat{q}) - I_m(\hat{q})\right]$$

$$A_{3j}, B_{3j}: \quad 2I'_m(q) \qquad \frac{\hat{q}^2 + q^2}{q^2} I'_m(\hat{q}) \qquad m\hat{q}^{-1} I_m(\hat{q}) .$$

$$(12.14)$$

In (12.13), (12.14) and in the following, the wavenumber q is normalized by means of the radius R of the column considered.

The determinants of A_{ij}, B_{ij} both depend on the axial wavenumber q, the azimuthal wavenumber m and the frequency $\omega R^2/\nu$ only. The surface tension σ is linked to the dynamic viscosity η. Splitting off the frequency according to $\omega\eta/\sigma = (\omega/\nu)(\rho\nu^2/\sigma)$, we find that the dispersion relation leads to the Ohnesorge number as a function of the axial and azimuthal wavenumbers q, m and the reduced frequency $\omega R^2/\nu$.

12.2.3 Natural Frequencies

The dispersion relation (12.13) is always satisfied for $\omega = 0, q^2 + m^2 = 1$. This reflects the well-known fact that a cylindrical liquid column becomes unstable if its length L exceeds its circumference; in this case $q = 2\pi R/L = 1$.

For numerical reasons, it is very convenient to divide the rows of $\det(A_{ij})$ and $\det(B_{ij})$ by $I'_m(q)$, $I'_m(\hat{q})$, and $I'_m(q)$, respectively. The dispersion relation then turns out to contain the ratios $I_m(q)/I'_m(q)$ and $I_m(\hat{q})/I'_m(\hat{q})$ only, which may be conveniently calculated by means of continued fractions (equation 9.1.73 in Abramowitz & Stegun [1965]). In the axisymmetric case $m = 0$, the dispersion relation reduces to

$$\frac{\rho\nu^2}{\sigma R}\left((\hat{q}^2+q^2)^2\frac{I_0(q)}{qI'_0(q)} - 4\hat{q}^2q^2\frac{I_0(\hat{q})}{qI'_0(\hat{q})} - (\hat{q}^2-q^2)\right) + (q^2-1) = 0. \quad (12.15)$$

Figure 12.1 depicts the real and imaginary parts of $\omega R^2/\nu$ as a function of the wavelength $\lambda/2\pi R = q^{-1}$ and the Ohnesorge number. The real part $\omega_1 R^2/\nu$ increases strongly with decreasing Ohnesorge number, i.e. with increasing surface tension. The imaginary part $\omega_2 R^2/\nu$ is primarily determined by the wavenumber. Short wavelengths are associated with strong damping. The curve limiting the existence of real frequencies is given by

$$\frac{\hat{q}^2+q^2}{q^2}\left(\frac{I_0(q)}{qI'_0(q)} - \frac{1}{\hat{q}^2}\right) + \left(\frac{I_0(\hat{q})}{I'_0(\hat{q})} - \frac{1}{\hat{q}}\right)^2 = 1. \quad (12.16)$$

12.2.4 Finite Liquid Columns

In the case of finite liquid columns supported and excited by coaxial circular disks, additional boundary conditions at the disks have to be satisfied:

- the axial flow velocity must match the velocity of the supporting disks
- the radial flow velocity must vanish.

At the same time, it is no longer the natural damped oscillations which are of interest, but rather the forced oscillations with real frequency ω.

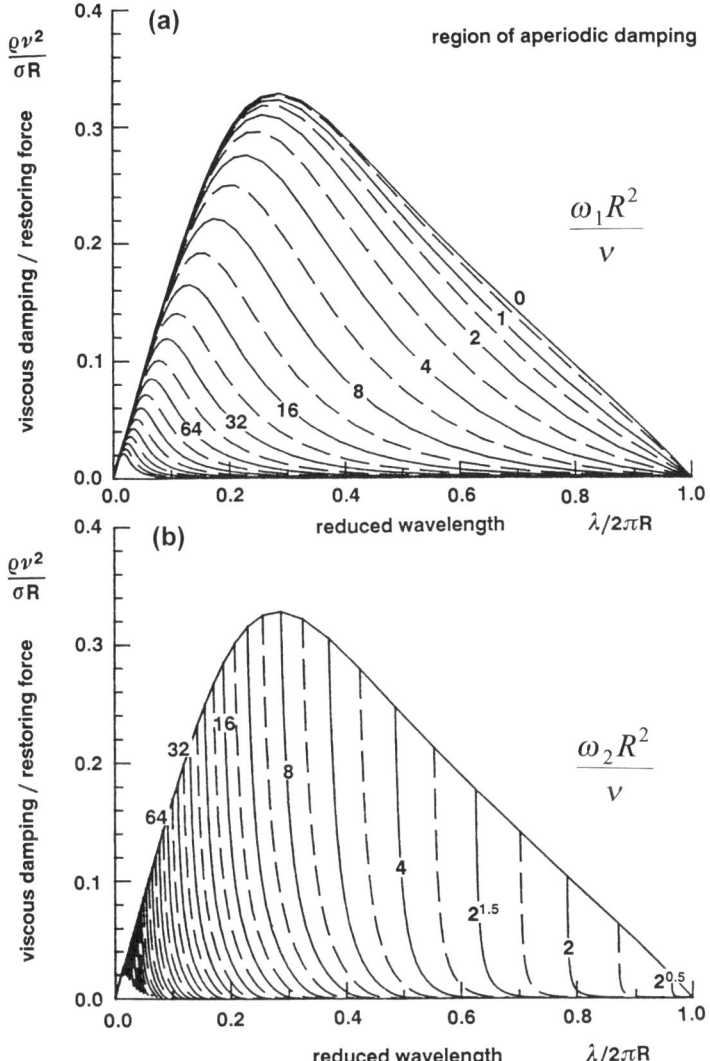

Fig. 12.1. Natural frequencies of an infinite cylindrical column versus wavelength $\lambda/2\pi R$ and Ohnesorge number $\sqrt{(\rho\nu^2/\sigma R)}$. (a) Reduced frequency $\omega_1 R^2/\nu$; (b) damping coefficient $\omega_2 R^2/\nu$

Let us consider for the moment a semi-infinite column supported by one solid disk only. Axial oscillations of the disk will cause axial oscillations of the column whose amplitude decreases with increasing distance from the supporting disk. We are thus dealing with spatially damped oscillations of the column. We have to consider the solutions of the dispersion relation for a given real frequency ω and complex wavenumber q. Once we have found the complete set of solutions, we may form linear combinations which satisfy, in addition to the boundary conditions along the free liquid surface, those at the solid disk.

Returning to finite columns supported by two solid disks, we have to satisfy two sets of boundary conditions. On the other hand, two sets of solutions of the dispersion relation are admitted: those which decrease axially and those which increase axially. Satisfying twice as many boundary conditions by means of twice as many functions seems more tedious and risky with respect to numerical convergence. However, this is not the case: simultaneous vibrations of the supporting disks with the same frequency but arbitrary amplitude and phase can always be decomposed into a symmetric oscillation of the column (corresponding to disk vibrations with opposite phases) and an antisymmetric oscillation of the column (corresponding to disk vibrations with equal phases).

12.2.5 Axially Damped Oscillations

The solutions of the dispersion relation (12.15) for real frequency ω and complex wavenumber q are generally governed by the fact that the dispersion function is an even function of q, such that if q is a solution of the dispersion relation, q is also a solution, and by the fact that the Bessel functions may be approximated by

$$\frac{I'_m(q)}{I_m(q)} = \frac{iJ'_m(q)}{J_m(q)} = -i\tan\left[iq - \left(m+\frac{1}{4}\right)\pi\right] \tag{12.17}$$

for large arguments q, such that an infinite set of solutions rather regularly spaced along the imaginary q axis arises. An exception is the first pair of solutions, to which (12.17) obviously does not apply.

Figure 12.2 shows the explicit dependence of the first six zeros $q_n, n = 1, 2, ..., 6$ of the dispersion relation (12.15) on $\rho\nu^2/(\sigma R)$ and $\omega R^2/\nu$. The full curves depict the dependence for constant $\rho\nu^2/(\sigma R)$, and the dash–dot curves depict the dependence for constant $\omega R^2/\nu$. The zeros q_{2n-1} and q_{2n} approach each other for low values of $\rho\nu^2/(\sigma R)$ and $\omega R^2/\nu$ at the nth zero of $J_0(iq)$. At large values of $\rho\nu^2/(\sigma R)$ all zeros of the dispersion relation (12.15) approach constant values proportional to $[\rho\nu^2/(\sigma R)]^{-1}$.

The wavenumbers $|\pm q'|$ of all solutions n obtained are in the range $|\pm q'| < 2\pi$, which means wavelengths $\lambda = 2\pi R/|\pm q'| > 1$ larger than the radius of the liquid column. The solutions, however, differ strongly with respect to spatial damping. The imaginary part q'' increases by π for each pair of solutions q_{2n-1}, q_{2n}. The solutions for high values of n decrease rapidly in the axial direction and change sign in the radial direction more frequently. This makes them well suited for satisfying the boundary conditions on the flow velocity at the supporting vibrating disks.

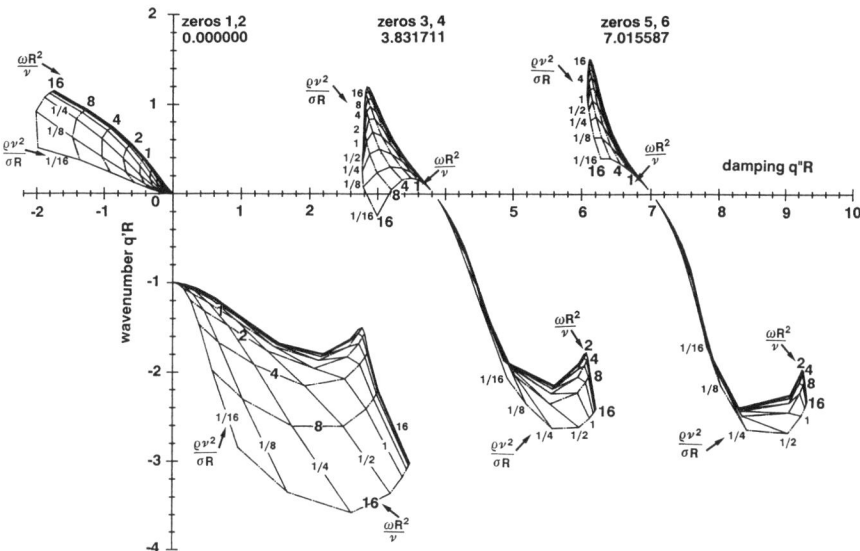

Fig. 12.2. First six zeros of the dispersion relation as a function of the Ohnesorge number Oh (*full lines*) and reduced frequency $\omega R^2/\nu$ (*dash-dot lines*)

12.2.6 Symmetric and Antimetric Oscillations

Let the liquid column under consideration have length L and let the positions of the supporting disks 1 and 2 be $z_1 = 0$ and $z_2 = L$. The flow velocity and the fluid-dynamic pressure within the oscillating column can be represented by an infinite sum over all solutions n of the dispersion relation for fixed real frequency ω and complex wavenumber q_n. Since both q_n and $-q_n$ solve the dispersion relation, we put

$$u_r(\mathbf{r},t) = e^{i\omega t} \sum_{n=1}^{\infty} [C(+n)e^{iq_n z} - C(-n)e^{iq_n(L-z)}] u_r(r,n) , \qquad (12.18)$$

$$u_z(\mathbf{r},t) = e^{i\omega t} \sum_{n=1}^{\infty} [C(+n)e^{iq_n z} + C(-n)e^{iq_n(L-z)}] u_z(r,n) , \qquad (12.19)$$

$$p(\mathbf{r},t) = e^{i\omega t} \sum_{n=1}^{\infty} [C(+n)e^{iq_n z} - C(-n)e^{iq_n(L-z)}] p(r,n) . \qquad (12.20)$$

The boundary conditions on the flow velocity at the supporting disks yield

$$\sum_{n=1}^{\infty} [C(+n) - C(-n)e^{iq_n L}] u_r(r,n) = 0 , \qquad (12.21)$$

$$\sum_{n=1}^{\infty} [C(+n) + C(-n)e^{iq_n L}] u_z(r,n) = a_1 e^{i\varphi_1} \qquad (12.22)$$

at $z_1 = 0$ and

$$\sum_{n=1}^{\infty}[C(+n)e^{iq_n L} - C(-n)]u_r(r,n) = 0 ,\qquad (12.23)$$

$$\sum_{n=1}^{\infty}[C(+n)e^{iq_n L} + C(-n)]u_z(r,n) = a_2 e^{i\varphi_2} \qquad (12.24)$$

at $z_2 = L$. Here a_1, φ_1 and a_2, φ_2 are the amplitude and phase of the vibrations of disks 1 and 2, respectively.

By forming even and odd combinations from the above conditions at the two disks, one obtains two independent sets of equations for the sums of coefficients $C(+n) + C(-n)$ and their differences $C(+n) - C(-n)$:

$$\sum_{n=1}^{\infty}[C(+n) \pm C(-n)](1 \mp e^{iq_n L})u_r(r,n) = 0 , \qquad (12.25)$$

$$\sum_{n=1}^{\infty}[C(+n) \pm C(-n)](1 \pm e^{iq_n L})u_z(r,n) = a_1 e^{i\varphi_1} \pm a_2 e^{i\varphi_2} . \qquad (12.26)$$

If the two disks vibrate with equal amplitude and phase, $a_1 e^{i\varphi_1} = a_2 e^{i\varphi_2}$, we find $C(+n) - C(-n) = 0$ and are left with antimetric oscillations, i.e. the two disks alternately initiate compression waves, which propagate along the column. If the disks vibrate with equal amplitude and opposite phase, $a_1 e^{i\varphi_1} = -a_2 e^{i\varphi_2}$, we find $C(+n) + C(-n) = 0$ and are left with symmetric oscillations, i.e. the disks cause compression waves synchronously, which propagate along the column in opposite directions.

The numerical solution of (12.25) and (12.26) for $C(+n) \pm C(-n)$ requires a truncation of the infinite sums over the axially damped solutions n. This means that on each disk, the boundary conditions for the flow velocity may be satisfied on a finite number of circles only. In order to satisfy the boundary conditions on the radial and axial flow velocities at N such circles, we have to take into account $2N$ solutions $u_r(r,n), u_z(r,n)$. The solution of (12.25) and (12.26) then requires the inversion of two $4N \times 4N$ matrices. It turns out that the use of just four circles is clearly insufficient, eight circles are sufficient, and the use of twelve circles provides a reasonable compromise between accuracy and the computer time required. The further improvements provided by satisfying the boundary conditions on 16 or 20 circles are only marginal.

Figure 12.3 demonstrates the convergence of the fluid-dynamic pressure. The pressure exerted by the liquid column on the vibrating disks is plotted versus the radius r. The different curves correspond to different numbers $N = 4, 8, 12, 16, 20$ of equidistant circles used for satisfying the boundary conditions. The deviation of the pressure from its exact value is found to decrease proportional to $1/N$. It decreases from the periphery to the center. The accuracy may be improved by selecting nonequidistant circles, for

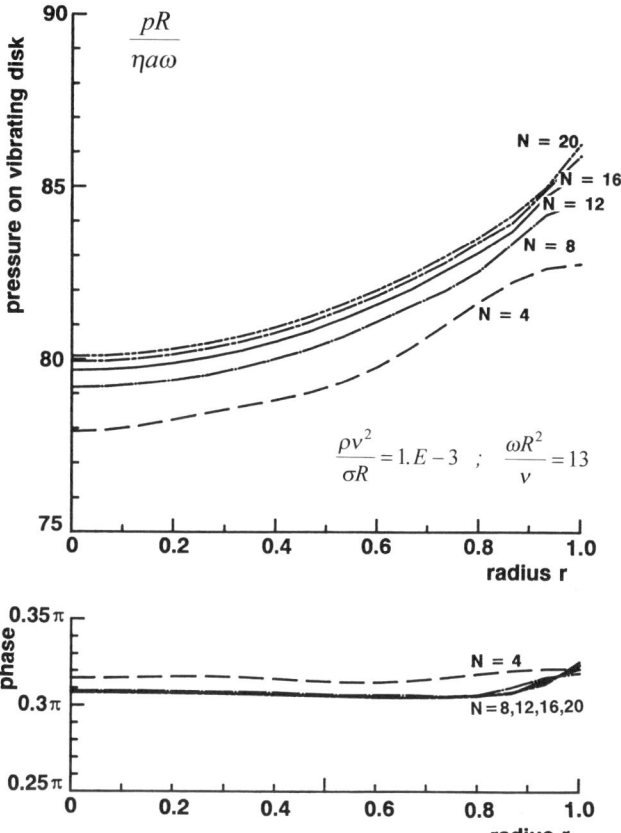

Fig. 12.3. Variation of the fluid-dynamic pressure on a vibrating disk when the boundary conditions on the flow velocity are exactly satisfied on an increasing number $N = 4, 8, 12, 16, 20$ of equidistant circles

instance circles enclosing equal areas, or by applying other arguments that lead to an increase of the density of circles towards the periphery.

12.2.7 Resonance Detection and Flow Patterns

Whether a given frequency ω of excitation represents a resonance frequency may be decided by direct observation or by photoelectric evaluation of a video or cine film. The most objective method, however, is to determine the fluid-dynamic pressure on the supporting vibrating disks. This method has the advantage of immediate control. At constant amplitude of disk vibration, the fluid-dynamic pressure reaches a maximum when a resonance frequency is applied. This may be used for the automatic detection of resonances.

There are various options for plotting the pressure. The dispersion relation (12.13), together with (12.10), suggests feeding the dimensionless parameters

$\rho\nu^2/\sigma R$ and $\omega R^2/\nu$ into the computer and calculating the fluid-dynamic pressure $pR/a\omega\eta$ at constant disk velocity $a\omega$. The results show that the pressure increases in inversely proportion to the Ohnesorge number Oh. It is therefore convenient to plot the product $(pR/a\omega\eta)(\rho\nu^2/\sigma R) = p\nu/a\omega\sigma$. Since measurements are usually based on constant disk amplitude rather than on constant disk velocity, we multiply with the reduced frequency, yielding $(p\nu/(a\omega\sigma))(\omega R^2/\nu) = pR^2/a\sigma$.

One might be tempted to reduce the frequency by means of the kinematic viscosity, i.e. to use $\omega R^2/\nu$. It is the competition between liquid inertia and viscosity which governs damping. However, it is the balance between inertia and surface deformation, between kinetic energy and potential energy, which determines the resonance frequencies. This is confirmed by the numerical calculations. The plots of pressure versus frequency in Figs. 12.4 and 12.5 were obtained by assuming that one disk was vibrating (the lower disk), whereas the other one was at rest (the upper disk). Figures 12.4 and 12.5 show the pressure on the vibrating disk and on the stationary disk versus the reduced frequency $(\rho\omega^2 R^3/\sigma)^{0.5}$. The different curves correspond to different Ohnesorge numbers Oh.

The influence of the Ohnesorge number is reduced to an effect on the sharpness of the pressure maxima and minima. For $\rho\nu^2/\sigma R > 10^{-2}$, detection of the resonances is hardly possible. For $\rho\nu^2/\sigma R = 10^{-4}$, the first four or five resonances may be detected with reasonable accuracy. As a rule, lowering $\rho\nu^2/\sigma R$ by an order of magnitude means that two or three additional resonances may be observed. However, at the same time fluid sloshing becomes stronger and the transient times required become longer.

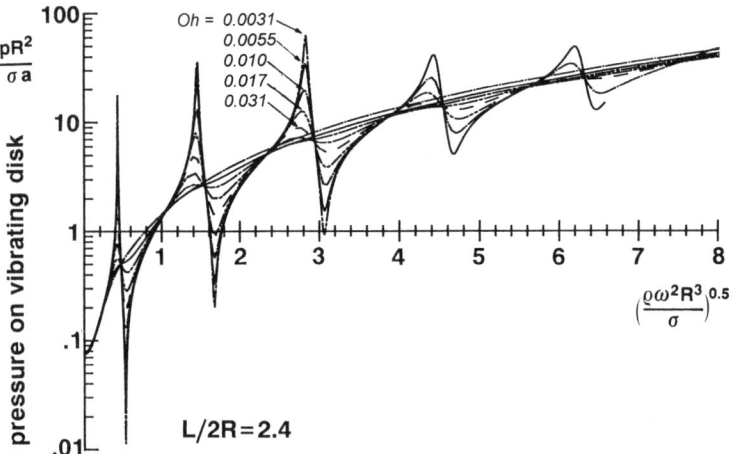

Fig. 12.4. Amplitude of the pressure $pR^2/\sigma a$ on the center of the vibrating disk versus frequency $(\rho\omega^2 R^3/\sigma)^{0.5}$ for varying Ohnesorge number Oh; $L_c = L/2R = 2.4$

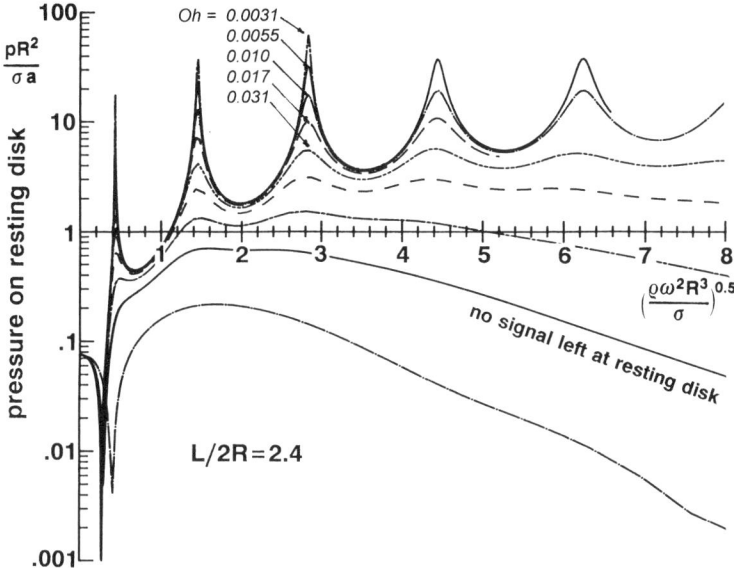

Fig. 12.5. Amplitude of the pressure $pR^2/\sigma a$ on the center of the stationary disk versus frequency $(\rho\omega^2 R^3/\sigma)^{0.5}$ for varying Ohnesorge number $Oh = \sqrt{(\rho\nu^2/\sigma R)}$; $L_c = L/2R = 2.4$

Independent of the aspect ratio and the Ohnesorge number, the pressure on the vibrating disk exhibits the same asymptotic behavior at high frequencies. This behavior is given by

$$p = \rho \times a\omega^2 \times 0.6R = \text{density} \times \text{acceleration} \times \text{length} \qquad (12.27)$$

This limit means that the liquid close to the vibrating disk is pushed forward and backward with acceleration $a\omega^2$ and that the actual length of the column affected by this acceleration corresponds to about 30 percent of its diameter $2R$. The liquid close to the vibrating disk has a standard flow field, which damps the vibration to the maximum possible extent. At the opposite disk, the stationary disk, no oscillation and therefore no pressure are left. An exception is the case of small aspect ratios $L_c = L/2R \leq 0.3$, when the flow field just discussed reaches the opposite disk.

Figure 12.6 shows the pressure at the first six resonance frequencies as a function of the Ohnesorge number Oh and the aspect ratio L_c. For each resonance, an existence limit arises at high Ohnesorge numbers, i.e. the resonances fade out when the oscillations are too strongly damped. From the slope of the curves in Fig. 12.6, it is obvious that the amplitude of the pressure at the resonance frequencies decreases in proportion to Oh^{-1}.

Figures 12.7 and 12.8 depict two characteristic flow patterns in a column with aspect ratio $L_c = 2.4$ and $Oh = 10^{-2}$. The lower disk vibrates, whereas the upper disk is at rest. Figure 12.7 shows the second resonance frequency,

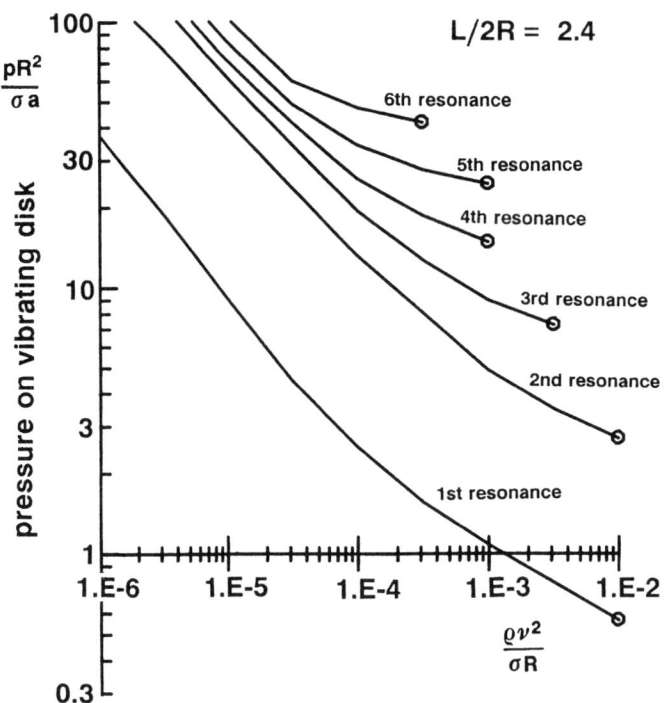

Fig. 12.6. Amplitude of the pressure on the vibrating disk at the first six resonance frequencies versus Ohnesorge number Oh for an aspect ratio $L_c = L/2R = 2.4$. The *circles* indicate those Ohnesorge numbers where the maximum in pressure is reduced to a slight rise of the pressure amplitude

and Fig. 12.8 a rather high frequency, where the resonances are fading out. Each oscillation is represented by the displacement of tracers between phases $(n-1)\pi/6$ and $n\pi/6$, with $n = 1$ to 6. The length of the tracers thus gives the local flow velocity.

In the case of resonance (see Fig. 12.7), it is very obvious that the flow velocity within the liquid becomes much higher than the velocity of the vibrating lower disk. This means that long transient times are required before the resonant flow pattern is reached. A considerable amount of kinetic energy has to be transferred from the vibrating disk to the liquid. The ratio of flow velocity to disk velocity increases with decreasing Ohnesorge number.

At high frequencies (see Fig. 12.8), only surface oscillations of the liquid column are left. The tracers near the axis of the column show hardly any motion. The surface oscillations do not reach the stationary disk for $Oh^2 \geq 10^{-3}$, but do so for $Oh^2 \leq 10^{-4}$. Owing to the lower viscosity, the damping length is longer in the latter case. The flow pattern depicted in Fig. 12.8 clearly shows the limitations of one-dimensional flow models. The flow velocity becomes

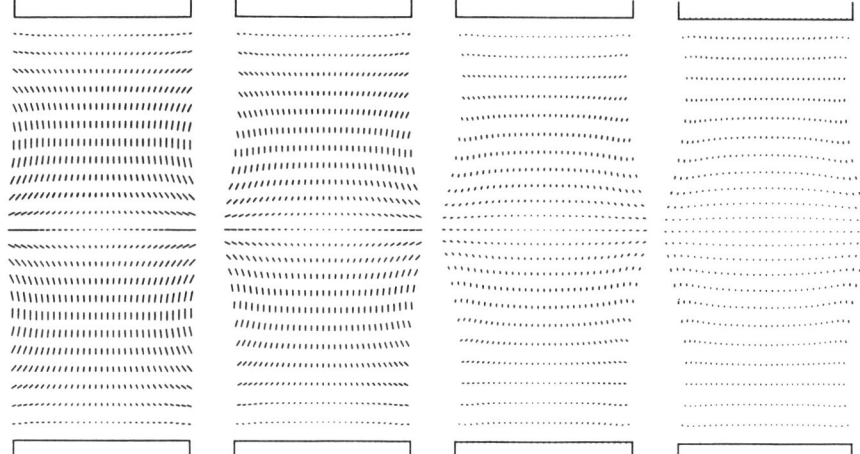

Fig. 12.7. Flow pattern of the second resonance in a column with aspect ratio $L_c = 2.4$ and Ohnesorge number $Oh = 10^{-2}$; $\omega R^2/\nu = 145$, $a/R = 0.025$, phase shift between patterns $= \pi/6$

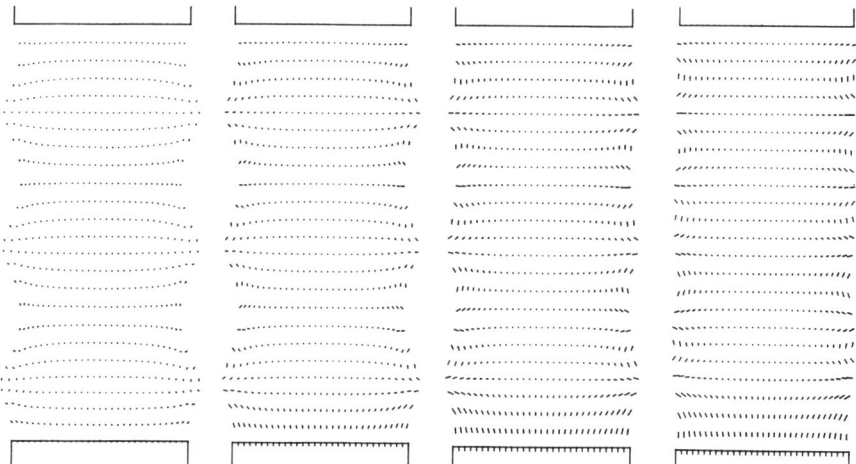

Fig. 12.8. Flow pattern of surface waves in a column with aspect ratio $L_c = 2.4$ and Ohnesorge number $Oh = 10^{-2}$; $\omega R^2/\nu = 750$, $a/R = 0.1$, phase shift between patterns $= \pi/6$

strongly dependent on the radial coordinate with increasing frequency and wavenumber.

Among the axially damped oscillations considered here, there are just two whose length of damping is longer than the column's diameter. Except for columns with small aspect ratio $L_c < 0.3$, the contribution of the other oscillations is limited to satisfying the boundary conditions at the disks and reshaping the surface deformation. Distant from the disks, only the two dominating oscillations are left. Among these, the first oscillation $D\{1,0\}$ either has a wavelength λ larger than the stable length $2\pi R$ of the bridge or else is strongly damped, whereas $D\{2,0\}$ has a wavelength $\lambda < 2\pi R$. Resonances, therefore, are primarily caused by the latter oscillation. For a given aspect ratio $L/2R$ and varying frequency, the observation of several resonances is possible for low values of $\rho\nu^2/\sigma R$ only. An increase in $\rho\nu^2/\sigma R$ involves an increase in damping, such that the possible resonances become less pronounced and eventually vanish.

12.3 Experiments

12.3.1 Short Liquid Columns

The resonance frequencies of liquid columns under weightlessness have been studied in the Spacelab D2 experiment LICOR (Liquid Columns' Resonances). In preparation for that experiment, the resonance frequencies of small liquid columns were observed [Ahrens et al. 1994] and investigations with density-matched liquids (silicone oil in a water/methanol mixture) were performed.

For observation of the resonance frequencies of small liquid columns the apparatus shown in Fig. 12.9 was developed. The small columns were established between disks made from stainless steel 3 mm, 4 mm and 5 mm in diameter. The upper disk could be positioned in three directions by means of micrometer screws. The lower disk was electrodynamically excited. Typical amplitudes a at frequencies between 15 Hz and 150 Hz were from 2 μm to $a = 5$ μm. It was possible to set the frequencies manually or to apply a frequency ramp with a variable rate.

The oscillations were observed by means of a telemicroscope and were video recorded, see Fig. 1.6. By using stroboscopic illumination, a stationary picture could be achieved at any desired phase. Applying a slow phase variation resulted in a slow-motion picture. This was of particular importance at high frequencies, where the oscillations become nonlinear.

For quantitative detection of the resonance frequencies, the pressure of the liquid column on the supporting disks was measured. The nominal sensitivity of the pressure sensors was 15 μV/Pa, such that voltages down to 30 μV needed to be observed. For calibration of the pressure sensors, the volume of the gas in a bottle was varied by 50 mm^3 by means of a microsyringe. The liquids used in the experiments were water and a water/glycerol mixture; their physical properties are listed in Table 12.1. With both liquids, systematic measurements with disks 5 mm in diameter and a frequency ramp from 15 Hz to 150 Hz were performed. The rate of rise in frequency during the ramp was 1 Hz per second. The aspect ratio L_c was varied systematically from 0.8 to 1.1.

Fig. 12.9. The device used for investigating the resonance frequencies of short liquid columns

Figure 12.10 depicts the first two resonance frequencies of a water column 5 mm in diameter and 3.5 mm in height. Figure 12.11 shows the resonance frequencies found versus the aspect ratio. The resonance frequencies obtained with water (x) are somewhat lower than those obtained with the water/glycerol mixture (+). This contradicts the theory, since the Bond number and also the Ohnesorge number for the mixture are slightly higher, which suggests slightly lower resonance frequencies. The full lines in Fig. 12.11 give the resonance frequencies according to the exact theory, and the dashed lines those according to the improved slice model, Sect. 12.2. The measured resonance frequencies agree excellently with the exact model.

12.3.2 Plateau Simulation

Plateau simulation of liquid oscillations, i.e. working with density-adjusted liquids, is often disqualified by the argument that the exterior fluid must also

304   12. Oscillations of Liquid Columns

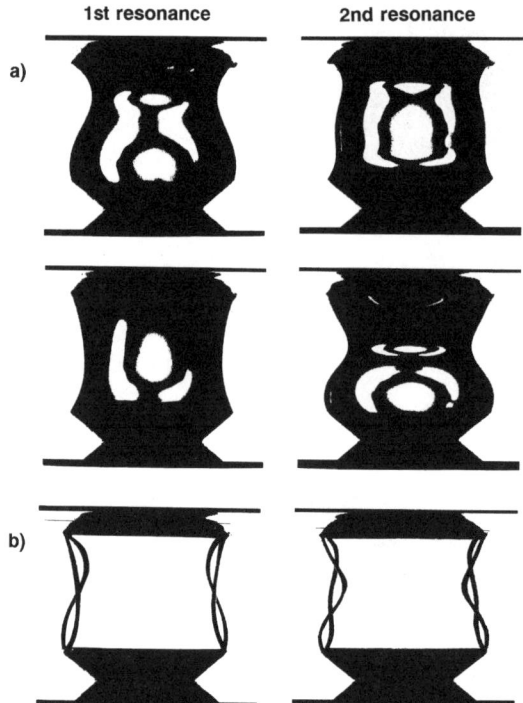

Fig. 12.10. The first and second resonant oscillations of a water column 5 mm in diameter and 3.5 mm in length. (**a**) Stroboscopic pictures of the extremal deformations; (**b**) superposition of the silhouettes of the extremal deformations

Table 12.1. Physical data for the liquids used

Liquid	ρ (g cm^{-3})	η (g cm^{-1} s^{-1})	σ(mN/m)
Water	1.00	0.01	50.3
Water/25% glycerol	1.06	0.021	48.5

oscillate, so that no clear comparison with theory is possible. In addition to the density and viscosity of the outer liquid bath, the shapes of the container used and of the leads to the supporting disks have to be taken into account. An exact calculation of the resonance frequencies, therefore, is possible by numerical methods only. However, considering that the viscosity affects strongly the sharpness of the resonance frequencies but only slightly their position, it appears possible to obtain useful results by means of Plateau simulation.

Plateau simulation was used to familiarize the science astronauts and the experimenters with the behavior of liquid columns and, in particular, with the detection of resonance frequencies. In a Plateau tank at the European Space Technology Centre (ESTEC) in Noordwijk (provided by ETSI Aeronáuticos,

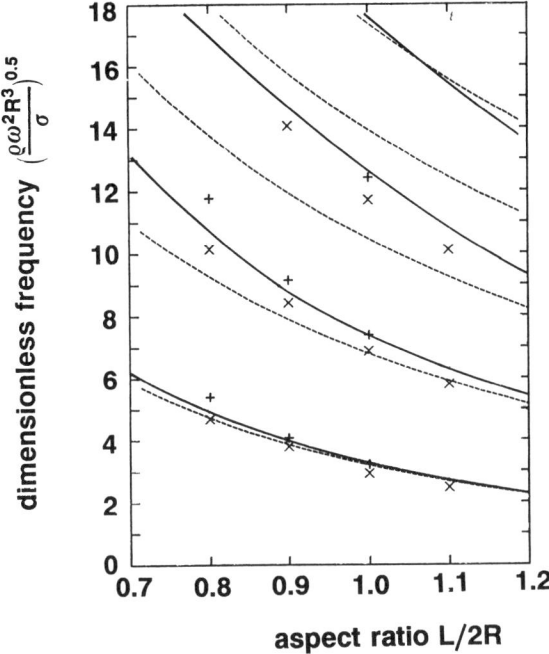

Fig. 12.11. The first three resonances of liquid columns 5 mm in diameter with different lengths: (x) water, (+) water/25% glycerol; *full lines*, exact theory; *dashed lines*, improved slice model

Madrid), a column of silicone oil (density 0.9 g/cm^3) was established in a bath of water and methanol (densities 1.0 g/cm^3 and 0.78 g/cm^3), see Fig. 1.3. After the first period of training of the science astronauts, a frequency ramp from 0 Hz to 5 Hz with a frequency rise of 1 Hz every 90 s was added. This proved particularly useful during the second training period.

By repeatedly studying the video recordings, i.e. after some visual practice, the science astronauts could determine resonance frequencies up to 0.02 Hz. The resonance frequencies shown in Table 12.2 were obtained from four columns 30 mm in diameter with different aspect ratios. Plotting the resonance frequencies versus the aspect ratio leads to Fig. 12.12. The scaling of the ordinate has been chosen such that optimum fitting of the theoretical results is achieved. The good agreement of the experimental with the theoretical curves reveals that the resonance frequencies in the Plateau tank show the same behavior as those of free liquid columns. The only adjustable parameter is the frequency scaling.

12.3.3 Automatic Resonance Detection

The Spacelab D2 experiment LICOR was performed in the Advanced Fluid Physics Module (AFPM), which was part of the Materials Research Labo-

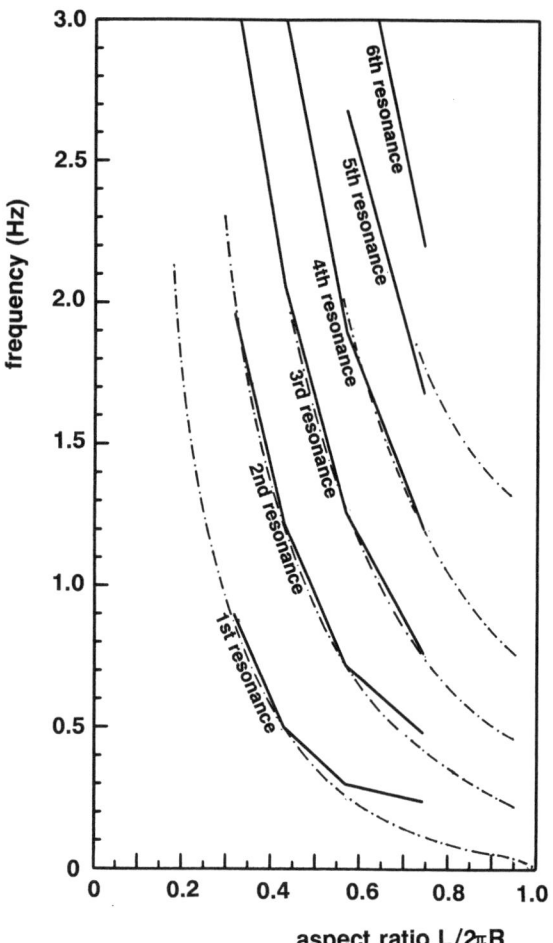

Fig. 12.12. Comparison of the resonance frequencies obtained in the Plateau tank with the exact theoretical model. The scaling of the frequency has been chosen according to the best fit. *Solid lines*, experiment; *dash-dot lines*, theory

ratory. For measuring and processing the pressure of the oscillating column, the special disks (front and rear disks) shown in Fig. 12.13 were designed. They were constructed and equipped with pressure sensors and the necessary electronics. The pressure exerted by the oscillating liquid column on the supporting disks was as low as 10 Pa. The hole for liquid injection is in the center of the front disk. The rear disk can be vibrated. Since the data downlink of the Materials Research Laboratory was just one signal per second per channel, it was necessary to determine the amplitude and phase of the pressure in the LICOR disks. The phase was measured relative to the vibration of the rear disk.

Table 12.2. The resonance frequencies (Hz) of a cylindrical liquid column 30 mm in diameter as determined in the Plateau tank in the COLUMBUS mock-up. Liquid column, silicone oil; liquid bath, water + methanol

Length (mm)	29.8	40.3	53.4	70.0
Amplitude (mm)	2.0	2.0	2.0	3.0
1st	0.90	0.50	0.30	0.24
2nd	1.96	1.22	0.72	0.48
3rd	3.10	2.06	1.26	0.76
4th		3.02	1.90	1.20
5th		>4.00	2.68	1.68
6th			3.52	2.20

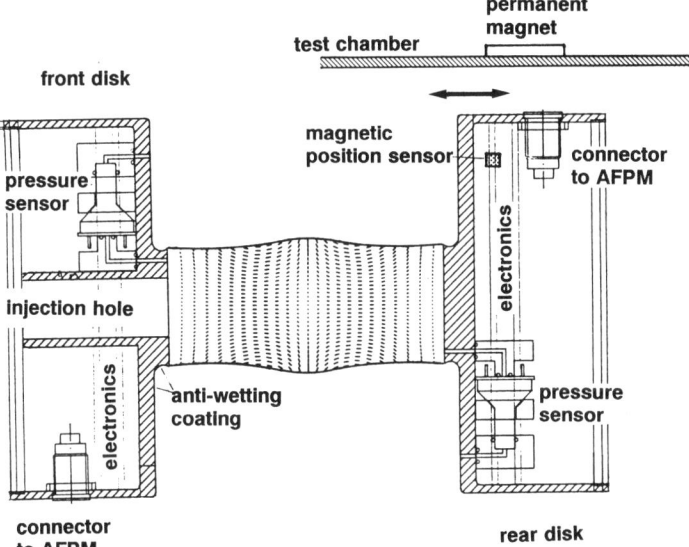

Fig. 12.13. The front and rear disks developed for use in the AFPM. The hole for liquid injection is in the front disk; the rear disk is excited with vibrations. The built-in electronics provide the analysis of the pressure signals into amplitude and phase

The data processing in the disks can evaluate pressure signals from 0.5 Pa to 10 Pa in a frequency range from 1 Hz to 5 Hz. By using this processing, it was possible to transmit the following to the ground at a rate of 4 Hz and with an accuracy of 16 bits:

- the amplitude of the pressure on the front disk
- the phase of the pressure on the front disk
- the amplitude of the pressure on the rear disk
- the phase of the pressure on the rear disk.

308   12. Oscillations of Liquid Columns

Three different methods were applied for resonance detection, namely visual observation, monitoring the pressure on the vibrating supporting disks and image processing of the video recordings.

12.3.4 The LICOR Runs

During the essential steps of the LICOR experiment, real-time TV was available. This experiment was performed by PS-1 Ulrich Walter. It was essential that the intended frequency ramps from 0 Hz to 5 Hz, each lasting 7 min 30 s, could be observed, recorded and processed in real-time TV. The video signal was transmitted via ITALSAT to the MARS Center, Naples, was image processed there and returned on line.

First, a column 30 mm in diameter and nominally 47.4 mm in length was created from 10 cSt silicone oil. In addition to the video signal of the oscillating columns, the pressure data were received on the ground. The amplitude of the pressure on the vibrating disk was plotted on one of the monitors at GSOC (German Space Operation Center, Oberpfaffenhofen) and agreed excellently with the previously determined curve. The pressure maxima, i.e. the resonance frequencies, could be easily identified from these data; see Fig. 12.14. The liquid column contained four unintended bubbles, which were not removed, owing to the stringent time limits. The bubbles did not disturb the pressure and frequency measurements. They generally tried to avoid the regions with high flow velocity and to move to the flow minima instead. After application of the frequency ramp, several frequencies close to the resonances noted by the payload specialist and principal investigator were set manually in steps of 0.04 Hz and the respective oscillations were observed in detail. This procedure proved less effective than the automatic detection by means of a frequency ramp. The resonance frequencies noted during the ramp are listed in Table 12.3.

Subsequently, the length of the column was reduced to a nominal value of 40 mm and the frequency ramp and the manual settings were repeated. As before, good agreement with theory was obtained. An axial line with equidistant radial ticks was placed between the background illumination and the liquid column. This line gave valuable additional information. The start of the frequency ramp can be seen from the motion of the ticks much earlier than from the surface deformation. Residual accelerations can be clearly observed, which were not detected at all by the Microgravity Measurement Assembly (MMA). Also, nonaxisymmetric surface deformations could be easily identified.

Figure 12.15 shows the amplitude of the pressure on the vibrating rear disk and on the stationary front disk during application of the frequency ramp to the first column, i.e. the one of nominal length 47.4 mm (actually 49.1 mm). The full lines give the theoretical pressure according to the improved slice model, and the dash–dot lines the pressure according to the exact theory.

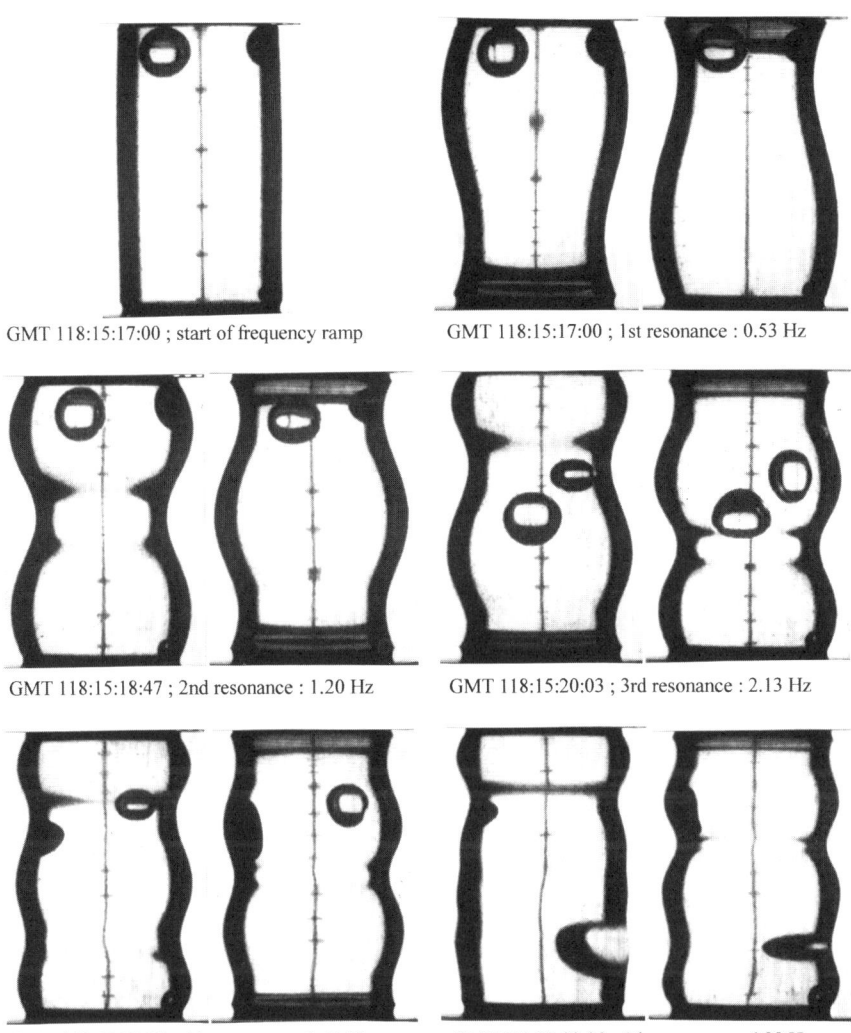

GMT 118:15:17:00 ; start of frequency ramp
GMT 118:15:17:00 ; 1st resonance : 0.53 Hz
GMT 118:15:18:47 ; 2nd resonance : 1.20 Hz
GMT 118:15:20:03 ; 3rd resonance : 2.13 Hz
GMT 118:15:22:06 ; 4th resonance : 3.12 Hz
GMT 118:15:23:28 ; 5th resonance : 4.28 Hz

Fig. 12.14. The first five resonant oscillations of a column of silicone oil 30 mm in diameter and 49.1 mm in length

Excellent agreement between the latter and the experimental values is obtained. The same is true for the phase of the pressure on the vibrating and stationary disks.

On the third day of the D2 mission, after a three-hour air-to-ground repair of the AFPM, a second run of LICOR was enabled by means of a replanning request. In order to save crew time, this run was performed with the 5 cSt silicone oil used in the preceding experiment, ONSET (onset of oscillatory Marangoni flows, R. Monti, Naples). In order to take account of the lower

Table 12.3. The resonance frequencies (Hz) of the column 30 mm in diameter and nominally 47.4 mm in length noted during the frequency ramp

Resonance	1st	2nd	3rd	4th	5th
Payload specialist	0.44	1.14	2.09	3.08	
Principal investigator	0.40	1.02–1.24	1.65–2.02		
MARS Center		1.55	1.99	3.82	4.7
Pressure maximum	0.53	1.20	2.14	3.12	4.28

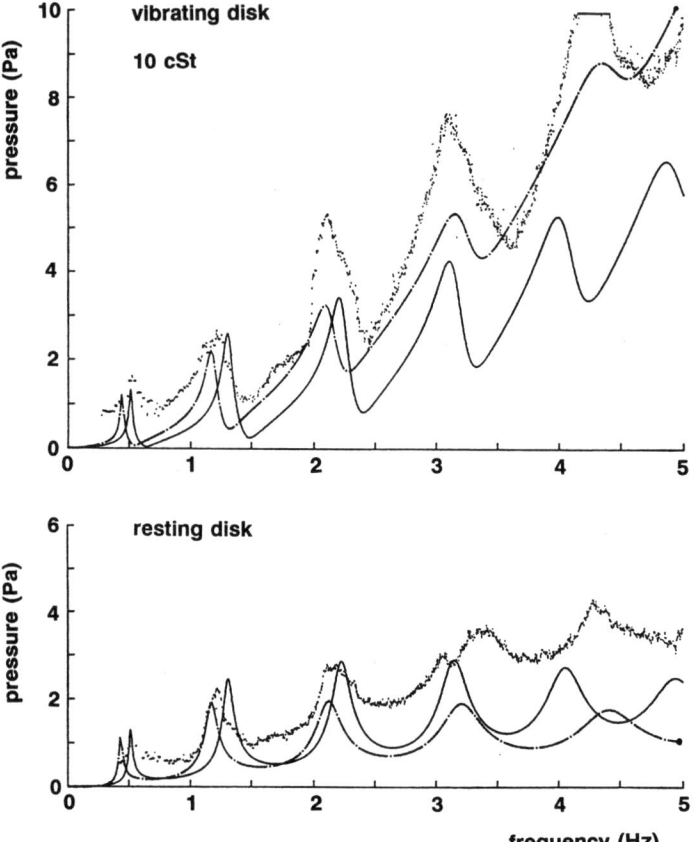

Fig. 12.15. The amplitude of the pressure on the vibrating disk (**a**) and on the stationary disk (**b**) during application of the frequency ramp to the column 49.1 mm in length: *full lines*, Cosserat model; *dash–dot lines*, exact model

viscosity, the amplitude of excitation of the rear disk was reduced from 1 mm to 0.5 mm. The front disk, owing to the preceding procedures, had a lateral shift of 0.7 mm. The request to rearrange it coaxially was declined, since this was the step at which the AFPM had broken down earlier.

Table 12.4. The resonance frequencies obtained from the pressure maxima and by visual evaluation of the video recordings

Silicone oil	Nominal length (mm)	Disk	Resonance frequency (Hz)				
			1st	2nd	3rd	4th	5th
10 cSt	47.4	Vibrating	0.53	1.20	2.13	3.12	4.28
		Stationary	0.53	1.21	2.19	3.38	4.30
		Video	0.48	1.20	2.04	3.40	4.31
10 cSt	40.0	Vibrating	0.69	1.53	2.70	3.96	
		Stationary	0.69	1.58			
		Video	0.64	1.53	2.70	3.96	4.98
5 cSt	38.0	Vibrating	0.80	1.77	3.00		
		Stationary	0.80	1.78			
		Video	0.78	1.77	300		

A column 40 mm in length was created and a frequency ramp from 0 Hz to 5 Hz was applied. As before, sharp maxima of pressure versus frequency appeared. The first resonance frequencies could be clearly identified; see Table 12.4. However, after the third resonance frequency had been passed, the oscillation became so intense that a strong lateral motion arose and the liquid spilled over the sharp edge of the rear disk. The column subsequently broke. Since no unusual accelerations of the Spacelab were recorded by the Microgravity Measurement Assembly, possible reasons for the spillage are

- the lateral shift of the front disk by 0.7 mm
- retardation of the full development of the resonant oscillations relative to the appearance of the pressure maximum.

The automatic detection of the resonances by means of pressure sensors proved to be very effective. The resonance frequencies found in the D2 experiment are shown in Fig. 12.16, together with the results from the short liquid columns and curves from theoretical modeling. The pressure on the supporting disks and the slope of the surface deformation follow the frequency of excitation without delay, whereas it requires several periods to fully develop the resonant oscillations, such that in the case of an increasing or decreasing frequency ramp, resonance frequencies that are too high or too low, respectively, will be detected.

The unintended lateral shift of the front disk, which was not corrected owing to the risk of breakdown of the AFPM, shows the sensitivity of liquid columns to nonaxial perturbations. This underlines the importance of nonaxial oscillations $D\{0,n\}$, i.e. oscillations with azimuthal wavenumbers $n \geq 1$. These oscillations show a much lower damping than the axial oscillations $D\{m,0\}$.

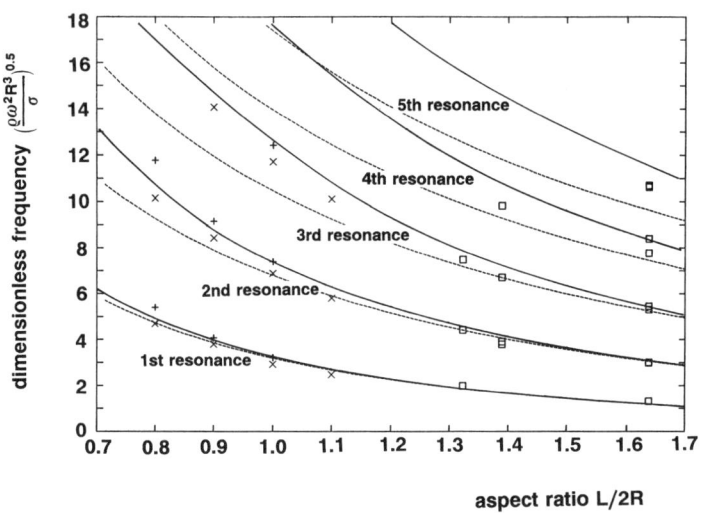

Fig. 12.16. The first five resonance frequencies obtained with columns of different length in the D2 experiment LICOR together with the resonance frequencies found in short liquid columns: (□) D2 experiments; (x) water, (+) water/25% glycerol; *full lines*, exact theory; *dashed lines*, Cosserat model; see also Fig. 12.12

12.4 Lateral Oscillations of Liquid Bridges

12.4.1 Damped Harmonic Oscillations

Turning to lateral oscillations of liquid columns, let us restrict ourselves to one-dimensional approaches. In this case, the lateral oscillation must induce neither a change in cross section of the column nor a flow in the axial direction. We assume that the lateral oscillation is superimposed independently on any simultaneous axial oscillation. As before, we consider liquid columns between circular disks of radius R, at a distance L apart. If the column is cylindrical, a sideways oscillation $D\{m,1\}$ with m nodes in the axial direction may be modeled by

$$r = r_0 + r_1 \cos\varphi \sin qz \sin \omega t , \tag{12.28}$$

where $r_0 = R$ and the axial wavenumber q is given by

$$q = \frac{(m+1)\pi}{L} , \tag{12.29}$$

where $m = 0, 1, 2$; see Fig. 12.17. In an initial step, we neglect damping and determine the frequency of oscillation from the balance between kinetic energy and surface energy. The kinetic energy of the oscillating column is given by

$$E_{\text{kin}} = \frac{1}{2}\rho\pi r_0^2 \int_0^L dz \, (r_1 \omega \sin qz \cos \omega t)^2 = \frac{1}{4}\rho\pi r_0^2 L r_1^2 \omega^2 \cos^2 \omega t, \quad (12.30)$$

whereas for the surface energy we obtain

$$E_{\text{sur}} = \sigma \int_0^L dz \int_0^{2\pi} r \, d\varphi \sqrt{1 + \left(\frac{\partial r}{\partial z}\right)^2} \approx \sigma 2\pi r_0 L \left(1 + \frac{1}{8}(qr_1 \sin \omega t)^2\right). \quad (12.31)$$

Comparison of the kinetic and surface energies leads to the dispersion relation

$$\frac{\rho \omega^2 R^3}{\sigma} = (qR)^2. \quad (12.32)$$

Now, considering the viscous energy losses, we find

$$\frac{d}{dt} E_{\text{kin}} = -\frac{1}{2} \int_0^L dz \, \pi r_0^2 \eta \left|\frac{\partial^2 \dot{r}}{\partial z^2}\right| \dot{r}$$

$$= -\frac{1}{2}\eta q^2 \pi r_0^2 \int_0^L dz \, r_1^2 \omega^2 \sin^2 qz \cos^2 \omega t = -\nu q^2 E_{\text{kin}}. \quad (12.33)$$

The kinetic energy and surface energy thus decrease as

$$E_{\text{kin}} \propto \exp\left(-\nu q^2 t\right). \quad (12.34)$$

For the equation of motion, we obtain

$$\rho R^3 \ddot{r}_1 \pm \eta q^2 R^3 \dot{r}_1 + \sigma (qR)^2 r_1 = 0. \quad (12.35)$$

By separating the frequency ω in (12.28) into its real part and its imaginary part, the frequency ω_1 and the damping decrement ω_2, i.e.

$$\omega = \omega_1 + i\omega_2, \quad (12.36)$$

we obtain

$$-\rho R^3 \omega^2 \pm i\omega \eta q^2 R^3 + \sigma (qR)^2 = 0, \quad (12.37)$$

$$\frac{\rho \omega_1^2 R^3}{\sigma} = (qR)^2 - \frac{Oh^2}{4}(qR)^4, \quad \omega_2 = \frac{\nu q^2}{2}. \quad (12.38)$$

The decrease in resonance frequency ω_1 with increasing viscous friction is proportional to the square of the Ohnesorge number Oh.

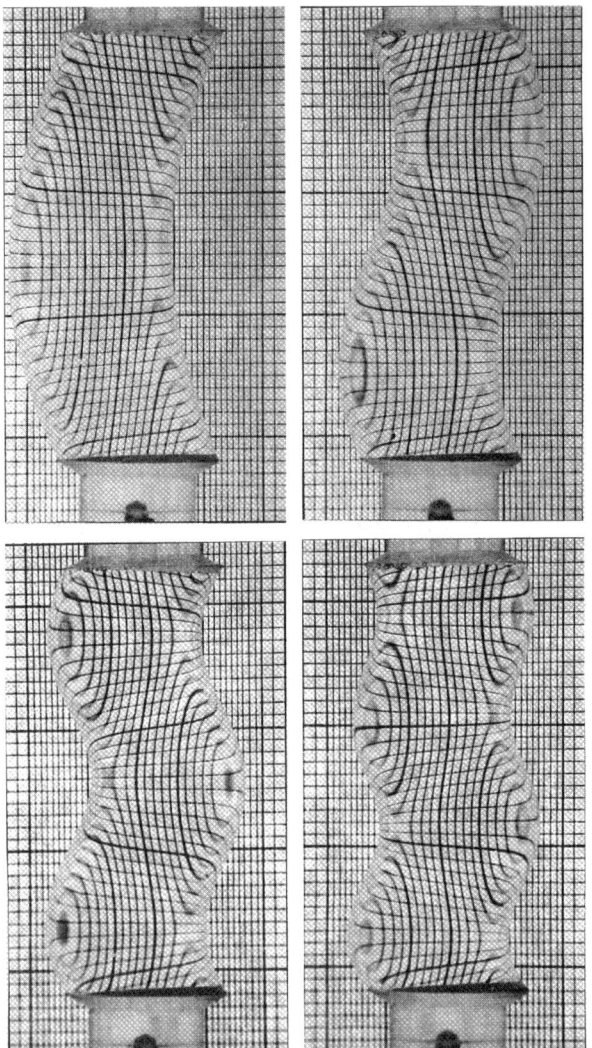

Fig. 12.17. The first four axial modes of lateral oscillation ($n = 1$) at the resonance frequencies of a liquid bridge surrounded by another liquid. The liquid in the bridge was a slightly dyed dimethyl silicone oil with a viscosity of $20 \times 10^{-6}\,\mathrm{m^2\,s^{-1}}$. The bridge was surrounded by a 1:2 methanol–water mixture, whose density was trimmed with the aid of observation of the deformation of the liquid bridge to match precisely that of the oil ($\rho = 945\,\mathrm{kg\,m^{-3}}$). The interfacial tension was measured as $\sigma = (9.7 \pm 0.5)\,\mathrm{\mu N/m}$ (from Fig. 9 of [Sanz & Lopez Diez 1989])

12.4.2 Periodic Lateral Deformations

Equation (12.28) is applicable to cylindrical columns only. In the case of noncylindrical shapes, the normal deformations $D\{m, n\}$ must be calculated

12.4 Lateral Oscillations of Liquid Bridges

from the capillary equation. In cylindrical coordinates $r(\varphi, z)$, the capillary equation reads (see (3.64))

$$\widehat{p} = \frac{1}{Sr} - \frac{1}{S^3} \left\{ \frac{\partial^2 r}{\partial z^2} \left[1 + \left(\frac{\partial r}{r\,\partial\varphi} \right)^2 \right] - \frac{\partial r}{r\,\partial\varphi} \frac{\partial r}{\partial z} \left(\frac{\partial}{r\,\partial\varphi} \frac{\partial r}{\partial z} + \frac{\partial}{\partial z} \frac{\partial r}{r\,\partial\varphi} \right) \right.$$

$$\left. + \frac{\partial}{r\,\partial\varphi} \left(\frac{\partial r}{r\,\partial\varphi} \right) \left[1 + \left(\frac{\partial r}{\partial z} \right)^2 \right] \right\}. \qquad (12.39)$$

As in Sect. 5.2.3, we assume a static axisymmetric shape $r_0(z)$ and a periodic lateral perturbation $r_1(z,t) \cos \varphi$, i.e. we replace the sinsoidal deformation given by (12.28) by the trial functions

$$r = r_0(z) + r_1(z,t) \cos \varphi, \quad \widehat{p} = \widehat{p}_0(z) + \widehat{p}_1(z,t) \cos \varphi. \qquad (12.40)$$

Substitution into (12.39) and comparison of the coefficients of $(\cos \varphi)^n$ for $n = 0, 1$ yields

$$\widehat{p}_0(z) = \frac{1}{S_0 r_0} - \frac{1}{S_0^3} \frac{d^2 r_0}{dz^2}, \quad \text{where } S_0 = \sqrt{1 + \left(\frac{dr_0}{dz}\right)^2}, \qquad (12.41)$$

$$\widehat{p}_1(z,t) = -\frac{1}{S_0^3 r_0} \frac{dr_0}{dz} \frac{\partial r_1}{\partial z} - \frac{1}{S_0^3} \frac{\partial^2 r_1}{\partial z^2} + \frac{3}{S_0^5} \frac{dr_0}{dz} \frac{d^2 r_0}{dz^2} \frac{\partial r_1}{\partial z}. \qquad (12.42)$$

The term $r^{-1} = (r_0 + r_1 \cos \varphi)^{-1}$ cancels with the term $\partial^2 (r_1 \cos \varphi)/\partial\varphi^2$; $p_1(z,t)$ drives the lateral motion with local velocity $u_x(z,t)$, i.e.

$$u_x(z,t) = \frac{\partial r_1(z,t)}{\partial t}. \qquad (12.43)$$

From the momentum equation, we have

$$\rho \frac{\partial u_x(x,z,t)}{\partial t} = -\frac{\partial p(x,z,t)}{\partial x} + \eta \frac{\partial^2 u_x(x,z,t)}{\partial z^2}. \qquad (12.44)$$

The lateral pressure gradient in the liquid equals

$$\frac{\partial p(x,z,t)}{\partial x} r_0(z) = p_1(z,t). \qquad (12.45)$$

Substitution of (12.42), (12.43) and (12.45) into (12.44) yields a partial differential equation for $r_1(z,t)$:

$$\left(\frac{3}{S_0^5} \frac{d^2 r_0}{dz^2} - \frac{1}{S_0^3 r_0} \right) \frac{dr_0}{dz} \frac{\partial r_1}{\partial z} - \frac{1}{S_0^3} \frac{\partial^2 r_1}{\partial z^2} = r_0(z) \left(\frac{\eta}{\sigma} \frac{\partial^2}{\partial z^2} \frac{\partial r_1}{\partial t} - \frac{\rho}{\sigma} \frac{\partial^2 r_1}{\partial t^2} \right). \qquad (12.46)$$

In the stationary case $\partial r_1/\partial t = 0$, (12.46) reduces to (5.8) for the lateral instability $D\{0, n\}$ with $n = 1$. Equation (12.46) describes a damped harmonic oscillator

$$\Gamma_2(z)\ddot{r}_1(z) + \Gamma_1(z)\dot{r}_1(z) + \Gamma_0(z)r_1(z) = 0 , \tag{12.47}$$

where

$$\Gamma_0(z) = \left(\frac{3}{S_0^5}\frac{\mathrm{d}^2 r_0}{\mathrm{d}z^2} - \frac{1}{S_0^3 r_0}\right)\frac{\mathrm{d}r_0}{\mathrm{d}z}\frac{\partial}{\partial z} - \frac{1}{S_0^3}\frac{\partial^2}{\partial z^2} , \tag{12.48}$$

$$\Gamma_1(z) = -\frac{\eta}{\sigma}r_0(z)\frac{\partial^2}{\partial z^2} , \quad \Gamma_2(z) = \frac{\rho}{\sigma}r_0(z) . \tag{12.49}$$

12.4.3 Coupled Damped Oscillations

As in Sect. 12.4.1, let us look for solutions of (12.47) in the form of damped harmonic oscillators. The general solution of that equation is given by

$$r_1(z,t) = [r_{11}(z) + ir_{12}(z)]\exp(i\omega t); \quad \omega = \omega_1 + i\omega_2 . \tag{12.50}$$

Substitution into (12.47) leads to

$$[\Gamma_2(z)\left(-\omega_1^2 + \omega_2^2 - 2i\omega_1\omega_2\right) + \Gamma_1(z)i(\omega_1 + i\omega_2) + \Gamma_0(z)](r_{11} + ir_{12})$$
$$= 0 . \tag{12.51}$$

These are two coupled differential equations, for the real part $r_{11}(z)$ and the imaginary part $r_{12}(z)$ of the perturbation $r_1(z,t)$, which differ only in the sign of ω_1. By nondimensionalizing (12.51) according to

$$\tilde{r} = \frac{r}{R} , \quad \tilde{z} = \frac{z}{R} , \quad \tilde{\omega}_1 = \omega_1\sqrt{\frac{\rho R^3}{\sigma}} , \quad \tilde{\omega}_2 = \omega_2\sqrt{\frac{\rho R^3}{\sigma}} , \quad Oh = \sqrt{\frac{\rho\nu^2}{\sigma R}} , \tag{12.52}$$

we reobtain (12.51) with all parameters now bearing the symbol $\tilde{}$. In components, (12.51) reads

$$\left[\left(\tilde{\omega}_1^2 - \tilde{\omega}_2^2\right)\tilde{\Gamma}_2(\tilde{z}) + \tilde{\omega}_2\tilde{\Gamma}_1(\tilde{z}) - \tilde{\Gamma}_0(\tilde{z})\right]\tilde{r}_{11}(\tilde{z}) - \tilde{\omega}_1\left[2\tilde{\omega}_2\tilde{\Gamma}_2(\tilde{z}) - \tilde{\Gamma}_1(\tilde{z})\right]\tilde{r}_{12}(\tilde{z})$$
$$= 0 , \tag{12.53}$$

$$\left[\left(\tilde{\omega}_1^2 - \tilde{\omega}_2^2\right)\tilde{\Gamma}_2(\tilde{z}) + \tilde{\omega}_2\tilde{\Gamma}_1(\tilde{z}) - \tilde{\Gamma}_0(\tilde{z})\right]\tilde{r}_{12}(\tilde{z}) + \tilde{\omega}_1\left[2\tilde{\omega}_2\tilde{\Gamma}_2(\tilde{z}) - \tilde{\Gamma}_1(\tilde{z})\right]\tilde{r}_{11}(\tilde{z})$$
$$= 0 , \tag{12.54}$$

where

$$\tilde{\Gamma}_1(\tilde{z}) = -Oh\,\tilde{r}_0(\tilde{z})\frac{\partial^2}{\partial\tilde{z}^2} , \quad \tilde{\Gamma}_2(\tilde{z}) = \tilde{r}_0(\tilde{z}) . \tag{12.55}$$

In the limit of small Ohnesorge numbers Oh, we conclude from (12.53), (12.54) that

$$\tilde{r}_{12}(\tilde{z}) \propto Oh , \quad \tilde{\omega}_2 \propto Oh , \quad \tilde{\omega}_1 = \tilde{\omega}_1(0) - \mathrm{const}(Oh)^2 . \tag{12.56}$$

12.4 Lateral Oscillations of Liquid Bridges

In the case of cylindrical columns, where $\tilde{r}_0(\tilde{z}) = 1$, $d\tilde{r}_0/d\tilde{z} = 0$ (12.53), (12.54) simplifies to give

$$\left((\tilde{\omega}_1^2 - \tilde{\omega}_2^2) + (1 - Oh\,\tilde{\omega}_2)\frac{d^2}{d\tilde{z}^2}\right)\tilde{r}_{11}(\tilde{z}) - \tilde{\omega}_1\left(2\tilde{\omega}_2 + Oh\frac{d^2}{d\tilde{z}^2}\right)\tilde{r}_{12}(\tilde{z}) = 0 \tag{12.57}$$

$$\left((\tilde{\omega}_1^2 - \tilde{\omega}_2^2) + (1 - Oh\,\tilde{\omega}_2)\frac{d^2}{d\tilde{z}^2}\right)\tilde{r}_{12}(\tilde{z}) + \tilde{\omega}_1\left(2\tilde{\omega}_2 + Oh\frac{d^2}{d\tilde{z}^2}\right)\tilde{r}_{11}(\tilde{z}) = 0. \tag{12.58}$$

Equations (12.57), (12.58) are solved by the sine and cosine circular functions. The boundary conditions at the supporting disks, together with volume conservation, require

$$\tilde{r}_{11}(\tilde{z}) = c_1 \sin \tilde{q}\tilde{z}, \quad \tilde{r}_{12}(\tilde{z}) = c_2 \sin \tilde{q}\tilde{z}, \quad \tilde{q} = \frac{(m+1)\pi R}{L}. \tag{12.59}$$

For (12.59) to solve (12.57) and (12.58), the terms in the large parentheses must vanish independently, yielding

$$\tilde{\omega}_1^2 + \tilde{\omega}_2^2 = \tilde{q}^2, \quad \tilde{\omega}_2 = \tfrac{1}{2}Oh\,\tilde{q}^2. \tag{12.60}$$

Equation (12.60) confirms the preliminary equation (12.38) obtained from the balance between kinetic energy and surface energy. The damping decrement $\tilde{\omega}_2$ may be generally calculated from the relative average loss in kinetic energy per period:

$$\tilde{\omega}_2 = \frac{Oh}{2}\frac{\int dz\, r_0^2(z) r_1(z)\left[-\partial^2 r_1(z)/\partial z^2\right]}{\int dz\, r_0^2(z) r_1^2(z)}. \tag{12.61}$$

Figure 12.18 shows the amplification of the disk amplitude versus the excitation frequency f for a liquid bridge of slenderness $L_c = 2.6$ surrounded by a density-matched liquid [Sanz & Lopez Diez 1989]. The similarity to Fig. 12.6 is obvious. The positions of the amplification maxima are given by $\tilde{\omega}_1$, and their sharpness by $\tilde{\omega}_2$. Figure 12.19 depicts the dimensionless frequency $\tilde{\omega}_1$ versus the aspect ratio L_c of the liquid bridges.

Figures 12.20–12.23 have been obtained by numerical integration of (12.53), (12.54). Figure 12.20 shows a noncylindrical column with aspect ratio $L_c = 1.5$. The liquid volume is slightly greater than that for a cylindrical shape; $V_c = 1.2$. The solid lines represent the static shape $r_0(z)$ of the liquid column, the dash–dot line the deformation $r_{11}(z)$, and the dashed line the deformation $-r_{12}(z)$. The latter follows $r_{11}(z)$ with a phase shift of $\pi/2$ according to

$$\text{Re}[r_1(z)] = [r_{11}(z)\cos(\omega_1 t) - r_{12}(z)\sin(\omega_1 t)]\exp(-\omega_2 t). \tag{12.62}$$

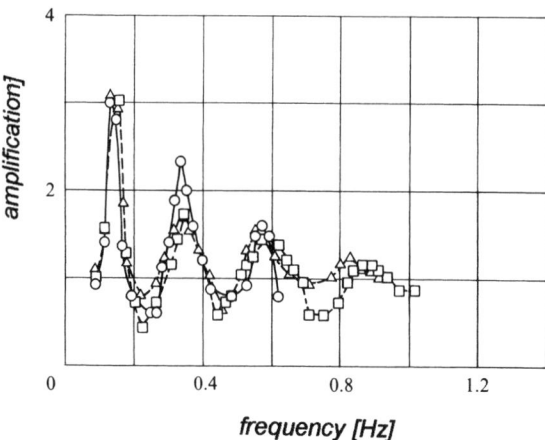

Fig. 12.18. Variation of the amplification of the disk amplitude with the excitation frequency f for a liquid bridge of slenderness $L_c = 2.6$ surrounded by a density-matched liquid (Fig. 8 of [Sanz & Lopez Diez 1989])

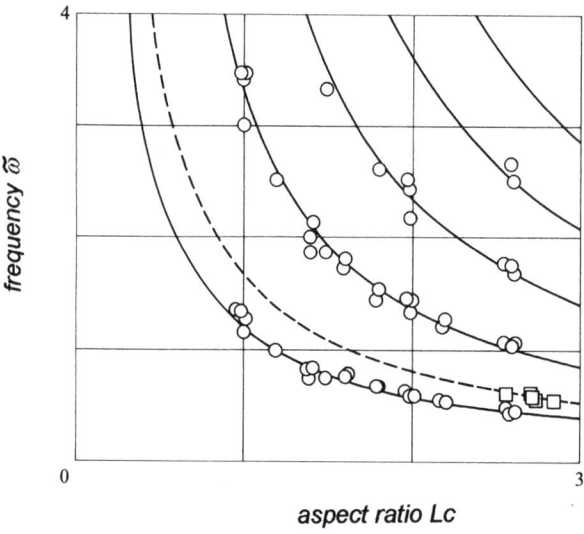

Fig. 12.19. Dimensionless frequency $\tilde{\omega}_1$ versus aspect ratio L_c of a liquid bridge. *Solid lines*, first five axial modes of a density-matched liquid bridge; *dashed line*, first axial mode of a nonsurrounded liquid bridge; *circles*, neutral buoyancy; *squares*, Spacelab D1 results [Fig. 10, Sanz & Lopez Diez 1989]

With another shift of $\pi/2$, the situation obtained by exchanging left and right in Fig. 12.20 appears, and so on. The points at which the maximum amplitude and an amplitude of zero occur, depend on the axial coordinate z. In the middle of the column, the maximum amplitude occurs first. The deformation in the middle of the column, where the strongest curvature (highest

12.4 Lateral Oscillations of Liquid Bridges   319

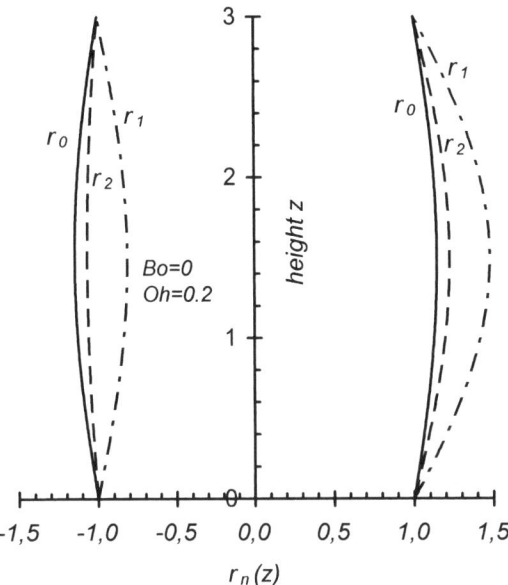

Fig. 12.20. Lateral oscillation of a noncylindrical column with aspect ratio $L_c = 1.5$. The liquid volume is slightly greater than that for a cylindrical shape; $V_c = 1.2$. The *solid lines* represent the static shape $r_0(z)$ of the liquid column, the *dash–dot line* the deformation $r_{11}(z)$ and the *dashed line* the deformation $r_{12}(z)$; $Bo = 0$, $Oh = 0.2$

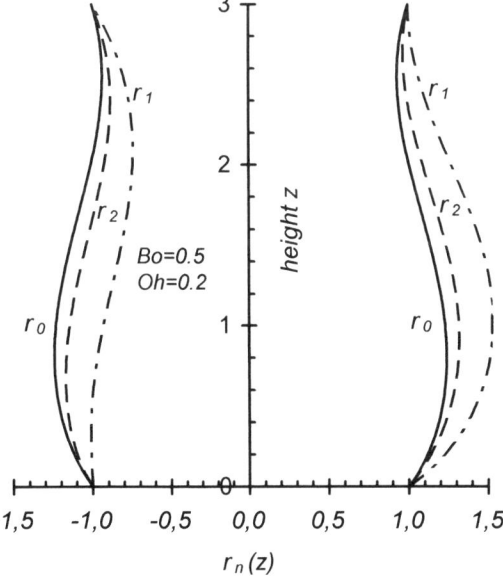

Fig. 12.21. The same oscillation as in Fig. 12.20 for a liquid column at Bond number $Bo = 0.5$ and $Oh = 0.2$

320   12. Oscillations of Liquid Columns

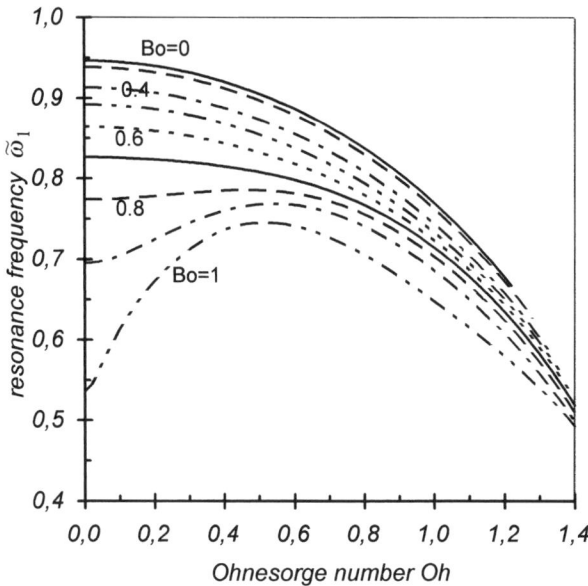

Fig. 12.22. The resonance frequency $\tilde{\omega}_1$ of the first lateral oscillation $D\{0,1\}$ as a function of the Ohnesorge number Oh and the Bond number Bo

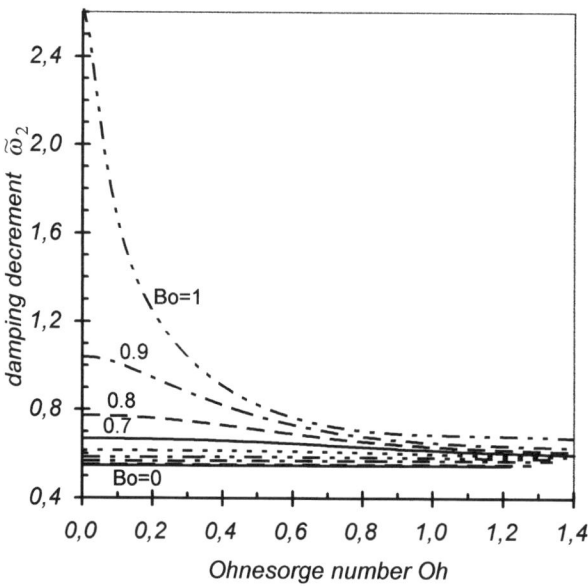

Fig. 12.23. The damping decrement $\tilde{\omega}_2$ of the first lateral oscillation $D\{0,1\}$ as a function of the Ohnesorge number Oh and the Bond number Bo

pressure) is driving the oscillation, goes ahead in phase and pulls behind the motion in the regions near the supporting disks.

Figure 12.21 depicts the same oscillation for a liquid column at a Bond number $Bo = g\rho R^2/\sigma = 0.5$.

The strongest curvature and pressure driving the oscillation arise in the lower part of the column. Consequently, the oscillation there is ahead of the rest of the column. The retardation of the oscillation near the disks increases with increasing Ohnesorge number Oh.

Figure 12.22 shows the resonance frequency $\tilde{\omega}_1$ of the first lateral oscillation $D\{0,1\}$ as a function of the Ohnesorge number Oh and the Bond number Bo. As in Fig. 12.20, the aspect ratio is $L_c = 1.5$ and the liquid volume is $V_c = 1.2$. At $Bo = 0$, the decrease in frequency with increasing Ohnesorge number is proportional to Oh^2. The resonance frequency also decreases with increasing Bond number. It approaches zero when the aspect ratio L_c approaches the maximum stable height of the liquid bridge, as discussed in Chap. 5. Figure 12.23 shows the corresponding behavior of the damping decrement $\tilde{\omega}_2$.

References

1. Abramowitz M, Stegun IA: *Handbook of Mathematical Functions*. Dover, New York (1965)
2. Ahrens S, Falk F, Großbach R, Langbein D: Experiments on oscillations of small liquid bridges. Microgravity Sci. Technol. **7** (1994) 2–5
3. Bauer HF: Natural damped frequencies of an infinitely long column of immiscible viscous liquids. Zeitschrift Angewandte Mathematik und Mechanik **64** (1984) 475–490
4. Bauer HF: Free surface and interface oscillations of an infinitely long viscoelastic liquid column. Acta Astronautica **13** (1986) 9–22
5. Bauer HF: Axial response of differently excited anchored viscous liquid bridges in zero-gravity. Arch. Appl. Mech. **63** (1992) 322–336
6. Langbein D: Oscillations of finite liquid columns. Microgravity Sci. Technol. **5** (1992) 73–85
7. Langbein D: Oscillations of finite liquid columns. ESA SP-353 (1992) 419–442
8. Lee HC: Drop formation in a liquid jet. IBM J. Res. Devel. **18** (1974) 364–369
9. Meseguer J: The breaking of axisymmetric slender liquid bridges. J. Fluid Mech. **130** (1983) 123–151
10. Meseguer J, Perales JM: Viscosity effects on the dynamics of long axisymmetric liquid bridges. In: *Microgravity Fluid Mechanics*. H.J. Rath (ed.). Springer, Berlin, Heidelberg (1992) 37–46
11. Meseguer J, Sanz A: Numerical and experimental study of the dynamics of axisymmetric slender liquid bridges. J. Fluid Mech. **153** (1985) 83–101
12. Sanz A, Lopez Diez J: Non-axisymmetric oscillations of liquid bridges. J. Fluid Mech. **205** (1989) 503–521
13. Sanz A: The influence of the outer bath on the dynamics of axisymmetric liquid bridges. J. Fluid Mech. **156** (1985) 101–140

14. Schilling U, Siekmann J: Gleichgewichtsformen und Eigenschwingungen von Flüssigkeitsbrücken unter Restschwere. Z. Flugwiss. Weltraumforsch. **13** (1989) 308–314
15. Schulkes RMSM: Eigenfrequencies of a rotating, viscous, incompressible fluid with a capillary free boundary. Proceedings of the 28th COSPAR (Annual Conference of the Committee of Space Research), The Hague (1990)

13. Microgravity Experiments in Sounding Rockets, Spacelab and EURECA

13.1 TEXUS 1–39

13.1.1 TEXUS 1

Launch date 13 Dec. 1977; number of experiments 11.

- Aluminum–Lead Alloy. E. Hodes, M. Steeg, Glyco-Wiesbaden
- Reaction Kinetics in Glass Melts. H.J. Barklage-Hilgefort, G.H. Frischat, Technical University of Clausthal
- Skin Technology. H. Sprenger, MAN, München, Schweitzer, MTU, München
- Powder Metallurgy. H.U. Walter, G. Ziegler, DFVLR, Köln
- Primary Crystallization. H. Ahlborn, Technical University of Hamburg
- Hydrogen Electrolysis. C.J. Raub, W. Wopersnow, FI Edelmetalle, Schwäbisch-Gmünd
- Dispersion Electrolysis. J. Ehrhardt, Battelle, Frankfurt
- Interface Convection. R. Brückner, Technical University of Berlin
- Dendrite Growth. H. Fredriksson, T. Carlberg, RIT, Stockholm
- Alloy with Oxide Inclusions. G. Neuschütz, Krupp FI, Essen
- Acoustic Positioning. E.G. Lierke, Battelle, Frankfurt

13.1.2 TEXUS 2

Launch date 16 Nov. 1978; number of experiments 10.

- Application of Skin Technology. H. Sprenger, MAN, München, K. Schweitzer, MTU, München
- Stability of Multicomponent Mixtures. H.U. Walter, G. Ziegler, DFVLR, Köln
- Fundamental Studies in the Manganese–Bismuth System. P. Pant, Krupp FI, Essen
- Brazing of Capillary Gaps. W. Bathke, BA Materialprüfung, Berlin
- Alloy Formation. F. Haeßner, K.J. Hettwer, Technical University of Braunschweig
- Interfacial Convection. R. Brückner, Technical University of Berlin
- Capillarity. P.J. Sell, D. Renzow, FHG Stuttgart

- Electrochemical Corrosion. G. Mix, Fachhochschule Kiel
- Solidification Studies. H. Fredriksson, T. Carlberg, RIT, Stockholm
- Solidification of an Alloy with Oxide Inclusions. J. Pötschke, D. Neuschütz, Krupp FI, Essen

13.1.3 TEXUS 3

Launch date 30 Apr. 1981; number of experiments 12.

- Skin Technology. H. Sprenger, MAN, München, K. Schweitzer, MTU, München
- Silver–Glas Powder Metallurgy. H.U. Walter, DFVLR, Köln
- Copper–Alumina Dispersion Alloy. J. Pötschke, K. Hohenstein, Krupp FI, Essen
- Aluminum–Hydrogen. H. Hoffmeister, J. Rüdiger, HdB Hamburg
- Boundary Layer Convection. R. Brückner, Technical University of Berlin
- Kinetics of Wetting. P.J. Sell, FHG Stuttgart
- Temperature Oscillation by Marangoni Convection. D. Schwabe, A. Scharmann, University of Giessen
- β-Galactosidase Diffusion Experiment. W. Littke, University of Freiburg
- Heat Transfer and Bubble Growth. A. Weinzierl, J. Straub, Technical University of München
- Marangoni Convection in Floating Zone. C.-H. Chun, DFVLR, Göttingen
- Halogen Lamp Experiments. A. Eyer, H. Walcher, University of Freiburg
- Coalescence in Immiscible Alloys. H. Fredriksson, A. Bergmann, RIT, Stockholm

13.1.4 TEXUS 4

Launch date 08 May 1981; number of experiments 11.

- Skin Technology. H. Sprenger, MAN, München, K. Schweitzer, MTU, München
- Metallic Foam 1 and 2. J. Pötschke, P. Pant, Krupp FI, Essen
- Reaction Kinetics in a Melt. G.H. Frischat, W. Beier, M. Breadt, Technical University of Clausthal
- Dotation Striations in Germanium 1, 2, 3 and 4. H.U. Walter, DFVLR, Köln
- β-Galactosidase Diffusion at Low Temperatures. W. Littke, University of Freiburg
- Dispersion Electrolysis. J. Ehrhardt, Battelle, Frankfurt
- Interfacial Convection. R. Brückner, Technical University of Berlin

13.1.5 TEXUS 5

Launch date 29 Apr. 1982; number of experiments 12.

- Skin Technology. H. Sprenger, MAN, München
- Dispersion Alloy. H.U. Walter, DFVLR, Köln
- Metal Foam. W. Schäfer, Dornier System Friedrichshafen
- Copper–Alumina Suspensions. J. Pötschke, Krupp FI, Essen
- Alloy Formation. F. Haeßner, P. Wehr, Technical University of Braunschweig
- Acoustic Mixing. W. Heide, D. Langbein, Battelle, Frankfurt
- Kinetics of Wetting. P.J. Sell, D. Renzow, E. Maisch, FHG Stuttgart
- Critical Marangoni Number. D. Schwabe, A. Scharmann, University of Giessen
- β-Galactosidase Diffusion Experiment. W. Littke, University of Freiburg
- Bubble Growth 1. J. Straub, A. Weinzierl, Technical University of München
- Coalescence in Immiscible Alloys. H. Fredriksson, A. Bergmann, RIT, Stockholm
- Soldering. T. Carlberg, RIT, Stockholm

13.1.6 TEXUS 6

Launch date 08 May 1982; number of experiments 11.

- Reaction Kinetics in Glass Melts. M. Breadt, G.H. Frischat, Technical University of Clausthal
- Metal Foam. J. Pötschke, P. Pant, Krupp FI, Essen
- Dispersion Alloy. A. Deruyttere, L. Froyen, University of Leuven
- Diffusion of Sulfur in Cast Iron. T. Luyendijk, H. Nieswaag, Technical University of Delft
- Aluminum–Nickel Eutectic. J.J. Favier, J. deGoer, CENG, Grenoble
- Aluminum–Copper Eutectic. J.J. Favier, J. deGoer, CENG, Grenoble
- Dotations Stripes in Germanium. H.U. Walter, DFVLR, Köln
- Dotations of Germanium. H.U. Walter, DFVLR, Köln
- Thermal Boundary Layer Convection. R. Brückner, Technical University of Berlin
- Phase Transformation. J. Straub. R. Lange, Technical University of München
- Bubble Electrolysis. C.J. Raub, FI Edelmetall and Metallchemie, Schwäbisch-Gmünd

13.1.7 TEXUS 7

Launch date 05 May 1983; number of experiments 13.

- Skin Technology. H. Sprenger, MAN, München, K. Schweitzer, MTU, München
- Dispersion Alloys 1. H.U. Walter, DFVLR, Köln
- Metallic Composite Materials with Particles. A. Deruyttere, L. Froyen, University of Leuven
- Metal Foam. J. Pötschke, P. Pant, Krupp FI, Essen
- Silicon Growth from Melting Zone. A. Eyer, H. Leiste, University of Freiburg
- Oscillation Modes of Marangoni Convection. C.-H. Chun, DFVLR, Göttingen
- Transparent Separation. D. Langbein, Battelle, Frankfurt
- Dispersion Electrolysis. J. Ehrhardt, Battelle, Frankfurt
- Single Crystals of $CeMg_3$. J. Pierre, J. Baruchel, M. Schlemker, E. Siaud, CNRS, Grenoble
- Collision of Small Crystals. H. Fredriksson, RIT, Stockholm
- Miscibility Gap Alloys. H. Fredriksson, RIT, Stockholm
- Floating Zone in a Mirror Furnace. T. Carlberg, RIT, Stockholm
- Immiscible Alloy System AlPb. P.D. Caton, W.G. Hopkins, Fulmer Research Institute, Slough

13.1.8 TEXUS 8

Launch date 05 May 1983; number of experiments 9.

- Liquid-Phase Sintering. F. Rositto, CESNEF, Milano
- Solidification of Nonmiscible Alloys. D. Langbein, W. Heide, Battelle, Frankfurt
- Boundary Layer Convection. R. Brückner, Technical University of Berlin
- Bubble Growth 2. J. Straub, A. Weinzierl, M. Zell, Technical University of München
- Critical Marangoni Convection. D. Schwabe, A. Scharmann, University of Giessen
- Influence on Marangoni Effect of a Surface Tension Minimum. J.C. Legros, G. Pétré, University of Bruxelles
- Gas–Liquid Phase Transformation. J. Straub, R. Lange, K. Nitsche, Technical University of München
- Liquid–Gas Phase Transformation. J. Straub, R. Lange, K. Nitsche, Technical University of München
- Interdiffusion. J. Richter, J. Hermanns, W. Merkens, RWTH Aachen

13.1.9 TEXUS 9

Launch date 03 May 1984; number of experiments 9.

- Skin Technology. H. Sprenger, MAN, München

- Copper–Molybdenum Suspension. J. Pötschke, P. Pant, Krupp FI, Essen
- Aluminum Foam. W. Schäfer, Dornier System Friedrichshafen
- Melting/Solidification of Particulate Metallic Composites. A. Deruyttere, L. Froyen, University of Leuven
- Acoustic Positioning. E.G. Lierke, R. Großbach, Battelle, Frankfurt, D.M. Herlach, DFVLR, Köln
- Thermal Marangoni Convection in a Floating Zone. R. Monti, L.G. Napolitano, University of Napoli
- Drops at a Solidification Front. D. Langbein, W. Heide, Battelle, Frankfurt
- Dispersion Electrolysis. J. Ehrhardt, Battelle, Frankfurt
- Influence on Marangoni Effect of a Surface Tension Minimum. J.C. Legros et al., University of Bruxelles

13.1.10 TEXUS 10

Launch date 15 May 1984; number of experiments 9.

- Directional Solidification of InSb–NiSb Eutectic. G. Müller, P. Kyr, University of Erlangen-Nürnberg
- Bubbles and Foams. P.J. Sell, D. Renzow, K.F. Gebhardt, FHG Stuttgart
- Bubble Growth. J. Straub, M. Zell, Technical University of München
- Boundary Layer in Solidifying Transparent Melts. P. R. Sahm, A. Ecker, RWTH Aachen
- Maximum Injection Rate in a Floating Zone. I. Martinez, A. Sanz, ETSIA, Madrid
- Ionic Solutions Near the Critical Point. D.J. Tuner, Central Electricity Research Lab., Leatherhead
- Tungsten Composites. L.B. Ekbom, National Defense Research Institute, H. Fredriksson, RIT, Stockholm
- Floating Zone Growth of Ga-Doped Germanium. T. Carlberg, RIT, Stockholm
- Retention of a Fine Precipitate Dispersion. P.J. Goodhew, T.W. Clyne, University of Surrey

13.1.11 TEXUS 11

Launch date 27 Apr. 1985; number of experiments 8.

- Skin Technology. H. Sprenger, MAN, München
- Interdiffusion 1 and 2. G.H. Frischat, Technical University of Clausthal
- Copper–Alumina Suspension. J. Pötschke, Krupp FI, Essen
- Boundary Layer Convection. R. Brückner, Technical University of Berlin
- Bubble Growth 2. J. Straub, Technical University of München
- Fusion of Yeast Cells. U. Zimmermann, University of Würzburg
- Spiniodal Decomposition. D. Beysens, CEN, Saclay

13.1.12 TEXUS 12

Launch date 06 May 1985; number of experiments 7.

- Interdiffusion. G.H. Frischat, Technical University of Clausthal
- Liquid-Phase Sintering 1 and 2. W. Graf, Krupp FI, Essen
- Skin Technology. H. Sprenger, MAN, München
- Silicon Growth from Melting Zone. A. Cröll, University of Freiburg
- Maximum Injection Rate in a Floating Zone. I. Martinez, ETSIA, Madrid
- Ionic Solutions Near the Critical Point. D.J. Tuner, Central Electricity Research Lab., Leatherhead

13.1.13 TEXUS 13

Launch date 19 Apr. 1986; number of experiments 10.

- Rare Earths 1 and 2. H. Bach, University of Bochum
- Fabrication of Superconducting Materials. K. Togano, Tsukuba
- Interdiffusion. G.H. Frischat, Technical University of Clausthal
- Electro Cell Fusion. U. Zimmermann, University of Würzburg
- Plant Protoplasts. R. Hampp, University of Tübingen
- Single Bubble. J. Straub, Technical University of München
- Surface Tension Minimum. J.C. Legros, M.C. Limbourg, G. Pétré, University of Bruxelles
- Crystal Growth from Solution. R. Vochten, University of Antwerp
- Spinoidal Decomposition. D. Beysens, P. Guenoun, F. Perrot, CEN, Saclay

13.1.14 TEXUS 14B

Launch date 30 Apr. 1987; number of experiments 9.

- Acoustic Positioning. E.G. Lierke, Battelle, Frankfurt
- Thermal Marangoni Convection in a Floating Zone. R. Monti, University of Napoli
- Ultrasonic Absorption in a Molten Salt. J. Richter, RWTH Aachen
- Fusion of Yeast Cells. U. Zimmermann, University of Würzburg
- Gene Transfer. U. Zimmermann, University of Würzburg
- Unidirectional Solidification of ZnBi. H. Fredriksson, RIT, Stockholm
- Solidification of Fine Dispersions. N.J.E. Adkins, P.J. Goodhew, University of Surrey
- Metal Matrix Composites. A. Deruyttere, L. Froyen, University of Leuven
- Floating Zone Experiments with Ge Crystals. T. Carlberg, RIT, Stockholm

13.1.15 TEXUS 15

Launch date 09 May 1987; number of experiments 8.

- Dispersion Alloys. H. Sprenger, MAN, München
- Liquid-Phase Sintering. V. Rogge, Krupp FI, Essen
- Separation Monotectic Ternary Alloys. B. Prinz, Metallgesellschaft Frankfurt
- Interdiffusion. G.H. Frischat, Technical University of Clausthal
- GaAs Crystal Growth. G. Müller, University of Erlangen-Nürnberg
- Plant Protoplasts. R. Hampp, University of Tübingen
- Surface Tension Minimum. J.C. Legros, University of Bruxelles
- Colloid Chemistry. H. Stein, University of Eindhoven

13.1.16 TEXUS 16

Launch date 23 Nov. 1987; number of experiments 10.

- Dispersion Alloys. H. Sprenger, MAN, München
- Liquid-Phase Sintering. V. Rogge, Krupp FI, Essen
- Separation Monotectic Ternary Alloys. B. Prinz, Metallgesellschaft Frankfurt
- Interdiffusion. G.H. Frischat, Technical University of Clausthal
- Liquid-Phase Sintering. L. Ekbohm, RIT, Stockholm
- Metal Matrix Composites. A. Deruyttere, University of Leuven
- Diffusion of Ni in CuAl. C. Köstler, University of Göttingen
- Refraction Gradient in Glasses. G.H. Frischat, Technical University of Clausthal
- Floating Zone Melting of GaAs Crystals. G. Müller, University of Erlangen-Nürnberg
- Influence of a Surface Tension Minimum on the Marangoni Effect. J.C. Legros, University of Bruxelles

13.1.17 TEXUS 17

Launch date 02 May 1988; number of experiments 5.

- Coagulation of Solid/Liquid Dispersions. H.N. Stein, University of Eindhoven
- Ultrasonic Absorption in a Molten Salt. J. Richter, E. Kirschbaum, RWTH Aachen
- Spinoidal Decomposition. H. Klein, DFVLR, D. Woermann, University of Köln
- Plant Protoplasts. R. Hampp, W. Mehrle, University of Tübingen
- Amphibian Embryos. G.A. Ubbels, S. Kerkvliet, Hubrecht Lab, Utrecht

13.1.18 TEXUS 18

Launch date 05 May 1988; number of experiments 5.

- Cress Growth. H.N. Stein, University of Eindhoven
- Electrophoresis. K. Hannig, MPI Martinsried
- Freezing of a Long Liquid Column. I. Martinez, A. Sanz, J.M. Perales, ETSIA, Madrid
- Electrofusion of Yeast Cells. U. Zimmermann, University of Würzburg
- Gene Transfer. U. Zimmermann, University of Würzburg

13.1.19 TEXUS 19

Launch date 28 Nov. 1988; number of experiments 10.

- Liquid-Phase Sintering. V. Rogge, Krupp FI, Essen
- Liquid-Phase Sintering. L. Ekbohm, RIT, Stockholm
- Dispersion Alloy. H. Sprenger, MAN, München
- Melt Pressing. T. Sato, STC, Tokyo
- Refraction Gradient in Glasses. G.H. Frischat, W. Beier, Technical University of Clausthal
- Solution Growth of GaAs. Y. Suzuki, STC, Tokyo
- Electrostatic Positioning. E.G. Lierke, R. Großbach, Battelle, Frankfurt
- Mobility of Spermatozoa. W.B. Schill, University of München
- Influence of a Surface Tension Minimum on the Marangoni Effect. J.C. Legros, University of Bruxelles
- Containerless Processing of Li-Silicate Glass. G.H. Frischat, Technical University of Clausthal

13.1.20 TEXUS 20

Launch date 02 Dec. 1988; number of experiments 8.

- Separation Monotectic Ternary Alloys. B. Prinz, A. Romero, Metallgesellschaft Frankfurt
- Self-Diffusion in Alkali-Silicate Glass Melts. G.H. Frischat, Technical University of Clausthal
- Metal Matrix Composites. A. Deruyttere, L. Froyen, University of Leuven
- Impurity Diffusion in Liquid III–V Compound Semiconductors. M. Watanabe, STC, Tokyo
- Diffusion of Ni in CuAl and CuAu Alloys. C. Köstler, University of Göttingen
- Vapor Growth of InP Crystal. Y. Yukimoto, STC, Tokyo
- Floating-Zone Melting GaAs. G. Müller, University of Erlangen-Nürnberg
- Electrophoresis. K. Hannig, MPI Martinsried

13.1.21 TEXUS 21

Launch date 30 Apr. 1989; number of experiments 5.

- Electrofusion of Plant Protoplasts. R. Hampp, W. Mehrle, University of Tübingen
- Single Growth. J. Straub, A. Vogel, Technical University of München
- Chararhizoid Cytoplasm. A. Sievers, B. Buchen, University of Bonn
- Bénard–Marangoni Instability in a Rectangular Cell. D. Schwabe, A. Scharmann, University of Giessen
- Bénard–Marangoni Instability in a Circular Cell. D. Schwabe, A. Scharmann, University of Giessen

13.1.22 TEXUS 22

Launch date 03 May 1989; number of experiments 5.

- Electromagnetic Positioning. Willnecker, DFVLR, Köln; J. Piller, Dornier Friedrichshafen
- Growth of a Si Crystal with Partially Free Melting Zone. K.W. Benz, A. Cröll, University of Freiburg
- Electrofusion of Yeast Cells. U. Zimmermann, University of Würzburg
- Transection of Genes. U. Zimmermann, University of Würzburg
- Telescience Experiment. E. Bennet, P. Kaul, H.P. Schmitt, K. Wittmann, DLR Köln

13.1.23 TEXUS 23

Launch date 05 Nov. 1989; number of experiments 6.

- Critical Marangoni Flow. R. Monti, University of Napoli
- Ultrasonic Absorption in a Molten Salt. J. Richter, RWTH Aachen
- Rotational Instability of a Long Liquid Column. I. Martinez, ETSIA, Madrid
- Colloid Chemistry. H.N. Stein, University of Eindhoven
- Growth of Cress. D. Volkmann, University of Bonn
- Motility of *Euglena*. P. Häder, University of Erlangen-Nürnberg

13.1.24 TEXUS 24

Launch date 06 Dec. 1989; number of experiments 3.

- Floating-Zone Melting GaAs. G. Müller, University of Erlangen-Nürnberg
- Thermal Conductivity Measurement in Fluid Sb. T. Hibiya, STC, Tokyo
- Electrophoresis. K. Hannig, MPI Martinsried

13.1.25 TEXUS 25

Launch date 13 May 1990; number of experiments 5.

- Electrofusion of Plant Protoplasts. R. Hampp, W. Mehrle, University of Tübingen
- Ultrasonic Absorption in a Molten Salt. J. Richter, RWTH Aachen
- Heat and Mass Transfer in Supercritical CO_2. D. Beysens, CEN, Saclay
- Function of Statoliths in the Rhizoids of *Chara*. A. Sievers, B. Buchen, University of Bonn
- Bénard Convection of Magnetic Colloids. K. Stierstadt, S. Odenbach, University of München

13.1.26 TEXUS 26

Launch date 15 May 1990; number of experiments 8.

- Single Bubble. J. Straub, Technical University of München
- Separation of Monotectic Alloys. H. Ahlborn, Technical University of Hamburg
- STC Experiment. Horitomi, STC, Tokyo
- Cast Iron. W. Amende, MAN, München
- Liquid-Phase Sintering. L. Ekbohm, RIT, Stockholm
- Electro Cell Fusion. U. Zimmermann, University of Würzburg
- Spinoidal Decomposition. H. Klein, DFVLR; D. Woermann, University of Köln
- Mobility of Spermatozoa. W.B. Schill, University of Giessen

13.1.27 TEXUS 27

Launch date 15 Nov. 1990; number of experiments 7.

- Heat Storage Module 1 and 2. F. Lindner, DLR Stuttgart
- Cast Iron. W. Amende. MAN, München
- Liquid-Phase Sintering. L. Ekbohm, RIT, Stockholm
- Pressure of Supercritical Salt Solution. D.J. Turner, Central Electricity Research Lab., Leatherhead
- Electrophoresis Visualization Experiment. H. Roux-de Balmann, University of Toulouse
- Role of Gravity in the Spatial Orientation of Cells. W. Briegleb, DLR Köln

13.1.28 TEXUS 28

Launch date 23 Nov. 1991; number of experiments 5.

- Crystal Growth of Ge:Ga. Hidmann, DLR Köln
- Electrophoresis Visualization Experiment. H. Roux-de Balmann, University of Toulouse
- Gravitaxis of Unicellular Algae. P. Häder, University of Erlangen-Nürnberg
- Role of Gravity in the Spatial Orientation of Cells. W. Briegleb, DLR Köln
- *Chara* Cytoskeleton. A. Sievers, B. Buchen, University of Bonn

13.1.29 TEXUS 29

Launch date 22 Nov. 1992; number of experiments 8.

- Separation of Monotectic Alloys. H. Ahlborn, Technical University of Hamburg
- Superconducting Materials. G. Krabbes, IFW Dresden
- Eutectic High-Temperature Alloys. W. Amende, MAN, München
- Solutal Diffusion in PbSnTe. Horitomi, STC, Tokyo
- Effects of Marangoni Convection on Separation. H. Klein, DFVLR; D. Woermann, University of Köln
- Gravitaxis and Phototaxis of Flagellates. P. Häder, University of Erlangen-Nürnberg
- A. Sievers, B. Buchen, University of Bonn
- Dopant Striations in Floating-Zone Silicon. K.W. Benz, A. Cröll, University of Freiburg

13.1.30 TEXUS 30

Launch date 01 May 1993; number of experiments 6.

- Segregation during Directional Solidification of GeSi. J. Schilz, DLR Köln
- Gravitaxis and Phototaxis of Flagellates. P. Häder, University of Erlangen-Nürnberg
- Cytoplasma Flows in *Chara*-Rhizoid. A. Sievers, B. Buchen, University of Bonn
- Convection of Magnetic Colloids. S. Odenbach, K. Stierstadt, University of München
- Protein Pattern in Mesophyll Protoplasts of *Vicia faba*. H. Schnabel, University of Bonn
- Effect of Changes in Gravity on Energy Metabolism of Plant Cells. R. Hampp, University of Tübingen

13.1.31 TEXUS 31

Launch date 26 Nov. 1993; number of experiments 7.

- Differential Wetting of GaInSb Melt (4 furnaces). T. Duffar, CENG, Grenoble
- Nucleation, Bubble Growth, Interfacial Multilayer. J. Straub, Technical University of München
- Coagulation of Suspensions. H.N. Stein, University of Eindhoven
- Marangoni–Bénard Instability. J.C. Legros, University of Bruxelles

13.1.32 TEXUS 32

Launch date 05 May 1994; number of experiments 7.

- Differential Wetting of GaInSb Melt (2 furnaces). T. Duffar, CENG, Grenoble
- Growth of Highly Doped GaSb. A. Danilewski, A. Cröll, University of Freiburg
- Coagulation during Separation of Binary Liquids. H. Klein, DFVLR, D. Woermann, University of Köln
- Protein Pattern in Mesophyll Protoplasts of *Vicia faba*. H. Schnabel, University of Bonn
- Effect of Changes in Gravity on Energy Metabolism of Plant Cells. R. Hampp, University of Tübingen
- Marangoni Convection. Moolenkamp, University of Groningen

13.1.33 TEXUS 33

Launch date 30 Nov. 1994; number of experiments 4.

- Controlled Acceleration of a Long Liquid Column. I. Martinez, ETSIA, Madrid
- Convectional Instabilities and Structures during Separation. G.M. Schneider, University of Bochum
- Thermocapillary Convection around Bubble. K. Wozniak, Mountain Academy of Freiberg, G. Wozniak, University of Essen
- Measurement of Interfacial Tension. A. Passerone, L. Liggieri, ICFAM-CNR, Genova

13.1.34 TEXUS 34

Launch date 02 Mar. 1996; number of experiments 6.

- Directional Solidification of Sn-Sb. Laakmann, RWTH Aachen
- Separation of Monotectic Alloys. H. Ahlborn, Technical University of Hamburg

- Oxygen Measurement. E. Messerschmid, University of Stuttgart
- Spinoidal Decomposition. H. Klein, DFVLR, D. Woermann, University of Köln
- Convectional Instabilities and Structures during Separation. G.M. Schneider, University of Bochum
- Interactive Experiment on Marangoni Migration. R. Fortezza, R. Monti, University of Napoli

13.1.35 TEXUS 35

Launch date 24 Nov. 1996; number of experiments 5.

- Plasma Crystals. Morfill, Rothermel, MPE München; Thomas, DLR Köln
- Fixation of Flagellates. P. Häder, University of Erlangen-Nürnberg
- Metabolism under Weightlessness. H. Schnabel, University of Bonn
- Gravitaxis of Flagellates on the Rotation Accelerator. P. Häder, University of Erlangen-Nürnberg
- Macrosteps in the Diffusion-Controlled Regime. A. Danilewski, University of Freiburg

13.1.36 TEXUS 36

Launch date 7 Feb. 1998; number of experiments 4.

- Plasma Crystals. Morfill, Rothermel, MPE München; Thomas, DLR Köln
- Signal Transduction in *Euglena*. P. Häder, Lebert, University of Erlangen-Nürnberg
- Development of Instabilities. S. Rex, Zimmermann, A. Weiß, Access Aachen
- Investigation of Microscopic Growth of Silicon. A. Danilewski, University of Freiburg

13.1.37 TEXUS 37

Launch date 27 Mar. 2000; number of experiments 4.

- Autonomous Directional Solidification of ZnSb. S. Rex, W. Huber, Access Aachen
- Critical Velocities in Open Capillary Flows. M. Dreyer, U. Rosendahl, ZARM Bremen
- Regulation of Metabolism under µg. R. Hampp, University of Tübingen
- In-Vivo Observation of Statolith Movement. B. Buchen, Hodieck, Braun, University of Bonn.

13.1.38 TEXUS 38

Launch date 2 Apr. 2000; number of experiments 4.

- Combustion Experiment. S. Tarifa, University of Madrid
- Investigation of Mechano-Transduction/Gravity Transducer Mechanism.
- Ca-Metabolism under µg. D. Jones, University of Marburg
- Droplet Evaporation. I. Gökalp, C. Chauveau, CNRS, Orléans

13.1.39 TEXUS 39

Launch date 8 May 2001; number of experiments 5.

- Second messenger levels in gravitactic ciliates at variable accelerations. R. Hemmersbach, DLR Köln
- Microgravity retardation of vesicle transport in plants during endocytosis and exocytosis. G. Scherer, University of Hannover
- Regulation of the primary metabolism of plant cells under microgravity. R. Hampp, University of Tübingen
- Investigation of Marangoni migration in monotectic alloys by electrical conductivity measurements. H. Neumann, Technical University of Chemnitz
- ARTEX – furnace with aerogels as cartridge material for the directional solidification of metallic alloys. L. Ratke, DLR Köln

13.2 MAXUS 1–4

13.2.1 MAXUS 1B

Launch date 8 Nov. 1992; number of experiments 7.

- Directional Solidification. Wallraffen, University of Bonn
- Lymphocytes in µg. M. Cogoli, ETH Zürich
- Floating Zone of GaAs. G. Müller, University of Erlangen-Nürnberg
- Floating Zone of Fluoride Glasses. Maze, Le Verre Fluore
- Oscillatory Marangoni Convection. R. Monti, University of Napoli
- Marangoni Convection. D. Schwabe, University of Giessen
- Electrophoretic Orientation of DNA. B. Norden, Chalmers University of Gothenburg

13.2.2 MAXUS 2

Launch date 28 Nov. 1995; number of experiments 8.

- Dynamics of Liquids at Edges and Corners. D. Langbein, A. de Lazzer, ZARM Bremen

- Gravitaxis of Protozoa. R. Hemmersbach, DLR Köln
- Monocytes. D. Schmitt, University of Toulouse
- Lymphocytes in Space. M. Cogoli, ETH Zürich
- Multilayer System. J.C. Legros, P. Géoris, University of Bruxelles
- Marangoni-Bénard Instability. J.C. Legros, O. Dupont, University of Bruxelles
- Instability in Liquid Containers. D. Schwabe, University of Giessen
- Electrophoresis. H. de Balmann, University of Toulouse

13.2.3 MAXUS 3

Launch date 23 Nov. 1998; number of experiments 5.

- Gravitactic Signal Perception and Transduction of *Euglena*. P. Häder, University of Erlangen-Nürnberg
- Transport of Statoliths in *Chara* Rhizoids. B. Buchen, Hodieck, Braun, University of Bonn
- Pulsating and Rotating Instabilities in Marangoni Flows. R. Monti, University of Napoli
- Electrohydrodynamic Sample Distortion during Electrophoresis. Clifton, Toulouse
- Perception and Signal Transduction in Microtubule Solutions. Tabony

13.2.4 MAXUS 4

Launch date 29 Apr. 2001; number of experiments 6.

- Pulsating and rotating instabilities in Marangoni flows 2. R. Monti, University of Napoli
- Physics of foams. M. Adler, University Marne-la-Vallée; B. Kronberg, YKI, Stockholm
- Multi-roll instability and thermocapillary flow and transition to oscillatory flow in long floating zones. D. Schwabe, University of Giessen
- Application of a rotating magnetic field for the suppression of time-dependent Marangoni convection. P. Dold, University of Freiburg
- Vibration induced convection in floating zone growth. A. Cröll, Mountain Academy of Freiberg
- Crystallization of MFI type zeolite from clear solution. J. Martens, University of Leuven

13.3 MiniTEXUS 1–6

- MiniTEXUS 1; launch date 29 Nov. 1993; J. Richter, G. Netter
- MiniTEXUS 2; launch date 3 May 1994; Utrehus, J.C. Legros

- MiniTEXUS 3; launch date 2 May 1995; S. Tarifa
- MiniTEXUS 4; launch date 29 Apr. 1995; J. Richter, G. Netter
- MiniTEXUS 5; launch date 10 Feb. 1998; D. Beysens, Garrabos, Evesque
- MiniTEXUS 6; launch date 3 Dec. 1998; Joulain, Torero

13.4 MASER 1–8

13.4.1 MASER 1

Launch date 19 Mar. 1987; number of experiments 6.

- Meniscus Stability in Immiscibles. A. Passerone, R. Sangiorgi, Genova; F. Rositto, Milano
- Coalescence of Immiscible Alloys. P.A. Sunnerkrantz, H. Fredriksson, A. Eliasson, RIT, Stockholm
- Three-Dimensional Marangoni Convection. J.H. Lichtenbelt, University of Groningen
- Thermocapillary Drop Motion. G. Wozniak, J. Siekmann, University of Essen
- Thermal Conductivity of Electrically Nonconducting Liquids. S. Engström, S. Aalto, Chalmers University, Gothenburg
- Macrosegregation in Ag–Sn alloys. H. Fredriksson, RIT, Stockholm

13.4.2 MASER 2

Launch date 29 Feb. 1988; number of experiments 8.

- Precipitation in the Zn–Bi system. P.A. Sunnerkrantz, H. Fredriksson, A. Eliasson, RIT, Stockholm
- Adhesion of Metals on Ceramic Substrates. L. Kozma, R. Warren, L. Ljungberg, I. Olefjord, Chalmers University, Gothenburg
- Gradient Solidification in Immiscible Alloys. H. Fredriksson, A. Eliasson, RIT, Stockholm
- Semiconfined Bridgman Growth of Ga doped Ge. D. Camel, P. Tison, J. de Goer, CENG, Grenoble
- Semiconfined Bridgman Growth of Ga Crystals. E. Tillberg, T. Carlberg, RIT, Stockholm
- Three-Dimensional Marangoni Convection. J.H. Lichtenbelt, University of Groningen
- Thermocapillary Drop Motion 2. G. Wozniak, J. Siekmann, University of Essen
- Directional solidification of Al–Cu alloys. Shahani

13.4.3 MASER 3

Launch date 10 Apr. 1989; number of experiments 9.

- Nucleation, Growth and Coalescence of Small Crystals. H. Fredriksson, RIT, Stockholm
- Metal Matrix Composites. A. Deruyttere, L. Froyen, University of Leuven
- Particle Agglomeration. M. Mathes, ACCESS Aachen
- Thermocapillary Drop Motion. G. Wozniak, J. Siekmann, University of Essen
- Protein Crystal Growth. L. Sjolin, Chalmers University, Gothenburg
- Membrane functions in green algae
- Regulation of cell growth and differentiation
- Binding of Concavalin A to lymphocytes
- Dorsoventral axis development in amphibian embryos

13.4.4 MASER 4

Launch date 29 Mar. 1990; number of experiments 8.

- Interfacial Tension between Immiscible Liquids. A. Passerone, ICFAM-CNR, Genova
- Orientation of DNA Molecules during Free-Solution Electrophoresis. B. Norden, Chalmers University, Gothenburg
- Laser Interferometric Observation of Unidirectional Solidification of Liquid Crystal Material. A. Iwasaki, ETL, Tsukuba
- Partial Melting Experiment of Y–Ba–Cu Oxide. N. Hamono, Hitachi
- Fertilization of urchin eggs and embryogenesis
- Embryogenesis lymphocytes in microgravity
- Regulation of cell growth and differentiation

13.4.5 MASER 5

Launch date 9 Apr. 1992; number of experiments 9.

- Wet Satellite Experiment. J.P.B. Vreeburg, NAL, Amsterdam
- Solutal Marangoni Effect. M. Adler, Laboratoire d'Aérothermique, Meudon
- Marangoni Convection due to Evaporation. L.P.B.M. Janssen, University of Groningen
- Marangoni in Space. L.P.B.M. Janssen, University of Groningen
- Laser Interferometric Observation of Solidification. S. Takei, Electrotechnical Laboratory, Tsukuba
- Oxide Superconductors Formation. M. Hamano, Hitachi Space Environment Utilization Center
- Fertilization of urchin eggs and embryogenesis
- Plasma–membrane fusion in human fibroblasts
- Nuclear response to protein

13.4.6 MASER 6

Launch date 04 Nov. 1993; number of experiments 5.

- Solutal Marangoni Effect 2. M. Adler, Laboratoire d'Aérothermique, Meudon
- Interactive Experiment on Marangoni Migration. R. Monti, R. Fortezza, University of Napoli
- Cleavage stage development of urchin eggs after microgravity exposure
- Development of xenopus eggs fertilized in space
- Cell growth regulation and differentiation

13.4.7 MASER 7

Launch date 03 May 1996; number of experiments 5.

- Pool Boiling With and Without Electrical Field. W. Grassi, University of Pisa
- Enzyme catalysis
- Liquid-phase epitaxial growth of SiC
- Signal transduction mechanisms in immobilized neuroendrine cells
- In-vivo culture of differentiated functional epithelial follicular cells from thyroid gland

13.4.8 MASER 8

Launch date 14 May 1999; number of experiments 4.

- Pool Boiling With and Without Electrical Field (Reflight). W. Grassi, University of Pisa
- Cosmic Dust Aggregation. J. Blum, University of Jena; A.C. Levasseur-Regourd, CNRS, Verrières
- Thermal Radiation Forces in Nonstationary Conditions. F.S. Gaeta, C. Albanese, F. Peluso, Napoli
- Jet Growth Motion in Aerosols. J.C. Legros, A. Vedernikov, ULB, Bruxelles

13.5 SPAR I–X

13.5.1 SPAR I

Launch date 11 Dec. 1975.

- Direct Observation of Solidification as a Function of Gravity Levels. M.H. Johnston, C.S. Griner, MSFC, Huntsville, AL

- Contained Polycrystalline Solidification in Low Gravity. J.M. Papazian, Grumman Aerospace Corporation, Bethpage, NY, W.R. Wilcox, Clarkson College, Potsdam, NY
- Thermal Migration of Bubbles and Their Interaction with Solidification Interfaces. J.M. Papazian, Grumman Aerospace Corporation, Bethpage, NY, T.Z. Kattamis, University of Connecticut
- Liquid Mixing Experiment. C.G. Schafer, G.H. Fichtl
- Bubble Behavior in Melts. J.M. Papazian, W.R. Wilcox, Clarkson College, Potsdam, NY
- Preparation of a Special Alloy for Manufacturing of Magnetic Hard Superconductors Under a Zero Gravity Environment. W. Heye, M. Klemm, Technical University of Clausthal

13.5.2 SPAR II

Launch date 17 May 1976.

- Immiscible Alloy AlIn. S.H. Gelles, A.J. Markworth, Battelle, Columbus, OH
- Immiscible Alloy AlIn. H. Ahlborn, Technical University of Hamburg
- Direct Observation of Dendrite Remelting and Macrosegregation in Casting. M.H. Johnston, C.S. Griner, MSFC, Huntsville, AL

13.5.3 SPAR III

Launch date 14 Dec. 1976.

- Interaction of Bubbles with Solidification Interfaces. J.M. Papazian, Grumman Aerospace Corporation, Bethpage, NY, W.R. Wilcox, Clarkson College, Potsdam, NY
- Contact and Coalescence of Viscous Bodies. D.R. Uhlman, MIT, Cambridge, MA

13.5.4 SPAR IV

Launch date 21 Jun. 1977.

- Immiscible Alloy AlIn. H. Ahlborn, Technical University of Hamburg
- Behavior of Second-Phase Particles at Solidification Front. D.R. Uhlman, MIT, Cambridge, MA
- Contained Polycrystalline Solidification in Low Gravity. J.M. Papazian, M. Kesselman, Grumman Aerospace Corporation, Bethpage, NY, T.Z. Kattamis, University of Connecticut
- Containerless Processing Technology Experiment. JPL, Pasadena, CA

13.5.5 SPAR V

Launch date 11 Sep. 1978; number of experiments 4.

- Uniform Dispersions by Crystallization Processing. D.R. Uhlman, MIT, Cambridge, MA
- Direct Observation of Dendrite Remelting and Macrosegregation in Casting. M.H. Johnston, C.S. Griner, MSFC, Huntsville, AL
- Contained Polycrystalline Solidification in Low Gravity. J.M. Papazian, Grumman Aerospace Corporation, Bethpage, NY, T.Z. Kattamis, University of Connecticut
- Agglomeration in Immiscible Liquids at Low-G. S.H. Gelles, A.J. Markworth, Battelle, Columbus, OH

13.5.6 SPAR VI

Launch date 17 Oct. 1979.

- Containerless Processing Technology Experiment. JPL, Pasadena, CA

13.5.7 SPAR VII

Launch date 14 May 1980.

- Dynamics of Liquid Drops. A.P. Croonquist, T.G. Wang, D.D. Elleman, P.H. Rayermann, JPL, Pasadena, CA

13.5.8 SPAR VIII

Launch date unknown.

- Dynamics of Liquid Bubbles. A.P. Croonquist, W.K. Rhim, D.D. Elleman, T.G. Wang, JPL, Pasadena, CA

13.5.9 SPAR IX

Launch date unknown.

13.5.10 SPAR X

Launch date 17 Jun. 1983.

- Containerless Processing Technology. T.G. Wang, JPL, Pasadena, CA

13.6 TR-IA 1–7

- TT-500A 8; launch date 14 Sep. 1980; number of experiments 4.
- TT-500A 9; launch date 15 Jan 1981; number of experiments 2.
- TT-500A 10; launch date 02 Aug 1981; number of experiments 2.
- TT-500A 11; launch date 16 Aug 1982; number of experiments 2.
- TT-500A 12; launch date 17 Jan 1983; number of experiments 3.
- TT-500A 13; launch date 19 Aug 1983; number of experiments 2.

13.6.1 TR-IA 1

Launch date 16 Sep. 1991; number of experiments 5.

- In-Situ Observation of Succinonitrile–Acetone Crystal. K. Tsukamoto, Tohoku University, Tokyo
- Observation of Marangoni Convection. H. Azuma, NAL, Tokyo
- Pool Boiling Experiment. Y. Abe, AIST, Tokyo
- Melting/Solidification of Particle-Dispersion-Strengthened Alloys. Y. Muramatsu, NRIM, Tsukuba
- Melting/Solidification of High-Temperature Oxide Superconductors. K. Togano, Tsukuba

13.6.2 TR-IA 2

Launch date 20 Aug. 1992; number of experiments 5.

- In-Situ Observation of Crystal Growth from Solution. K. Kuribayashi, ISAS, Tokyo
- Observation of Marangoni Convection. A. Hirata, Waseda University
- Bubble Injection and Oscillation. M. Ishikawa, MRI, Tokyo
- Semiconductor Solution Growth. T. Nishinaga, University of Tokyo
- Melting and Solidification of Glass. J. Hayakawa, AIST, Osaka

13.6.3 TR-IA 3

Launch date 08 Sep. 1993; number of experiments 7.

- In-Situ Observation of Interface Morphology during Crystal Growth. T. Sawada
- Generation and Control of Marangoni Flow. K. Kuwabara, IH Heavy Industry
- Nucleation, Growth and Movement of Bubbles. S. Ishii, MHI, Takasago
- Ceramics Synthesis. O. Odawara, Tokyo Institute of Technology
- Shape of Solid/Solution Interface. K. Kinoshita, NTT Research Laboratory
- Solidification Visualization Experiment. Japan Association of Air Companies
- Thermal–Hydraulic Performance Experiment. Japan Association of Air Companies

13.6.4 TR-IA 4

Launch date 25 Aug. 1995; number of experiments 5.

- Morphological Stability of Polyhedral Crystal. K. Tsukamoto, University of Tohoku
- Solidification of SnPb Eutectic Alloy. T. Motegi, Chiba Institute of Technology
- Liquid Diffusion in PbSnTe. M. Uchida, IH Heavy Industry
- 3D Velocity of Marangoni Convection in Liquid Column. H. Kwamura, University of Tokyo
- Temperature Oscillation of Marangoni Flow in Molten Silicone. T. Hibiya, NEC Corporation

13.6.5 TR-IA 5

Launch date 25 Sep. 1996; number of experiments 6.

- Growth of Colloidal Crystals under Microgravity. M. Ishikawa, MRI Tokyo
- Heat Transfer in Microgravity Nucleate Boiling. H. Ohta, University of Kyushu
- Formation and Combustion of Homogeneous Liquid Sprays. J. Sato, IH Heavy Industry
- Capillary Gap Penetration of Molten Alloy during Brazing. K. Sasabe, NRIM, Tsukuba
- Self-Diffusion in Germanium Semiconductor Melt. T. Itami, University of Hokkaido
- Diffusion Measurement by Shear Cell Method in Ge. S. Yoda, NASDA, Tokyo

13.6.6 TR-IA 6

Launch date Sep. 1997; number of experiments 5.

- 3D-Fluid Flow and Liquid Column Surface Temperature. K. Nishino, University of Yokohama
- Study of the Isotope Effect on the Self-Diffusion in Liquid Metals. T. Itami, University of Hokkaido
- Solidified Structures of AlTi Peritectic System Alloy. T. Motegi, Chiba Institute of Technology
- Cell Growth-Related Gene Expression in Osteoplast. A. Sato, NASDA, Tokyo
- Visualization of Marangoni Flow in Molten Silicone Column. T. Hibiya, NEC Corporation, Tsukuba

13.6.7 TR-IA 7

Launch date summer 1998; number of experiments 6.

- Dendritic Ice Crystals during Free Growth in Supercooled Water. Y. Furukawa, University of Hokkaido
- Diffusion of Metallic Complex Melts with High Melting Points. T. Itami, University of Hokkaido
- Measurement of InAs–GaAs Mutual Diffusion Coefficients. K. Kinoshita, NTT Research Laboratory
- Physiology and Function of Cultured Cells. H. Osada, Institute of Physical and Chemical Research
- Studies on Flame Propagation in Monodispersed Fuel Spray. H. Nomura. University of Nihon
- Evaluation of Position Control Function of Electrostatic Levitation Furnace. S. Yoda, NASDA, Tokyo

13.7 Skylab, May 1973

Skylab 2, Skylab 3, Skylab 4, see Table 1.1

13.7.1 Material Processing Facility

- 2   Metals-Melting Experiment. R.M. Poorman, MSFC Astronautics Laboratory, Huntsville, AL
- 2   Exothermic Brazing Experiment. J.R. Williams, MSFC Product Engineering Laboratory, Huntsville, AL
- 2   Sphere-Forming Experiment. E.A. Hasemeyer, MSFC Product Engineering Laboratory, Huntsville, AL
- 2   GaAs Crystal Growth Experiment. R.S. Seidensticker, Westinghouse Research Laboratory

13.7.2 Multipurpose Furnace System

- 3 + 4 Vapor Growth of IV–VI Compounds. H. Wiedemeier, Rensselaer Polytechnic Institute, Troy, NY
- 3 + 4 Immiscible Alloy Compositions. J.L. Reger, TRW Systems
- 3   Radioactive Tracer Diffusion. A.O. Ukanwa, MSFC Space Sciences Laboratory, Huntsville, AL
- 3   Microsegregation in Ge. F.A. Padovani, Texas Instruments
- 3 + 4 Growth of Spherical Crystals, H.U. Walter, University of Alabama, Huntsville, AL
- 3 + 4 Whisker-Reinforced Composites. T. Kawada, Tsukuba
- 3 + 4 InSb Crystals, H.C. Gatos, MIT, Cambridge, MA

- 3 + 4   Mixed III–V Crystal Growth. W.R. Wilcox, USC
- 3   Alkali Halide Eutectics. A.S. Yue, UCLA
- 3   Silver Grids Melted in Space. A. Deruyttere, University of Leuven
- 3 + 4   CuAl Eutectic. E.A. Hasemeyer, MSFC Product Engineering Laboratory, Huntsville, AL

13.7.3 Skylab Science Demonstrations

- 3   Diffusion in Liquids
- 3   Ice Melting
- 4   Liquid Floating Zone. J. Carruthers, Bell Laboratories; E.G. Gibson, NASA
- 4   Stability of Liquid Dispersions in Low Gravity. L.L. Lacy, G.H. Otto, MSFC, Huntsville, AL
- 4   Liquid Film Demonstration Experiment. W. Darbro, MSFC, Huntsville, AL
- 4   Rochelle Salt Growth
- 4   Deposition of Silver Crystals
- 4   Fluid Mechanics Demonstration. O.H. Vaughan, B. Facemire, MSFC, Huntsville, AL
- 4   Charged Particle Mobility

13.8 Apollo–Soyuz Test Project (ASTP)

Rendezvous of a Saturn-1B rocket with Soyuz 19
Launch date of both spacecrafts 15 July 1975

- Static Electrophoresis. G.V.F. Seaman, R.E. Allen, G.H. Barlow, M. Bier
- Free-Flow Electrophoresis. K. Hanning, R. Schindler, MPI for Biochemistry, München
- Crystal Growth from Solution. M.D. Lind, Science Center Division of Rockwell International
- Germanium Crystal Growth. H.C. Gatos, A.F. Witt, MIT, Cambridge, MA
- Crystal Growth from the Vapour Phase. H. Wiedemeier, Rensselaer Polytechnic Institute, Troy, NY
- Surface-Tension-Induced Convection in Metals
- Growth of LiF-NaCl Eutectica
- Processing of Monotectic (PbZn) and Synthetic (AlSb) Alloys
- Processing of Permanent-Magnet Materials

13.9 Spacelab 1 (STS-9)

13.9.1 Materials Science Double Rack (MSDR)

- Isothermal Heating Facility (IHF)
- Solidification of Immiscible Alloys. H. Ahlborn, University of Hamburg
- Solidification of Technical Alloys. J. Pötschke, Krupp Essen
- Skin Technology. H. Sprenger, MAN, München
- Vacuum Brazing. W. Schönherr, Bundesanstalt für Materialprüfung, Berlin
  Vacuum Brazing. R. Stickler, University of Wien
- Emulsions and Dispersion Alloys. H. Ahlborn, University of Hamburg (also Battele, Frankfurt)
- Reaction Kinetics in Glass. H.G. Frischat, Technical University of Clausthal
- AlPb Metallic Emulsions. P.D. Caton, Fulmer Research Institute, Slough
- Bubble Reinforced Materials. P. Gondi, University of Bologna
- Nucleation Behavior of Ag–Ge. Y. Malméjac, CEA, Grenoble
- Solidification of Near Monotectic ZnPb Alloys. H. Fischmeister, Moutain University of Leoben
- Dendrite Growth and Microsegregation. H. Fredriksson, RIT, Stockholm
- Composites with Short Fibers and Particles. A. Deruyttere, University of Leuven
- Unidirectional Solidification of Cast Iron. T. Luyendijk, Lab. Metaalkunde, University of Delft

13.9.2 Low Temperature Gradient Furnace (TGF)

- Unidirectional Solidification of Al–Zn Emulsions. C. Potard, CEA-CEN, Grenoble
- Unidirectional Solidification of Al–Al_2Cu, Ag–Ge Eutectics. Y. Malméjac, CEA, Grenoble
- Growth of Lead Telluride. H. Rodot, CNRS, Meudon-Bellevue
- Unidirectional Solidification of Eutectics. G. Müller, University of Erlangen
- Thermodiffusion in Tin Alloys. Y. Malméjac, CEA, Grenoble

13.9.3 Mirror Furnace

- Zone Crystallization of Silicon. R. Nitsche, University of Freiburg
- Traveling Solvent Growth of CdTe. H. Jäger, Battelle, Frankfurt
- Traveling Heater Method of II–V Compounds (InSb). K.W. Benz, University of Stuttgart
- Crystallization of Silicon Spheres. H. Kölker, Consortium Elektrochemische Industrie, München

13.9.4 Fluid Physics Module (FPM)

- Oscillation Damping of a Liquid in Natural Levitation. H. Rodot, CNRS, Meudon-Bellevue
- Kinetics of Spreading of Liquids on Solids. J.M. Haynes, University of Bristol
- Free Convection in Low Gravity. L.G. Napolitano, University of Napoli
- Capillary Surfaces in Low Gravity. J.F. Padday, Kodak, Harrow
- Coupled Motion of Liquid–Solid Systems in Near Zero Gravity. J.P.B. Vreeburg, NAL, Amsterdam
- Floating Zone Stability in Zero Gravity. I. DaRiva, University of Madrid
- Interfacial Instability and Capillary Hysteresis. J.M. Haynes, University of Bristol

13.9.5 Single Experiments

- Crystal Growth of Proteins. W. Littke, University of Freiburg
- Self-Diffusion and Interdiffusion in Liquid Metals. K.H. Kraatz, Technical University of Berlin
- Adhesion of Metals, UHV Chamber. G. Ghersini, Centro Informationi Studi Esperienze, Milano
- Tribological Experiments in Zero Gravity. C.H.T. Pan, University of Columbia
- Crystal Growth from Solution under Microgravity Conditions. K.F. Nielsen, Technical University of Denmark, Lyngby
- Crystal Growth of Mercury Iodide by Physical Vapor Transport. C. Belouet, Laboratoires d'Electronic et de Physique Appliqueé, Limeil-Brévannes
- Astronomy and solar physics; 6 experiments
- Space plasma physics; 6 experiments
- Atmospheric physics and earth observation; 6 experiments
- Life sciences; 15 experiments

13.10 Spacelab 3 (STS-51B)

- Solution Growth of Triglycine Sulfate. R. Lal, University of Alabama, Huntsville, AL
- Vapor Crystal Growth of Mercury Iodide. W. Schnepple, EG&G, Goleta, CA
- Mercury Iodide Solution Growth. R. Cadoret, Lab. Phys. Milieux Condense's
- Dynamics of Rotating and Oscillating Drops. T.G. Wang, JPL, Pasadena, CA
- Geophysical Fluid Flow Cell. J.E. Hart, University of Colorado, Boulder, CO

13.11 Spacelab D-1 (STS-61A)

13.11.1 Materials Science Double Rack (MSDR)

- Separation of Immiscible Melts. H. Ahlborn, University of Hamburg
- Suspended Particles at a Cu Solidification Front. J. Pötschke, Krupp FI, Essen
- Skin Technology. H. Sprenger, MAN, München
- Ostwald Ripening in Metallic Melts. H.F. Fischmeister, MPI Stuttgart
- Homogeneity of Glasses. G.H. Frischat, Technical University of Clausthal
- Particles at Melting and Solidification Fronts. D. Langbein, Battelle, Frankfurt
- Skin Casting of Cast Iron. H. Sprenger, MAN, München
- Melting and Solidification of Metallic Composites. A. Deruyttere, L. Froyen, University of Leuven
- Nucleation of Eutectic Alloys. Y. Malmejac, CEN, Grenoble
- Silicon Growth from Melting Zone. A. Cröll, University of Freiburg
- THM Growth of GaSb. K.W. Benz, University of Stuttgart
- THM Growth of CdTe. R. Schönholz, University of Stuttgart
- Crystallization of a Silicon Sphere. H. Kölker, Wacker München
- Thermal Diffusion, Soret Effect. J. Dupuy, University of Lyon
- Cellular Morphologies in PbTl Alloys. B.B. Billia, University of Marseille
- Dendritic Solidification of AlCu Alloys. C. Potard, CEN, Grenoble
- Ge/GeI Chemical Growth. J.C. Launay, University of Bordeaux
- GeI Vapor Phase Growth. J.C. Launay, University of Bordeaux
- Thermomigration of Co in Sn. J.P. Praizey, CEN, Grenoble
- Marangoni Convection in Gas/Liquid Mass Transfer. A.A.H. Drinkenburg, University of Groningen
- Separation of Liquid Phases and Bubble Dynamics. R. Nähle, DFVLR, Köln
- Mixing and Demixing of Transparent Liquids. D. Langbein, Battelle, Frankfurt
- Floating-Zone Hydrodynamics. J. Da Riva, University of Madrid
- Surface-Tension-Induced Convection. J.C. Legros, G. Pétré, University of Bruxelles
- Adhesion Forces in Liquid Films. J.F. Padday, Kodak, Harrow
- Marangoni Flows. L.G. Napolitano, University of Napoli
- Liquid Motions in Partially Filled Containers. J.P.B. Vreeburg, NAL, Amsterdam
- Interdiffusion in Metallic Melts. K.H. Kraatz, Technical University of Berlin
- Protein Crystal Growth. W. Littke, University of Freiburg

13.11.2 Process Chamber

- Bubble Transport by Chemical Waves. A. Bewersdorff, DFVLR, Köln
- Density Distribution and Phase Separation at the Critical Point. H. Klein, DFVLR; D. Woermann, University of Köln
- Bubble Migration Induced by a Temperature Gradient. D. Neuhaus, DFVLR, Köln
- Boundary Layers of Solidifying Transparent Melts. A. Ecker, RWTH Aachen
- Marangoni Convection in an Open Boat. D. Schwabe, A. Scharmann, University of Giessen
- Interdiffusion in Salt Melts. W. Merkens, RWTH Aachen

13.11.3 MEDEA

- Growth of III–V Semiconductors. K.W. Benz, University of Paderborn
- THM Growth of PbSnTe. M. Harr, Battelle, Frankfurt
- Vapor Zone Crystallization of CdTe. M. Bruder, University of Freiburg
- Directional Solidification of the InSb–NiSb Eutectic. G. Müller, University of Erlangen-Nürnberg
- Diffusion at Phase Boundaries. H.M. Tensi, Technical University of München
- Crystallization Front Convection of AlMg Alloy. S. Rex, RWTH Aachen
- Heat Capacity. H. Straub, Technical University of München

13.11.4 Materials Science Experiment Assembly (MEA)

- Diffusion of Liquid Zn and Pb. R.B. Pond, Marvalaud Inc., USA
- Liquid Phase Miscibility Gap Materials. H.S. Gelles, Columbus, OH
- Vapor Growth of Alloy-Type Semiconductor Crystals. H. Wiedemeier, Troy, USA
- Semiconductor Materials Growth in Low Gravity. R. Crouch, NASA Langley Research Center
- Containerless Melting of Glass. D.E. Day, University of Missouri-Rolla, USA

13.11.5 Miscellaneous Experiments

- Biorack; 13 experiments
- Biosciences; 7 experiments
- Vestibular Sled; 2 experiments
- NAVEX; 2 experiments

13.11.6 Science Demonstration Experiments

- Mass Discrimination. H.E. Ross, University of Stirling
- Spatial Description in Space. A.D. Friderici, MPI, Nijmegen
- Gesture and Speech in µg. A.D. Friderici, MPI, Nijmegen
- Reaction Time Measurement. M. Hoscheck, J. Hund, Mühltal

13.12 Spacelab D-2 (STS-55)

13.12.1 MEDEA

- Floating-Zone Growth of GaAs. K.W. Benz, University of Freiburg
- THM Growth of GaAs from Gallium Solutions. K.W. Benz, University of Freiburg
- Floating Zone Crystal Growth of Gallium-Doped Germanium. D. Camel, CENG, Grenoble
- Crystal Growth of GaAs by the Floating Zone Method. G. Müller, University of Erlangen-Nürnberg
- Directional Solidification of Ge/GaAs Eutectic Composites. Y. Furuhuta, Tokyo
- Cellular–Dendritic Solidification with Quenching of AlLi Alloys. P. Morgand, Grenoble
- Directional Solidification of CuMn Alloy. M. Stehle, RWTH Aachen
- Thermoconvection in Dendritic–Eutectic Solidification of AlSi11 Alloy. H.M. Tensi, Technical University of München
- Diffusion of Ni in Liquid Cu–Al and Cu–Au Alloys. J. Urbanek, Göttingen
- Hysteresis of the Specific Heat During Cooling Through the Critical Point. J. Straub, Technical University of München

13.12.2 Materials Science Double Rack (MSDR)

- Crystallization of Nucleic Acids and Nucleic Acid–Protein Complexes. V.A. Erdmann, University of Giessen
- Crystallization of Ribosomal Particles. A. Yonath, University of Hamburg
- Higher Modes of Oscillatory Marangoni Convection and their Instabilities. C.-H. Chun, DLR Göttingen
- Liquid Columns' Resonances. D. Langbein, Battelle, Frankfurt
- Marangoni–Bénard Instability. J.C. Legros, University of Bruxelles
- Stability of Long Liquid Columns. I. Martinez, ETSIA, Madrid
- Onset of Oscillatory Marangoni Flows. R. Monti, University of Napoli
- Convective Effects on the Growth of GaInSb Crystals. T. Duffar, CENG, Grenoble
- Stationary Interdiffusion in a Nonisothermal Molten Salts Mixture. J. Dupuy, J. Bert, University of Lyon

- Cellular Dendritic Solidification at Low Rate of Al–Li Alloys. H. Nguyen Thi, University of Marseille
- Vapor Growth of InP Crystal with Halogen Transport in a Closed Ampoule. H. Ono, Tokyo
- Directional Solidification of the LiF–LiBaF$_3$ Eutectic. H. Wallrafen, University of Bonn
- Oxide Dispersion-Strengthened Single-Crystal Alloys. W. Amende, MAN, München
- Self-Diffusion in Pure Metals. G. Frohberg, Technical University of Berlin
- Impurity Diffusion and Interdiffusion in Different Systems. G. Frohberg, Technical University of Berlin
- Separation Behavior of Monotectic Alloys. L. Ratke, DLR Köln
- Heating and Remelting of an Allotropic FeCSi Alloy in a Ceramic Skin and Volume Change. W. Amende, MAN, München
- Interfacial Tension and Heterogeneous Nucleation in Immiscible Liquid Metal Systems. R. Sangiorgi, ICFAM-CNR, Genova
- Solution Growth of GaAs Crystals. Y. Suzuki, STC, Tokyo
- Impurity Transport and Diffusion in InSb Melt. M. Watanabe, STC, Tokyo
- Nucleation and Phase Selection during Solidification of Undercooled Alloys. K. Wittmann, DLR Köln

13.12.3 HOLOP

- Phase Separation in Liquid Mixtures with Miscibility Gap. H. Klein, DLR Köln
- Interferometric Determination of the Differential Interdiffusion Coefficient of Binary Molten Salts. W. Merkens, RWTH Aachen
- Measurement of Diffusion Coefficient in Aqueous Solution. M.C. Robert, Paris
- Marangoni Convection in a Rectangular Cavity. D. Schwabe, University of Giessen

13.12.4 MAUS

- Gas Bubbles in Glass Melts. G.H. Frischat, Technical University of Clausthal
- Reaction Kinetics in Glass Melts. G.H. Frischat, Technical University of Clausthal
- Pool Boiling. J. Straub, Technical University of München

13.12.5 Miscellaneous Experiments

- Anthrorack; 19 experiments
- Baroreflex; 1 experiment
- Biolabor; 16 experiments

- Radiation Complex; 5 experiments
- AOET. 1 experiment
- Microgravity Measurement Assembly; 2 experiments
- GAUSS; 1 experiment
- MOMS; 3 experiments

13.13 IML-1 (STS-42)

- Microgravity Vestibular Investigations (MVI). M.F. Reschke, JSC Houston, TX
- Space Physiology Experiments. 6 experiments
- Mental Workload and Performance Experiment. H.L. Alexander, MIT, Cambridge, MA
- Gravitational Plant Physiology Facility. 2 experiments
- Biorack. 17 experiments
- Biostack. H. Bücker, DLR Köln
- Radiation Monitoring Container Device (RMCD). S. Nagaoka, NASDA, Tokyo
- Space Acceleration Measurement System (SAMS). H. Hamacher, DLR Köln
- Protein Crystal Growth. C.E. Bugg, University of Alabama, Huntsville, AL
- Single-Crystal Growth of β-Galactosidase and β-Galactosidase/Inhibitor Complex. W. Littke, University of Freiburg
- Crystal Growth of Electrogenic Membrane Protein Bacteriorhodopsin. G. Wagner, University of Giessen
- Crystallization of Proteins and Viruses in Microgravity by Liquid–Liquid Diffusion. A. McPershon. University of Riverside, CA
- Vapor Crystal Growth Studies of Single Mercury Iodide Crystals. L. Van den Berg, EG&G, Goleta, CA
- Mercury Iodide Nucleations and Crystal Growth in Vapor Phase. R. Cadoret, University of Clermont-Ferrand, Aubière
- Organic Crystal Growth Facility. A. Kanbayashi, NASDA, Tokyo

13.13.1 Critical Point Facility

- Study of Density Distribution in a Near-Critical Sample Fluid. A.C. Michels, Van der Waals Laboratory, Amsterdam
- Heat and Mass Transport in a Pure Fluid in the Vicinity of the Critical Point. D. Beysens, CENS, Saclay
- Phase Separation in an Off-Critical Binary Mixture. D. Beysens, CENS, Saclay
- Investigation of the Thermal Equilibrium Dynamics of SF_6 Near the Liquid–Vapor Critical Point. A. Wilkinson, LRC, Cleveland, OH

13.13.2 Fluids Experiment System

- Study of Solution Crystal Growth in Low Gravity. R.B. Lal, University of Alabama, Huntsville, AL
- An Optical Study of Grain Formation: Casting and Solidification Technology (CAST). M.H. McCay, University of Tennessee

13.14 Spacelab J (STS-47)

- Growth Experiment on Narrow-Band-Gap Semiconductor PbSnTe. T. Yamada, NTT
- Growth of PbSnTe Single Crystal by Traveling-Zone Method. S. Iwai, Inst. of Physical & Chemical Res.
- Growth of Semiconductor Compound Single Crystal with a Floating-Zone Technique. I. Nakatani, NRIM, Tsukuba
- Casting of Superconducting Composite Materials. K. Togano, NRIM, Tsukuba
- Formation Mechanism of Deoxidation Products in Iron Ingots Deoxidized with Two or Three Components. A. Fukuzawa, NRIM, Tsukuba
- Preparation of Ni-Base Dispersion Strengthened Alloys. Y. Murumatsu, NRIM, Tsukuba
- Diffusion in Liquid State and Solidification of Binary System. T. Dan, NRIM, Tsukuba
- Behavior of Glass at High Temperatures. N. Soga, Kyoto University
- Growth of Spherical Silicon Crystals and Surface Oxidation. T. Nishinaga, University of Tokyo
- Fabrication of Ultra-Low Density, High-Stiffness Carbon Fiber and Aluminum Composite. T. Suzuki, Tokyo Institute of Technology
- Study on Liquid Phase Sintering. S. Kohara, Science University of Tokyo
- Fabrication of SiAsTe Semiconductor in Microgravity Environment. Y. Hamakawa, Osaka University
- Gas Evaporation in Low Gravity. N. Wada, Nagoya University
- Drop Dynamics in an Acoustic Resonant Chamber and Interference with the Acoustic Field. T. Yamanaka, NAL, Tokyo
- Bubble Behavior in a Thermal Gradient and Stationary Acoustic Wave. H. Azuma, NAL, Tokyo
- Preparation of Optical Materials for Use in Nonvisible Region. J. Hajakwa, Government Industrial Institute
- Marangoni-Effect Induced Convection in Material Processing under Microgravity. S. Enya, IHI, Tokyo
- Solidification of Eutectic System Alloys in Space. A. Ohno, Chiba Institute of Technology
- Growth of Samarskite Crystal under Microgravity Conditions. S. Takekawa, National Inst. for Res. in Inorganic Mat.

- Crystal-Growth Experiment on Organic Superconductors in Low Gravity. H. Anzai, NEL, Tsukuba
- Crystal Growth of Compound Semiconductors (InGaAs) in a Low-Gravity Environment. M. Tatsumi, Sumitomo Electric Industries

13.15 IML-2 (STS-65)

13.15.1 Large Isothermal Furnace

- Gravitational Role in Liquid Phase Sintering. R.M. German, Pennslyvania State University, PA
- Mixing of a Melt of Multicompound Semiconductor. A. Hirata, Waseda University, Tokyo
- Effect of Weightlessness on Microstructure and Strength of Ordered TiAl
- Intermetallic Alloys. M. Takeyama, A. Sato, STA, Tokyo

13.15.2 TEMPUS

- Effects of Nucleation by Containerless Processing in Low Gravity. R.J. Bayuzik, Vanderbilt University, Nashville, TN
- Alloy Undercooling Experiments. M.C. Flemings, MIT, Cambridge, MA
- Nonequilibrium Solidification of Largely Undercooled Melts. D.M. Herlach, DLR Köln
- Structure and Solidification of Largely Undercooled Melts of Quasicrystal-Forming Alloys. K. Urban, Institut für Festkörperforschung, Jülich
- Thermodynamics and Glass Formation in Undercooled Liquid Alloys. H.J. Fecht, University of Augsburg
- Metallic-Glass Research in Space: Thermophysical Properties of Metallic Glasses and Undercooled Alloys. W.L. Johnson, CALTECH, Pasadena, CA
- Viscosity and Surface Tension of Undercooled Melts. I. Egry, DLR Köln
- Measurement of the Viscosity and Surface Tension of Undercooled Melts under Microgravity Conditions and Supporting MHD Calculations. J. Szekely, MIT, Cambridge, MA

13.15.3 Bubble, Drop, and Particle Unit (BDPU)

- Bubble Migration, Coalescence and Interaction with Melting and Solidification Fronts. R. Monti, University of Napoli
- Thermocapillary Migration and Interactions of Bubbles and Drops. R.S. Subramanian, Clarkson University, Potsdam, NY
- Bubble Behavior under Low Gravity. A. Viviani, University of Napoli
- Interfacial Phenomena in a Multilayered Fluid System. J.N. Koster, University of Boulder, CO

- Thermocapillary Instability in a Three-Layer-System. J.C. Legros, University of Bruxelles
- Nucleation, Bubble Growth, Interfacial Microlayer, Evaporation and Condensation Kinetics. J. Straub, Technical University of München
- Static and Dynamic Behavior of Liquids in Edges and Corners. D. Langbein, ZARM Bremen

13.15.4 Critical Point Facility (CPF)

- The Piston Effect. D. Beysens, CEN, SACLAY, Gif Sur-Ivette
- Thermal Equilibrium in a One-Component Fluid. R.A. Ferrell, University of Maryland, MD
- Density Equilibrium Time Scale. H. Klein, DLR Köln
- Heat Transport and Density Fluctuations in a Critical Fluid. A.C. Michels, Van der Waals–Zeeman Lab., Amsterdam

13.15.5 Vibration Isolation Box Experiment System (VIBES)

- Influence of G. Jitter on Natural Convection and Diffusive Transport. H. Azuma, NAL, Tokyo
- Study on Thermally Driven Flow under Microgravity. M. Furukawa, NASD, Tsukuba

13.15.6 Free Flow Electrophoresis Unit (FFEU)

- Gravitational Role in Electrophoretic Separations of Pituitary Cells and Granules. W.C. Hymer, Pennsylvania State University, PA
- Separation of Chromosome DNA of a Nematode, *C. Elegans*, by Electrophoresis. H. Kobayashi, Josai University, Saitama
- Experiments Separating the Culture Solution of Animal Cells in High Concentration under Microgravity. T. Okusawa, Hitachi, Ibaraki
- Optimization of Protein Separation. V. Sanchez, University of Toulouse
- Electrohydrodynamic Sample Distortion. R. Snyder, MSFC, Huntsville, AL

13.15.7 Miscellaneous Experiments

- Space Acceleration Measurement System (SAMS). C. Baugher, MSFC, Huntsville, AL
- Quasi-Steady Acceleration Measurement (QSAM). H. Hamacher, DLR Köln
- Advanced Protein Crystallization Facility; 10 experiments
- Aquatic Animal Experiment Unit; 4 experiments
- Biorack; 28 experiments
- Thermoelectric Incubator; 3 experiments
- Extended Duration Orbiter Medical Project; 2 experiments
- Spinal Changes in Microgravity. J.R. Ledsome, UBC, Vancouver

13.16 EURECA

- Gas Phase Growth of II–VI Semiconductors. R. Nitsche, University of Freiburg
- Solution Zone Growth of II–VI Semiconductors. R. Nitsche, University of Freiburg
- Solution Zone Growth of III–V Semiconductors. K.W. Benz, University of Freiburg
- Solution Zone Growth of PbSnTe. M. Harr, Battelle, Frankfurt
- Zone Growth of Ternary Sulfides. V. Kramer, University of Freiburg
- Growth of TSeF–TCNQ, M. Harr, Battelle, Frankfurt
- Growth of Lysozyme and Beta-Galactosidase. W. Littke, University of Freiburg
- Particles before a Solidification Front. F. Nilmen, Battelle, Frankfurt
- Fluid Phase Sintering. W. Graf, KFI, Essen
- Ostwald Ripening and Diffusion. H.F. Fischmeister, MPI Stuttgart
- PV-Diagram at the Critical Point. W. Wagner, University of Bochum
- Adsorption at the Critical Point. G.H. Findenegg, University of Bochum
- Exobiological Radiation Measurement. G. Horneck, DFVLR, Köln
- Biostack. H. Bücker, DFVLR, Köln
- "Spares" and Organic Molecules. K. Dose, University of Mainz
- Dosimetric Mapping. G. Reitz, DFVLR, Köln
- Inactivation and Changes due to Solar Radiation. J. Kiefer, University of Giessen
- Radiofrequency Ion Thruster Assembly. Bassner, MBB, München

13.17 MIR and FOTON

13.17.1 MIR

- Measurement of Radiation in Space. K. Fujitaka
- MIR 92 – 17 Mar–25 Mar 1992; 14 experiments in life sciences
- MIR 92E. 14 experiments in life sciences

13.17.2 MIR 97

Launch date 06 Feb. 1997, landing 24 Feb. 1997

- Homogeneity of Heavy Metal Fluoride Glasses. G.H. Frischat, Technical University of Clausthal
- Heterogeneous Nucleation of Glass Forming Melts. H. Reiß, M. Müller, University of Jena, I. Gutzow, Bulgarian Academy of Science, Sofia
- Microstructure Evolution in Immiscible Aluminium Alloys. L. Ratke, G. Korrekt, DLR Köln, B. Prinz, ID-Ingenieurdienste Friedberg

- Investigation of Thermosolutally Caused Convection during Bridgman-Growth of BiSbTe$_3$-mixed Crystals. P. Reinshaus, University of Halle-Wittenberg
- Macro- and Microsegregation of Different Components in BiSbTe$_3$ Crystals Grown by the Zone Melting Technique. F. König, G. Bärwolff, G. Seifert, Technical University of Berlin
- Interaction Between Semiconductor Melts and the Corresponding Gas Phases in a µg Crystal Growth System. E. Buhrig, U. Wunderwald, Mountain Academy of Freiberg
- Kinetics of Phase Transition Around the Critical Point of Pure Fluids. J. Straub, Technical University of München
- Smart Gas Sensor. H. Wessels, ESTEC Noordwijk, H.-J. König, RST Raumfahrt und Umweltschutz, Rostock
- Microgravity Measurement on MIR. U. Merbold, ESA, H. Hamacher, DLR Köln
- 20 experiments in life sciences

13.17.3 FOTON 9

Mid Dec. 1993.

- Crystallization of CdTeSe. University of Freiburg
- Crystallization of CdZnTe. HU Berlin

13.17.4 FOTON 12

Launch date 09 Sep. 1999.

- Diffusion coefficient of Te and In in GaSb melts. J.P. Praizey, CENG, Grenoble
- Diffusion Measurements in Metallic Systems. G. Frohberg, Technical University of Berlin
- GeSi (10 at% Si max). K.W. Benz, University of Freiburg
- Zn-Doped Ge and InAs. Buhrig, Mountain University of Freiberg

Bibliography

Proceedings of the 3rd Space Processing Symposium – SKYLAB Results. 30 Apr–1 May 1974. MSFC, Huntsville, AL

J.P.B. Vreeburg: *Summary Review of Microgravity Fluid Science Experiments.* NAL Amsterdam, ESA Nov. 1986

R.S. Snyder: Summary of Pre-ASTP Results. *Proceedings of 2nd European Symposium on Material Sciences in Space.* Frascati 1976, ESA, SP-114

Long Term Planning in Materials Science under Microgravity, MSWG. ESA, Paris, January 1980

Experiment Reports from Flight Opportunities on MASER 1 and 2. Compiled by J.P.B. Vreeburg. ESA February 1989

Summary Report of Materials Science and Fluid Science Experiments, TEXUS 1–10. H. Ahlborn, O. Minster, H.U. Walter (eds.), ESA February 1989

G. Frohberg (ed.): *Summary Report of ESA-Experiments on TEXUS 13–20* ESA February 1989

Summary Report of Sounding Rocket Experiments in Fluid Science and Materials Sciences. TEXUS 1 to 20 and MASER 1 to 2. ESA SP 1132, Vol. 1

Final Reports of Sounding Rocket Experiments in Fluid Science and Materials Sciences. TEXUS 21 to 24 and MASER 3. ESA SP 1132, Vol. 2

Final Reports of Sounding Rocket Experiments in Fluid Science and Materials Sciences. TEXUS 25 to 27 and MASER 4. ESA SP 1132, Vol. 3

Final Reports of Sounding Rocket Experiments in Fluid Science and Materials Sciences. TEXUS 28 to 30, MASER 5 and MAXUS 1. ESA SP 1132, Vol. 4

H.U. Walter (ed.): *Focused Science in ESA's Microgravity Programme: Physics, Fluids and Material Science.* ESA, May 1993

J.P.B. Vreeburg: *Summary Review of Microgravity Fluid Science Experiments.* NAL Amsterdam, ESA, Nov. 1986

I. Meyer: *Mission Documentation DASA-Nord.* 16 Jun 1999

G. Seibert: *A World Without Gravity.* ESA SP-1251, Noordwijk (2001)

H. Ahlborn (ed.): *TEXUS 1–10 Abschlußbericht.* BMFT-DLR

H. Ahlborn (ed.): *TEXUS 11/12 Abschlußbericht.* BMFT-DLR 1985

H. Ahlborn (ed.): *TEXUS 17–20 Abschlußbericht.* BMFT-DARA 1990

H. Ahlborn (ed.): *TEXUS 21–30 Abschlußbericht.* BMFT-DARA 1996

STS-9/SPACELAB-1: Press kit. ESA/NASA, Oct. 1983

TR-IA-2 Launch Information for Press. NASDA, August 1992

TR-IA-3 Launch Information for Press. NASDA, September 1993

TAKESAKI-7, Microgravity Experiments using Sounding Rocket TR-IA-7. NASDA, September 1998

Subject Index

aircraft 14, 17
angular momentum 113
Apollo program 2
Archimedes 15
axially decreasing (damped) waves 288, 294
axially increasing waves 288
axisymmetry 64, 72

balance 13, 261
Bessel function 78, 287, 289
bifurcation 72, 84, 95, 115, 123, 125, 140, 144, 198
Bond number 15, 25, 59, 70, 104, 129
boundary condition
– Neumann 30, 58
– no-slip 264
– of forces 41, 135
– Young 30, 31, 46
breakage, wavenumber of 204
bubble motion 308
bungee jumping 13

canthotaxis 90, 149
– interval of 160
capillary
– equation 61, 67, 84, 200
– number 32
– pressure 16, 45, 57
– surface 199
catapult 13
catenoid 37, 52, 54, 120, 169
centrifugal force 13
circular cylinder 76, 155, 179, 246
Columbus 2, 3, 307
column
– axisymmetric 89
– finite 292
– infinite 290
– rotation 89

complex axial wavenumber 287
cone 158
contact angle 121
– advancing 12, 21, 119, 242, 261
– dynamic 31
– hysteresis 274
– receding 12, 21, 119, 242, 276
– static 30
continuity equation 260
convection
– candle 28
– Cognac 28
– free 6
– interface 6, 10, 28
– Marangoni 12, 27
– solutocapillary 12
– thermocapillary 8, 12, 27
coordinates
– Cartesian 62
– cylindrical 63
– polar 62
corner volume 225, 228
coupled damped oscillation 316
crystal growth 6, 35
cube 49
cylinder
– regular 190
– rhombic 188

damped harmonic oscillation 312
damping decrement 317
decrement 77
deformation
– antimetric 77, 90, 94, 102
– asymmetric 115
– axially periodic 200
– critical 86, 117
– cross section 79
– lateral 79, 89, 96, 103, 110
– local 74
– longitudinal 199

Subject Index

– nonaxisymmetric 79, 82
normal 69, 77
– symmetric 77, 90, 102
Delaunay curve 52
diffusion 24
– interdiffusion 9
– mass 6
– self-diffusion 9
– temperature 6
dispersion relation 292
double float zone 119, 135
– nonsymmetric 139
– symmetric 139
drop 49, 59
– dynamics module 106
– free 47
– hanging 26, 58, 61, 133
– merging 36
– rotation 106, 112
– sessile 26, 58, 61, 133, 157
– surface energy 28
drop tower 1, 17
– CNES, Grenoble 4
– JAMIC, Sapporo 4
– NASA Lewis, Cleveland, OH 4
– ZARM, Bremen 4

electric charging 38
energy balance 261, 272, 312
engulfment 8
Euler–Lagrange equation 46, 67, 114
exotic container 246, 249
– mismatch of contact angle 253
– mismatch of volume angle 253
– residual gravity 254
experiments
– adhesion forces in liquid films 37
– double float zone 141
– DYLCO (dynamics of liquids in edges and corners) 205, 272
– exotic container 256
– floating liquid zones 69
– ICE (interface configuration experiment) 241
– LICOR (liquid column resonances) 302
– ONSET (onset of oscillatory Marangoni flows) 309
– STACO (stability of rotating columns) 87
exponential 240

exponential fading out 222
exponential piling-up 223

facilities
– AFPM (Fluid Physics Module) 305
– BDPU (Bubble, Drop and Particle Unit) 273
– DDM (Drop Dynamics Module) 106
– FPM (Fluid Physics Module) 348
– MSDR (Materials Science Double Rack) 347
– TEMPUS (Tiegelfreies elektromagnetisches Prozessieren unter Schwerelosigkeit) 106
fingering 35
finite-element method 46, 114
– SURFACE EVOLVER 151, 225
flow in parallelogram 283
flow in rectangular tube 278
fluid dynamics 6
fluid-static pressure 1, 6
forced oscillation 289
Fourier expansion 101
Fourier series 199
friction coefficient 278

g-level
– tolerable 5
– attainable 10
glider 11
GSOC (German Space Operation Center) 308

Hagen–Poiseuille flow 259
Hamaker constant 25
helicoid 162, 169
hyperbola 41
hysteresis 82, 151, 237, 245

ice cream cone 186
inflection point 55, 120, 195
integral theorem 86, 162, 179, 221, 240
interface energy 27
interface tension 24, 25
intermolecular attraction 8, 21
International Space Station 4
isopleth 34
isopyknic 15

Kepler 13

Subject Index

Lagrange multiplier, parameter 45, 74, 84
Landen transformation 56
lateral oscillation 312, 314
lead pellets 1
linear stability analysis 67, 86, 105
liquid ring 81, 98

Marangoni convection 6
MASER 338
MAXUS 336
meniscus height, central 226
microgravity 14
– attainable 5
– availability 1
– conditions 1
– cost 1, 5, 17
– quality 1, 17
– repeatability 1
– time 1
– tolerable 11
minimization of energy 44, 183
minimum volume condition 67, 83, 104
MMA (Microgravity Measurement Assembly) 308
modeling
– analytical 12, 46
– numerical 15
molecular attraction 33
moment of inertia 113
momentum equation 259, 288
monotectic alloys 9

N-pod, regular 217
natural frequency 292
natural oscillation 289
Navier–Stokes equation 259
Newton 33
Newton's law of acceleration 12
Newton's method 47
Newton's no-slip condition 33
nodoid 52, 57, 120, 137
nucleation 6, 23

Ohnesorge number 68, 298, 313
oscillation
– antimetric 295
– symmetric 295
Ostwald ripening 35

parabolic flights 1
parallel plates 119
payload specialists 3

penetration into wedge 262, 264
pinning line 149
Plateau simulation 17, 106, 303
pole radius 61
polygonal cylinder 184
– regular 184
– rhombic 181, 188
– square 161, 184
polyhedron 51, 213, 222, 232
principal axis 73
principal curvature 42
proboscis 235, 238, 241

raindrop 1
resonance detection 287, 297
resonance frequency 288, 300, 303, 307
rise in capillary 263
rotating column 81
rotating drop 17, 107
rotation number 94, 98, 106, 110, 199
Runge–Kutta method 47, 75, 89, 105, 193

Scherk function 164
Scotch tape 35
sedimentation 1, 6
shooting method 92
Shuttle 3
similarity solution 266
skipping-rope mode 71
Skylab 2, 92, 345
slice model 288, 289
small liquid columns 302
snail 154
soap bubble 1, 68
sounding rockets
– MASER (Material Science Experiments Rockets) 3, 4
– MAXUS (formed from MASER and TEXUS) 4
– SPAR (Space Processing Application Rockets) 1, 4
– TEXUS (Technologische Experimente unter Schwerelosigkeit) 3, 26
– TR-IA 3, 4
space agencies
– ESA (European Space Agency) 15
– ESRO (European Space Research Organisation) 15

- NASA (National Aeronautics and Space Administration)   12
- NASDA (National Space Development Agency of Japan)   344

space platforms
- (ISS) International Space Station   11
- (MIR) International Space Station (mir = wonder)   1

Space Shuttle   2
- Challenger   2
- Columbia   2
- Hermes   2
- STS (Space Transportation System)   3

Spacelab   2
- FMPT (First Material Processing Test, Spacelab J)   354
- FSLP (First Spacelab Payload, STS-9)   9

SPAR   2, 340
spherical cap   213
square cylinders   164
stability   125, 181, 206
- amphora mode   69
- antisymmetric   72
- diagram   94, 95, 97, 98, 105, 143, 144, 197
- Kelvin   36, 68
- lateral   100, 195
- linear analysis   85
- linear stability analysis   67, 96, 105
- longitudinal   195
- minimum-volume condition   67, 83, 123, 127, 131
- Rayleigh   53, 67, 69, 75, 146, 154, 208, 237, 290
- second variation of the energy   73, 80
- symmetric   72

stress tensor   24, 291

surface oscillation   300
surface tension   42
  see also interface tension
surface tension tank   179
sword-like shape   264

Taylor expansion   162
tentacle   157
TEXUS   2, 323
tip position, location   267, 271
toroid   109
trajectory
- tide   14
tripod   213, 217, 221
truncation parameter   168

unduloid   52, 57, 75, 120, 137

van der Waals
- attraction   21, 22, 24
- repulsion   8
variation
- calculus of   45, 73
- second   73
vector potential   260
vorticity   260

Weber number   67
wedge   158
- circular   183, 185
- contribution   185
- rotation   192, 195
- spherical   216
- straight   49
- volume   222
wetting   8, 31, 34, 57, 180, 264
- barrier   149, 150
- cross   154
- limit   233, 274
- stripe   153
- tile   151
winding rate   170

Springer Tracts in Modern Physics

170 **d–d Excitations in Transition-Metal Oxides**
A Spin-Polarized Electron Energy-Loss Spectroscopy (SPEELS) Study
By B. Fromme 2001, 53 figs. XII, 143 pages

171 **High-T_c Superconductors for Magnet and Energy Technology**
By B. R. Lehndorff 2001, 139 figs. XII, 209 pages

172 **Dissipative Quantum Chaos and Decoherence**
By D. Braun 2001, 22 figs. XI, 132 pages

173 **Quantum Information**
An Introduction to Basic Theoretical Concepts and Experiments
By G. Alber, T. Beth, M. Horodecki, P. Horodecki, R. Horodecki, M. Rötteler, H. Weinfurter, R. Werner, and A. Zeilinger 2001, 60 figs. XI, 216 pages

174 **Superconductor/Semiconductor Junctions**
By Thomas Schäpers 2001, 91 figs. IX, 145 pages

175 **Ion-Induced Electron Emission from Crystalline Solids**
By Hiroshi Kudo 2002, 85 figs. IX, 161 pages

176 **Infrared Spectroscopy of Molecular Clusters**
An Introduction to Intermolecular Forces
By Martina Havenith 2002, 33 figs. VIII, 120 pages

177 **Applied Asymptotic Expansions in Momenta and Masses**
By Vladimir A. Smirnov 2002, 52 figs. IX, 263 pages

178 **Capillary Surfaces**
Shape – Stability – Dynamics, in Particular Under Weightlessnes
By Dieter Langbein 2002, 182 figs. XVIII, 364 pages